Proceedings of the 29th International Geological Congress Part B

Also available from VSP

PROCEEDINGS OF THE 29TH INTERNATIONAL
GEOLOGICAL CONGRESS - PART A:
Metamorphic Reaction: Kinetics and Mass Transfer
Edited by T. Nishiyama and G.W. Fisher
Sandstone Petrology in Relation to Tectonics
Edited by F. Kumon and K.M. Yu
Evaporite and Desert Environment
Edited by Y. Watanabe and A. Motamed

PROCEEDINGS OF THE 29TH INTERNATIONAL
GEOLOGICAL CONGRESS - PART C:
Siliceous, Phosphatic and Glauconitic Sediments of the
Tertiary and Mesozoic
Edited by A. Iijima, A.M. Abed and R.E. Garrison

PROCEEDINGS OF THE 29TH INTERNATIONAL
GEOLOGICAL CONGRESS - PART D:
Circum-Pacific Ophiolites
Edited by A. Ishiwatari, J. Malpas and H. Ishizuka

Related titles

Facies Models in Exploration and Development of
Hydrocarbon and Ore Deposits
Edited by A.H. Bouma and R.M. Carter

Regional Metamorphism of Ore Deposits and
Genetic Implications
Edited by P.G. Spry and L.T. Bryndzia

Tectonics of Circum-Pacific Continental Margins
Edited by J. Aubouin and J. Bourgois

Proceedings of the 29th International Geological Congress
Part B

Kyoto, Japan, 24 August - 3 September, 1992

Reconstruction of the Paleo-Asian Ocean
Editor: R.G. Coleman

Quaternary Environmental Changes
Editor: E.H. Juvigné

Utrecht, The Netherlands, 1994

VSP BV
P.O. Box 346
3700 AH Zeist
The Netherlands

© VSP BV 1994

First published in 1994

ISBN 90-6764-174-X

All rights reserved. No part of this publication may be reproduced, stored in a retrieval system, or transmitted in any form or by any means, electronic, mechanical, photocopying, recording or otherwise, without the prior permission of the copyright owner.

CIP-DATA KONINKLIJKE BIBLIOTHEEK, DEN HAAG

Proceedings

Proceedings of the 29th International Geological Congress.
- Utrecht : VSP
Pt. B / ed.: R.G. Coleman, E.H. Juvigné.
ISBN 90-6764-174-X bound
NUGI 816
Subject headings: geology.

Printed in The Netherlands by A-D Druk, Zeist.

CONTENTS

RECONSTRUCTION OF THE PALEO-ASIAN OCEAN
Editor: R.G. Coleman

Preface	5
The remnants of the Paleo-Asian Ocean within Central Kazakhstan *A.E. Yakubchuk and K.E. Degtyarev*	7
Tectonic evolution of Northern Xinjiang, N.W. China: An introduction to the tectonics of the southern part of the Paleo-Asian Ocean *Xiao Xuchang, Tang Yaoqing, Zhao Min, and Wang Jing*	25
Evidences for the West China Craton and its evolution *Yuan Xuecheng, Zuo Yu, and Zhang Chaowen*	39
Geodynamic evolution of Southern Siberia in Late Precambrian - Early Paleozoic time *N.A. Berzin and N.L. Dobretsov*	53
The geodynamic evolution of the mobile fold belts of the territory of Mongolia *D. Dorjnamjaa, G. Badarch, and D. Orolmaa*	71
Ophiolites of the southern Siberia and northern Mongolia *E.V. Sklyarov, V.A. Simonov, and M.M. Buslov*	85
Blueschist belts of North Asia and models of subduction-accretion wedge *N.L. Dobretsov and A.G Kirkdyashkin*	99
Blueschists discovered in Kumishi, South Tianshan and their tectonic significance *Gao Jun, Tang Yaoqing, Zhao Min, Wang Jun, and Wu Hanquan*	115
Early arc plutonic rocks in the Olyutor Range, Northeastern Kamchatka, Russia *H. Tanaka, P.K. Kepezhinskas, S. Miyashita, and I. Rueber*	119
Geodynamic evolution of continental margins in Eastern Asia and tectonic setting of East China *E.Z. Chang, X. Ying, D. Zhou, and L. Wang*	133
Comparisons of arc-trench systems in the early Paleozoic Gorny Altai and the Mesozoic-Cenozoic of Japan *T. Watanabe, M.M. Buslov, and S. Koitabashi*	169

QUATERNARY ENVIRONMENTAL CHANGES
Editor: E.H. Juvigné

Preface	191
Peculiarities of formation of flat glaciolacustrine hills *A. Bitinas*	193
What controls the calcium carbonate content in Quaternary sediment cores raised from below the carbonate lysocline depth in the Equatorial Indian Ocean? *P. Divakar Naidu, B.A. Malmgren, and L. Bornmalm*	201
Rates of erosion by the Colorado River in the Grand Canyon, Arizona *W.K. Hamblin*	211
Method of estimating mean annual temperatures from plant fossils in the Pliocene and Pleistocene periods *Y. Hase and A. Iwauchi*	219
Spatial variation of CO_2^- and SO_3^- radicals in massive coral from Ishigaki Island, Japan and its implications *S. Ikeda, M. Furusawa, and M. Ikeya*	225
Late Quaternary paleoceanography of the Japan Sea: a tephrochronological and sedimentological study *K. Ikehara, K. Kikkawa, H. Katayama, and K. Seto*	229
A Pleistocene stratotype at Alzenau (Vorspessart, Germany) *E. Juvigné, R. Geeraerts, F. Geissert, H.-J. Gregor, M. Hottenrott, J.J. Hus, G. Seidenschwann, and R.C. Walter*	237
Analysis of paleo-environments based on electric conductivity of the STICS-water - On five drill cores in Kanto-plain, Central Japan *M. Koarai and Research Group for Alluvium Deposit*	251
Holocene dinoflagellate cyst assemblage in lagoonal lakes along the east coast of Japan *N. Kojima*	263
The late Quaternary environment around Lake Nojiri in Central Japan *Nojiri-ko Excavation Research Group*	269
Vegetation and climate since Last Glacial Maximum in Darjeeling (Mirik Lake), Eastern Himalaya *C. Sharma and M.S. Chauhan*	279

Quaternary environmental changes in the Northeastern Japan
S. Takeuti and K. Manabe   289

Beginning of a new period - the Technogene
G. Ter-Stepanian   299

The essential characteristics and facies model of windblown sand deposits during the Last Glacial period in Eastern China
M. Zhang and J. Liu   309

RECONSTRUCTION OF THE PALEO-ASIAN OCEAN

Editor
R.G. Coleman

CONTENTS

Preface   5

The remnants of the Paleo-Asian Ocean within Central Kazakhstan
A.E. Yakubchuk and K.E. Degtyarev   7

Tectonic evolution of Northern Xinjiang, N.W. China: An introduction to the tectonics of the southern part of the Paleo-Asian Ocean
Xiao Xuchang, Tang Yaoqing, Zhao Min, and Wang Jing   25

Evidences for the West China Craton and its evolution
Yuan Xuecheng, Zuo Yu, and Zhang Chaowen   39

Geodynamic evolution of Southern Siberia in Late Precambrian - Early Paleozoic time
N.A. Berzin and N.L. Dobretsov   53

The geodynamic evolution of the mobile fold belts of the territory of Mongolia
D. Dorjnamjaa, G. Badarch, and D. Orolmaa   71

Ophiolites of the southern Siberia and northern Mongolia
E.V. Sklyarov, V.A. Simonov, and M.M. Buslov   85

Blueschist belts of North Asia and models of subduction-accretion wedge
N.L. Dobretsov and A.G Kirkdyashkin   99

Blueschists discovered in Kumishi, South Tianshan and their tectonic significance
Gao Jun, Tang Yaoqing, Zhao Min, Wang Jun, and Wu Hanquan   115

Early arc plutonic rocks in the Olyutor Range, Northeastern Kamchatka, Russia
H. Tanaka, P.K. Kepezhinskas, S. Miyashita, and I. Rueber   119

Geodynamic evolution of continental margins in Eastern Asia and tectonic setting of East China
E.Z. Chang, X. Ying, D. Zhou, and L. Wang   133

Comparisons of arc-trench systems in the early Paleozoic Gorny Altai and the Mesozoic-Cenozoic of Japan
T. Watanabe, M.M. Buslov, and S. Koitabashi   169

PREFACE

The Eurasian continent is a landmass that contains numerous orogenic belts of different ages and origins. The geologic map of Eurasia is like a patchwork quilt because the continent consists of many irregular blocks welded together, and it contains imbricated ophiolites and folded nappe-belts within the suture zones of accretion. These features are evidence that Eurasia is a composite continent amalgamated from many continents and microcontinents of Precambrian to Mesozoic in age. The understanding of this mosaic has become more evident in recent times by application of the plate tectonic theory allowing us to make preliminary paleogeographic reconstructions of this area. It is apparent that some of these continental blocks have traveled great distances across ancient oceans, now vanished, until they eventually collided together. Other blocks in this mosaic may be local and have undergone only rotation or post amalgamation translation along major strike-slip faults. The Paleozoic orogenic belts bordering these blocks occupy vast areas of central Asia and contain the record of accretion and amalgamation within in the Paleoasian Ocean.

This book contains new information on attempts to reconstruct Paleoasian Ocean as a result of cooperative studies under the sponsorship of the International Geological Correlation Program (IGCP). This project called: "Reconstruction of the Paleoasian Ocean (Project #283) was initiated in 1989 and the main goals were to collect and evaluate petrologic and tectonic data within specific terranes so as to understand the geodynamic processes leading to the amalgamation of Asia. Paleomagnetic data from individual terranes were obtained to provide new data for paleogeographic reconstructions. Compilation of tectonic, terrane, and metamorphic maps along strips have been completed and new seismic refraction and reflection traverses have also been completed or initiated.

The significant aspect of this project is that we have a working group that consists of geologists from Japan, China, Russia, USA and the Peoples Republic of Mongolia. Geologic maps of various scales had been published in these different countries, but this project has now produced a multinational geodynamic map. The geodynamic maps encompasses ideas related to terrane maps but will show actual distribution of magmatic, sedimentary, and metamorphic rock assemblages that are indicative of plate boundaries. These rock assemblages will be classified as: (1) divergent, (2) convergent, (3) intraplate. Main suture zones and the outlines of terranes are drawn on the maps and their boundaries may be strike-slip faults (transform) or thrust faults. These collated maps will produce new approaches to geodynamic interpretations which can assist in exploration for hydrocarbons and metallic deposits.

In the process of this project we have had three separate meetings in order to exchange ideas and data. Associated field trips in Northwestern China, Western Mongolia, Southern Siberia and Eastern China near Shenyang has provided multi-national interpretations that have produced some surprising new revelations. The most recent meeting of IGCP Project 283 took place in Kyoto, Japan where a symposium entitled "Reconstruction of the Paleoasian Ocean" was convened. The papers contained in this volume represent an important sampling of our international cooperation and the progress we have made in reconstructing the Paleo-Asian Ocean. These papers represent vital geologic discussions for interpreting the geodynamic maps being produced by Project 283.

Editor- Professor *ROBERT G. COLEMAN, Geology Department, Stanford University, Stanford, California, 94305 USA*

The remnants of the Paleo-Asian Ocean within Central Kazakhstan

A.S. YAKUBCHUK[1] and K.E. DEGTYAREV[2]

[1]*Geological Faculty, Moscow State University, Lenin Hills, Moscow, 119899 Russia*
[2]*Geological Institute, Russian Academy of Sciences, Pyzhevskii per. 7, Moscow, 109017 Russia*

ABSTRACT - The remnants of the Paleo-Asian ocean belong to the Caledonian and Variscan accretionary systems of Central Kazakhstan. Caledonain systems composing the western and eastern domains of Central Kazakhstan contain mainly continental and arc terranes, ophiolite sutures and accretionary complexes with dislocated flysch units. Three stages of Caledonian evolution are distinguished according to time-space distribution of terranes and sutures: (1) Vendian-Early Cambrian pre-island arc stage when rift complexes and ophiolites are formed during destruction of the 1100 Ma continental crust; (2) Cambrian-Middle Ordovician stage, when three synchronous pairs of arc-back arc structures were formed during successive developments; (3) Caradocian-Early Silurian non-ophiolite stage characterized by predominant accretional-collisional processes corresponding to amalgamation of terranes within the Western domain (Taconian Kokchetav-North Tianshan system and Early Silurian Erementau-Chu-Ili system) and the Eastern domain (outer Early Silurian Boshekul-Chinghiz system) which were juxtaposed by a possible transform-like fault. So only the inner part of Central Kazakhstan (Junggar-Balkhash system) was preserved as a relict basin surrounded by earlier amalgamated terranes. Total re-organization of the structural plan occurred after this stage and orogenic (Andean type) volcanism took place along the boundary of the Relict basin and the previously accreted terranes. Intermediate latest Caledonian stage (Middle Devonian) is characterized by coexistence of accretionary within the northwestern part of Junggar-Balkhash system. Commencement of differentiated volcanism within the new Late Paleozoic volcanic belt disrupting the Junggar-Balkhash system into two parts with ophiolite formation on its southeastern part marking the beginning of the Variscan epoch. Further evolution took place during the successive amalgamation of terranes within the southeastern part of the system during the Late Visean and Middle Carboniferous. The last amalgamation occurred synchronously with collision within the Ural-South Tianshan and Irtysh-Zaissan megasutures that formed during the final stages of accretion of the Kazakhstan block to Northern Eurasia contemporaneous with a Late Paleozoic granitization of the whole area. Early Mesozoic strike-slip dislocations were developed during gradual cooling of the Kazakhstan block. The proposed model is supported by paleomagnetic data.

Key Words - Kazakhstan, tectonics, ophiolites, sutures, paleogeography

INTRODUCTION

The problems of the Caledonian-Variscan evolution of Central Kazakhstan have been discussed since the 1930's. The main reason for such continuing discussion is the role of ophiolites and their basalt-chert suites.

Traditionally the evolution of this region has been based on the older geosynclinal theory. According to this view the present structure of Kazakhstan represents the conjunction of the Caledonian eugeosynclinal and miogeosynclinal anticlinorium and synclinorium structures composing the basement of the Kazakhstan block (median massif) which is overlain by orogenic volcanic and molasse formations [7, 5, 24]. Since the 1970's this point of view was also re-established from the mobilistic approach [2, 3, 21, 38]. These authors regarded the structure of the Central Kazakhstan mosaic fold region to have formed as a result of its evolution through oceanic (Vendian- Early Cambrian), transitional or island-arc (Middle Cambrian-Silurian), and continental (Devonian-Quaternary) stages. Such an evolutionary succession was based on ideas about ophiolites and their chert cover as the most ancient complexes of the Caledonian due to the absence of any macrofaunal remnants. The rare K-Ar data on highly altered rocks (gabbros and basalts) supported ideas about the Late Proterozoic age of the ophiolites. The concept of a former great Paleo-Asian ocean or perhaps a Kazakhstan-Mongolian ocean was

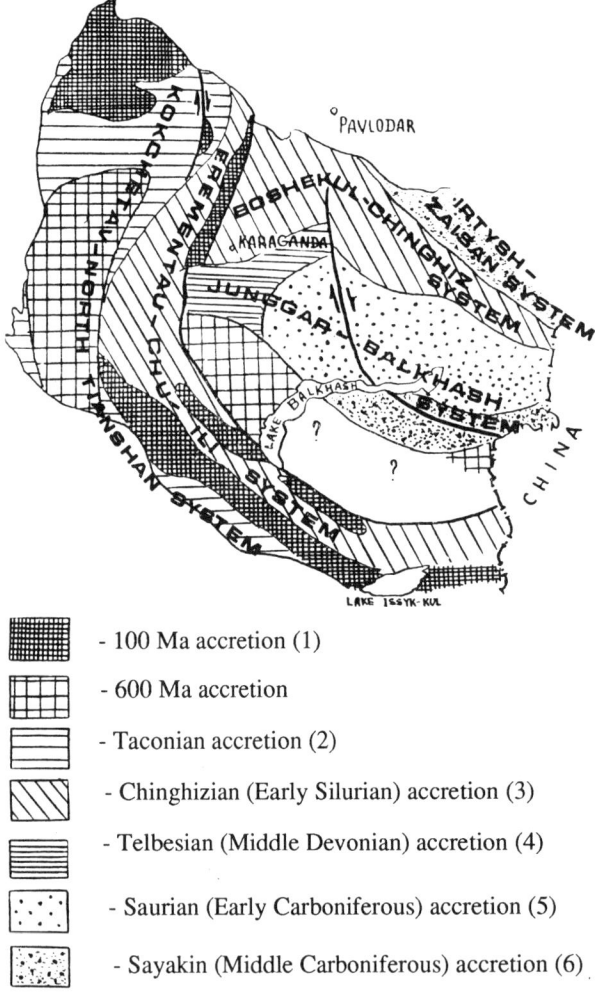

- 100 Ma accretion (1)
- 600 Ma accretion
- Taconian accretion (2)
- Chinghizian (Early Silurian) accretion (3)
- Telbesian (Middle Devonian) accretion (4)
- Saurian (Early Carboniferous) accretion (5)
- Sayakin (Middle Carboniferous) accretion (6)

Figure 1. Sketch map of polychronous accretional structure of Central Kazakhstan (slightly modified after Peive and Mossakovskii [21]; Zaitsev [34]. Only pre-orogenic complexes are shown. Numbers in brackests correspond to numbers of accretional events (see Fig. 2)

formulated in the above cited papers. Furthermore the process of lithosphere thickening by imbrication produced the upper granite crust. Tectonic structures of the pre-Devonian complexes were subdivided into Western and Eastern domains (geoblocks)[3, 21] distinguished by the presence of Riphean continental massifs and Caledonian ophiolites. A line going along the eastern rim of Niyaz and Aktau-Mointy Precambrian massifs was selected as the border of these domains due to the prevailing Riphean blocks in the Western domain and their absence in the Eastern domain where ophiolites and island-arc volcanics are widely developed.

Zaitsev [34, 35] modernized simultaneously the classical ideas of geosynclinal evolution of Central Kazakhstan, which he regarded this as a structure of concentric-zonal type where successive centripetal migration of folding events from the outer parts in towards the inner parts took place (Fig. 1). Zaitsev accounts for the Early Paleozoic ophiolites and chert-basalt complexes from one side and the differentiated volcanics and flysch from the other side as being synchronous in many cases. This point of view was based on the age of the chert-basalt complexes containing Arenigian conodonts [12]. These complexes were previously regarded as Vendian (Latest Proterozoic).

Later analysis of the data was done by Kheraskova [15]. The synchroneity of the calc-alkaline and chert-basalt sequences (upper part of the ophiolites) was interpreted as an indication of an Early Paleozoic arc-back arc basin and not a mid-ocean basin. This interpretation was supported also by petrochemistry of the various volcanic units [16]. The remnants of the real Kazakhstan-Mongol or Paleo-Asian ocean is probably found only within the Altai-Sayan region to the north. However, not all of the paleontological evidence has not yet been obtained for all of these basalt-chert units in the ophiolites.

Most of these age gaps were filled at the end of the 1980's [18, 6] which gave us the idea to consider these as remnants of arc or back-arc systems and not a single basin [31].

The aim of this paper is to consider the evolution of the Paleo-Asian oceanic fragments in Central Kazakhstan through the Late Proterozic to Paleozoic (Caledonian to Variscan epochs) using for the first time in this region a terrane-suture terminology.

TECTONIC STRUCTURE OF CENTRAL KAZAKHSTAN

The triangular Kazakhstan block occupies a central part of the Ural-Mongolian fold belt. The block is surrounded by Irtysh-Zaissan and Ural-South Tianshan Variscan megasutures. Outer parts of Kazakhstan block were accreted during Caledonian tectonic events and range in age from Late Ordovician up to the Middle Devonian; the smaller inner part of this area was accreted during Variscan times (Early Carboniferous to Middle Carboniferous) synchronously with events in the surrounding megasutures.

It is obvious that modern tectonic structure of Central Kazakhstan is the result of multiple deformations. According to our point of view there are two tendencies in the younging of the accretional-collisional events where the Eastern domain reveals the younging from north to south and the Western domain shows a younging from west to east (regardless of the earlier Precambrian events)(Fig.1). We regard these domains as two major parts of the pre-Late Llandoverian (Early Silurian) structure of the Kazakhstan block. Two domains are conjugated along the boundary which is interpreted as the trace of the former pre-Late Llandoverian (or Late Llandoverian strike-slip or perhaps even a transform fault. Establishing lithologies of both domains reveals the presence of Riphean continental terranes, ophiolitic sutures and Early Paleozoic arc terranes in both domains. However, the dominance of continental terranes in the west domain and ophiolites in the east domain is evident (Fig. 2 & 3). The presence of these contrasting complexes in the opposite domains marks each domain main structural boundary along which facing structural directions abut at high angles.

Previously designated anticlinorium and synclinorium structures now reveal their allochthonous nature and provide reasons to use formational zones or belts for these regional structures [15,21,31]. However, such terms also used for orogenic structures [21] that have undergone different types of evolution and deformation. Therefore the terrane-suture terminology is most useful in designating basement that consists of ancient
continental blocks, island arcs, flysch suites or ophiolites. Orogenic complexes will be regarded here as later volcanic belts and molasse depressions.

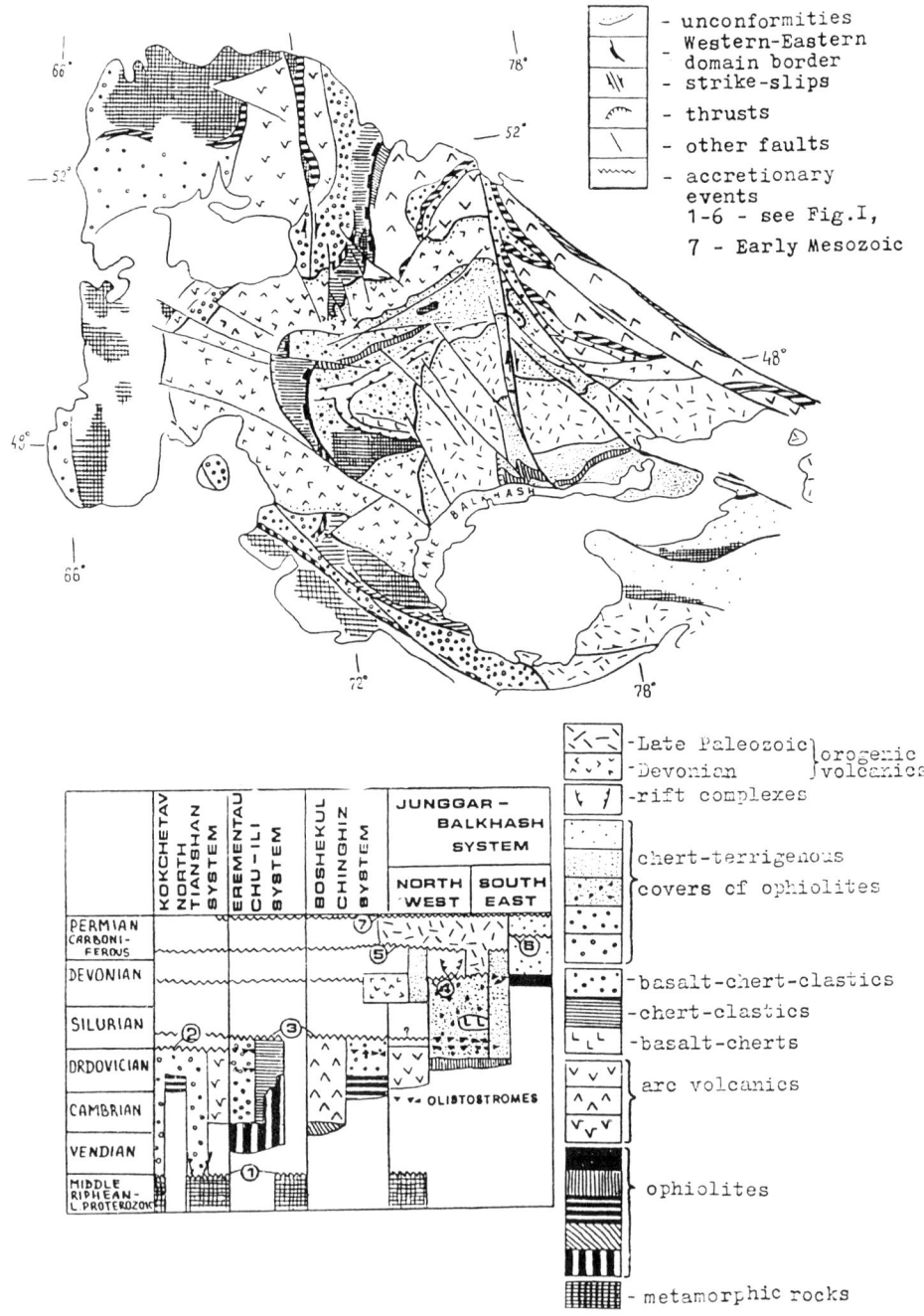

FIGURE 2 Modern stucture and time-space distribution of major Pre-cambrian-Paleozoic complexes.

CALEDONIAN COMPLEXES

Inner structures of these domains form as a result of amalgamation of continental (1100 Ma) and Early Paleozoic island-arc terranes during earlier tectonic events. Ophiolitic and non-ophiolitic (where chert-basalt complexes are present only in East Erementau and Agadyr) sutures divide the different terranes. Thick accretionary complexes, such as the Selety-Konskii and Zhaman-Sarysuiskii, are developed and are composed mainly of thick flysch and olistostrome units.

Western Domain
Consisting of Kokchetav-North Tianshan and Erementau-Chu-Ili accretionary-collisional systems.

The Kokchetav-North Tianshan system is composed of the Kokchetav and Ulutau-Chu-Ili continental terranes separated by Chistopol ophiolitic suture. The northern and western rims of Ulutau-Chu-Ili composite terrane are overprinted by Late Riphean-Early Paleozoic Baikonur-Jarkaingach rift. The eastern edge of the system is marked by Middle Cambrian-Ordovician Stepnyak arc terrane. Aksu-Iradyr and Chistopol sutures represent disrupted former single suture. The main accretionary event took place in the Late Ordovician (Taconian event) [34].

Terrane stratigraphy. Kokchetav and Ulutau-Chu-Ili terranes reveal metamorphic basement composed of Early Proterozoic rocks and in Kokchetav ultra high pressure metamorphic minerals are known [25]. The cover of the basement is shallow-marine terrigenous quartz-arkose clastics (Vendian) which change gradually to shallow marine carbonates (Cambrian-Early Ordovician) followed by terrigenous turbidites (Middle-Late Ordovician). The Baikonour Jarkainagach rift system consists of Late Riphean and Vendian alkaline basalts and tilloids at the base. Cambrian terrigenous-siliceous-tuffaceous sequences and Late Cambrian-Early Ordovician carbonate-shale-chert sequences changed to turbidites in the Middle-Late Ordovician. Stepnyak island-arc terrane consists of Middle Cambrian-Ordovician Calc-alkaline volcanics and associated tuffaceous sediments gradually differeniate with time. Tremadocian marianite-boninite volcanics are known in the northern part of the terrane[26]. The eastward decrease of total alkalinity is established for the Ordovician calc-alkaline volcanics and the Late Ordovician granodiorites of the Stepnyak terrane indicating that the former subduction zone dips westward from the modern Erementau-Chu-Ili system and suggests a back-arc position for ophiolites from the Chistopol and Aksu-Iradyr sutures composed of melange-type ophiolites. However in some cases the upper parts of the ophiolite sections are preserved [26] with isotropic gabbro, sheeted dyke complex and pillow-basalts of Arenigian age according to conodonts [14]. So these ophiolites were formed immediately after marianite-boninite volcanism in the Tepnyak terrane. However, there are some scientists who consider these volcanics to be Early Cambrian basalts [6, 27] . Blocks in serpentinite melange have dunite-harzburgite composition of residual peridotites associated with dunite-wehrlite-lherzolite-pyroxenite-gabbro-norite belonging to the lower sequence of the cumulates [20, 26]. Petrochemistry of the basalts indicates a normal tholeiite affinity [26].

Ermentau-Chu-Ili system is 1500 km long and its inner structure is different from the Kokchetav-North Tianshan system. It is divided into two main parts separated by the Jalair-Naiman ophiolite suture and its northern part is masked by cover. The western part is the giant Selety-Konskii accretionary complex of Central Kazakhstan and the eastern part is a chain of continental terranes (Niyaz and Buruntau) that are overthrusted by complexly imbricated basalt-chert-terrigenous nappes separated by lenses of serpentinite melange within the accretionary complex. The direction of thrusting and imbrication is not clear but appears to be eastward. The thrusting and imbrication occured twice: (1) In
the Middle Ordovician where nappes are underlain by Middle Ordovician olistostromes [6]. (2) In the Early Silurian when the main accretional event took place [34].

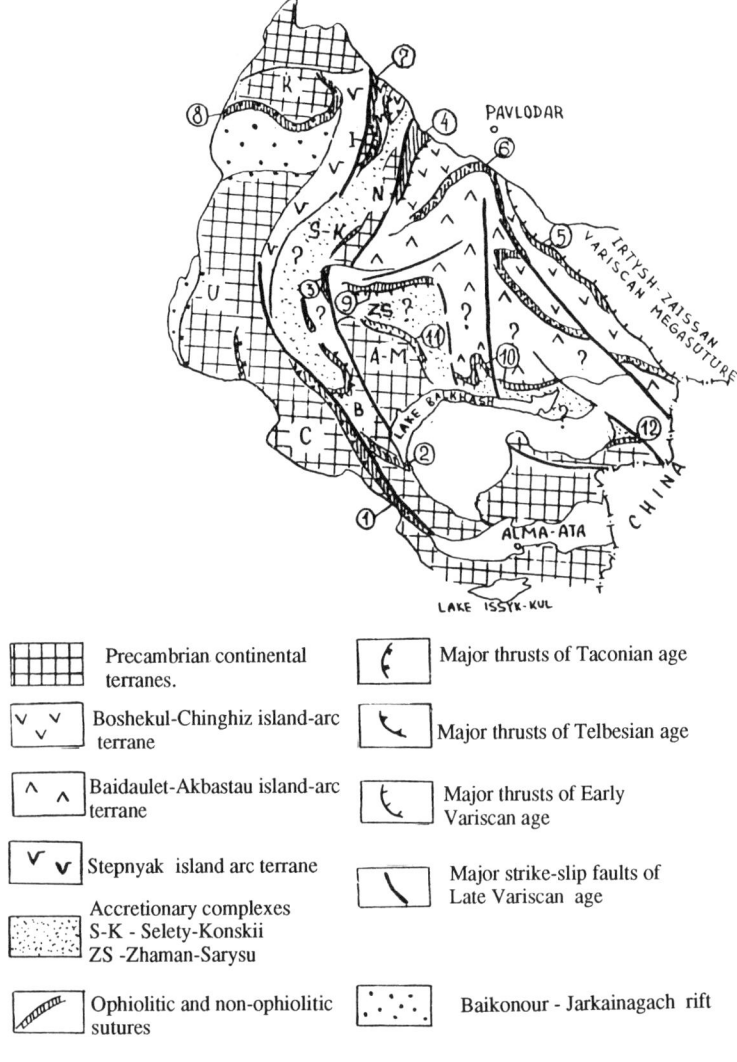

Figure 3 Terrane structure of Central Kazakhstan. Letters: K-Kokchetav terrane, I-Ishkeolmes terrane, N-Niyaz terrane, U-Ulutau terrane, B-Buruntau terrane, C-Chu-Ili terrane, A-M-Aktau-Mointy terrane. Numbers in circles: 1- Jalair-Naiman, 2- Sarytun, 3- Satybad, 4- Boshekul, 5- Arkalyk, 6- Maikain-Balkybek, 7- Aksu-Iradyr, 8- Chistopol, 9- Tekturmas, 10- North Balkhash, 11- Agadyr, 12- North Junggar sutures.

Terrane stratigraphy The Jalair-Naiman ophiolitic suture and its possible northern extension contain numerous melange-type ophiolites. Comparatively well preserved "oceanic sequences" consisting of conformably layered depleted dunite and harzburgite overlain by wehrlite, lherzolite, pyroxenite, hornblende-gabbro, and norite cumulates are found within the Jalair-Naiman suture [8]. Composition of the cumulates and their petrochemistry reveal a subalkaline affinity. The age of ophiolites is Vendian to Early Cambrian estimated by their geologic position. However, there are Early Cambrian basalts overlain in some slices by Middle(?)-Late Cambrian-Llanvirnian conodont-bearing cherts and Early Ordovician basalts (with conodonts) overlain by conodont-bearing Arenigian cherts. Ophiolites and their fragments indicate a post-Ordovician gliding age [11]. These facts show that ophiolites were formed during a long term period (both before and synchronously with arc volcanism in the Stepnyak terrane).

The Selety-Konskii accretionary complex has a different composition and can be divided into eastern and western parts along an ophiolite suture. The western part contains Late Cambrian-Ordovician terrigenous clastics which overlie (?) the ophiolites of the Jalair-Naiman suture and only at the Middle Ordovician times do these andesite units contain clastic sequences. The eastern part of the accretionary complex is overthrust eastward onto continental terranes and underlain by Middle Ordovician (Caradocian or Llandeilian) olistostrome bodies of Late Ordovician level.

Continental terranes have a different stratigraphy. The Buruntau terrane is divided by the Sarytum rift which is composed of Vendian-Early Cambrian dolomite-carbonate unit at the base alternating with Early Cambrian terrigenous clastics and shales containing interlayered alkaline basalts. The final stage of the Sarytum rift evolution is marked by a Late Cambrian-Early Ordovician gabbro and plagiogranite bodies. The Buruntau terrane itself has an unexposed Precambrian metamorphic basement. The cover of the Buruntau terrane is composed of Late Cambrian-Llanvirnian cherts which are overlain by Middle Ordovician terrigenous clastics. The Niyaz continental terrane has a cover of Vendian arenites, younger layers are represented only by Middle Ordovician olistostromes as mentioned above

Eastern Domain
The eastern domain is composed of Caledonian Boshekul-Chingiz and the Caledonian-Variscan Junggar-Balkhash accretionary systems.

Boshekul-Chingiz system The inner structure consists of the Boshekul-Chinghiz arc terrane and the northern rim of the Baidaulet-Chingiz arc terrane is separated by the Maikain-Baikybek ophiolitic suture. The northern boundary of the system is the Boshekul and Arkalyk sutures with Arkalyk taking an intermediate position among the Boshekul-Chingiz and Irtysh-Zaissan systems. The main accretion took place during the Early Silurian (Chingiz event), however, synsedimentation continued into the Late Ordovician [6].

Terrane stratigraphy. Boshekul-Chingiz arc terrane reveals three stages of volcanism: (1) Early-Middle Cambrian, (2) Late Cambrian-Early Ordovician, and (3) Middle-Late Ordovician. This terrane contains Early Cambrian Boshekul ophiolites at the base [6,17]. These ophiolites have bi-modal composition due to the presence of tholeiitic basalts (50 percent) and rhyolites (50 percent) indicating an ensimatic nature for this arc terrane [21]. Cumulates consist of olivine-orthopyroxene-clinopyroxene-(hornblende)-plagioclase sequences suggesting an origin of these ophiolites during the initial stages of arc magmatism. Middle Cambrian-Early Ordovician (with interruption at the beginning of the Late Cambrian) calc-alkaline and alkaline volcanics fractionate upward, typical for island-arc evolution. Middle-Late Ordovician arc volcanics repeat this petrochemical sequence [21]. Almost the same situation is observed within the Baidaulet-Akbastau arc terrane, but here the volcanic activity took place only during the Early-Middle Ordovician. Sialic basement is suggested for this arc terrane due to the presence of metamorphic rock of undetermined age at the base of the volcanic pile (the Matak horst) and numerous granite bodies [34]. The Maikain-Balkybek suture contains Late Cambrian-Early Ordovician melange-type dismembered ophiolite and in two cases complete ophiolite sequences [32]. Covering cherts and ash-bearing cherts reveal gliding age by conodonts at their base as will as within the ophiolite themselves [33]. Ophiolites change their petrochemical signatures from the earliest subalkaline type to arc-tholeiites and normal tholeiites in the latest stages reflecting a gradual opening of the paleobasin [19]. The cumulate mineralogy reveals olivine-orthopyroxene-clinopyroxene-hornblende-plagioclase assemblage typical for supra-subduction zone ophiolites [13, 23]. Arkalyk suture contains only serpentinite melange lenses and blocks of Cambrian basalt in olistostromes. Younger Ordovician and Devonian ophiolites are known from the adjacent Irtysh-Zaissan system. Older Vendian-Early Cambrian basalt complexes are also known from the East Erementau zone of the Boshekul suture [6]. Terrigenous clastics cover both chert suites of ophiolites and arc volcanics during the intermediate stage of arc terrane evolution. These intermediate stages of Boshekul-Chingiz arc terrane are synchronized with synsedimentation and olistostrome development [21]. So, ophiolites of Maikain-Balkybek suture are synchronous with volcanics of Boshekul-Chinghiz terrane and have to be regarded as remnants of intra-arc basin. Ophiolite remnants of Arkalyk suture and East Erementau in regard to the adjacent Irtysh-Zaissan system should be considered to be remnants of the Paleo-Asian ocean basin because of their wide age spectrum.

Junggar-Balkhash system. The inner structure is composed of Late-Caledonian-Variscan accretionary complexes and the Aktau-Mointy continental terrane. Accretionary complexes range in age from Middle Devonian to Late Visean and Middle Carboniferous [34]. The southern part of the Baidaulet-Akbastau arc terrane is included in the Junggar-Balkhash system. Accretionary complexes contain Tekturmas and North Balkhash sutures containing Middle Ordovician ophiolites. Probable Precambrian basement lies to the south of Tekturmas suture [31, 33, 34], supported by the presence of large granite bodies derived from ensialic continental crust rather than ensimatic greywackes.

Terrane stratigraphy. Aktau-Mointy continental terrane has 1100 Ma basement overlain by Vendian-Ordovician carbonate-terrigenous suites. Tekturmas and North Balkhash sutures reveal the synchroniety of Llanvirnian-Llandeilian ophiolites in both [37, 31]. The ophiolites have a gliding age synchronous with volcanics of the Baidaulet-Akbastau arc terrane. Olivine-clinopyroxene-plagioclase cumulates indicate a primitive subalkaline magmatic composition similar to Llanvirnian ophioilites. The arc-tholeiite affinity of the Llandeilian ophiolites indicates that they formed in a supra-subduction zone. Llanvirnian-Caradocian chert and ash-bearing chert covering the basalts are in turn overlain by thick Silurian-Early Devonian flysch complexes in the northwestern part and by Silurian-Early Carboniferous clastics in the southeastern part. Agadyr suture of the northwestern part is also characterized by the presence of Early Silurian basalt-chert suites synchronous with olistostromes of the Tekturmas suture. The middle Devonian (Telbessian) accretion completes the back-arc evolution of the northwestern part of the system. This stage is characterized by the sedimentation of Middle Devonian olistostromes in the southeast. This accretionary event took place only during the Late Visean.

CALEDONIAN EVOLUTION

Using the terrane stratigraphy it is possible to select sub-synchronous two pairs of arc terranes and ophiolitic sutures:(1) Stepnyak arc terrane and the Chistopol-Aksu-Iradyr sutures; (2) Boshekul-Chingiz arc terrane and the sutures of Tekturmas and North Balkhash. We regard these pairs as former arc-back-arc basins. Continental terranes with their cover were developed in close connection to adjacent structures. They manifest carbonate formations during stages of basin opening and during stages of basin closing terrigenous clastics occur on the continental crust Two ophiolitic sutures (Jalair-Naiman and Arkalyk with regard to the adjacent Irtysh-Zaissan system) reveal a wide age spectrum of ophiolite formation without apparent dependence on arc magmatism. We regard these ophiolites as indicators of an autonomous spreading basin (Jalair-Naiman) or the remnants of a larger true Paleo-Asian ocean (East Erementau, Arkalyk, Irtysh-Zaissan)

These relationships determine three stages of Caledonian evolution of Central Kazakhstan: (1) pre-island arc Vendian-Early Cambrian Stage, when destruction and extension of the ancient (1100 Ma) continental crust took place with formation of ophiolites and rift complexes; (2) syn-island arc Cambrian-Middle Ordovician stage, when island arc magmatism and back-arc spreading developed and supra-subduction

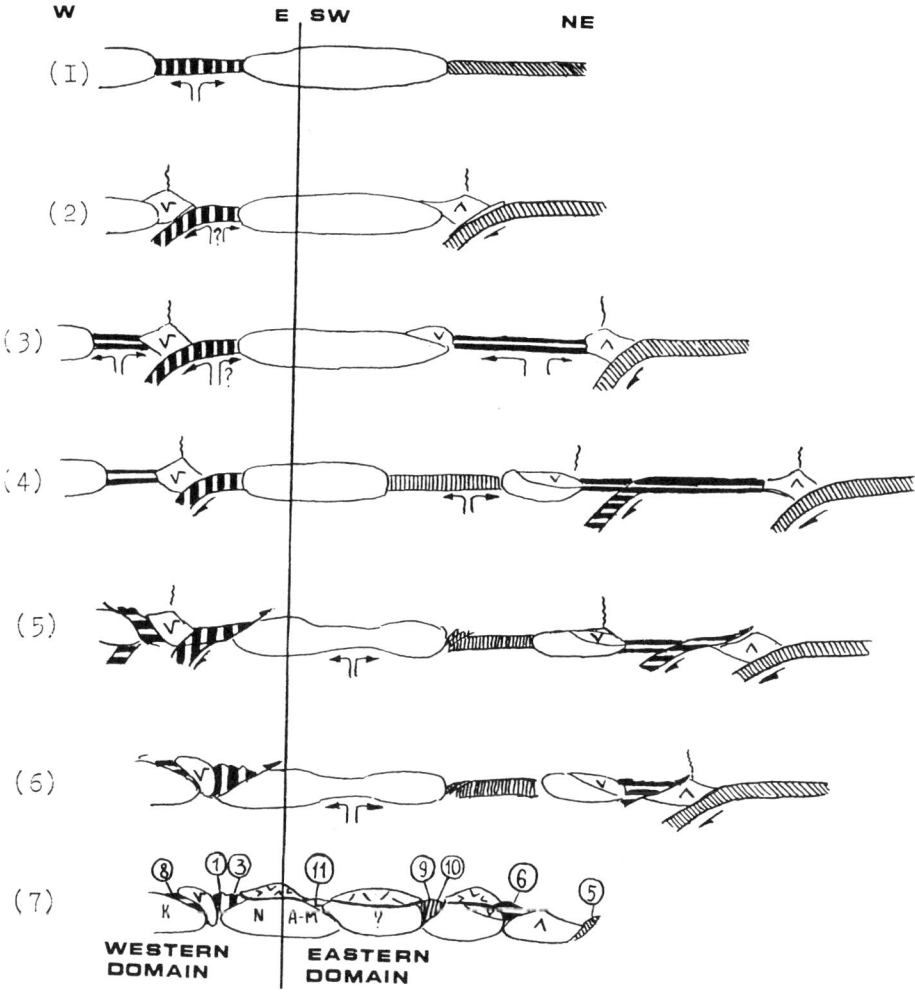

Figure 4 Schemes of Early Paleozoic evolution of Central Kazakhstan. Arrows show occurence of spreading and subduction. Numbers in brackets: (1)- Vendian-Early Cambrian, (2)- Early-Middle Cambrian, (3)- Late Cambrian-Early Ordovician, (4)- Middle Ordovician (without Caradocian), (5)- Caradocian-Ashgillian, (6)- Early Silurian, (7)- Modern structures. Other symbols correspond to those shown in Fig. 2 & 3.

ophiolites formed, and (3) Caradocian-Llandoverian stage marks the peak of accretion and collision with interrupted ophiolite formation and only weak rifting manifested in the Agadyr suture. The transition between the last two stages is not sharp beginning diachronously in different basin as a result of their variable openings and their interdependence. The opening in the younger basins is correlated with closing and subduction in the adjacent older basins since the Middle Ordovician. Figure 4 shows our ideas on the Vendian-Ordovician evolution. It is obvious that there are no signs of evident interdependence among Western and Eastern domains until the Late Ordovician. These facts will be discussed in the paleomagnetic section. The idea of such independence is supported by distribution of Archaeocytes which are found only within the Boshekul-Chingiz arc terrane and are completely absent in other terranes where Early Cambrian sedimentary units are present. It is obvious also that short-term development is also characteristic of these back-arc basin such as observed in modern back-arc basins. Furthermore there is no evidence of long-term telescopic spreading evolution within any of these back arc basins. However, southward (towards the continent) younging of supra-subduction zone ophiolites in the Eastern domain is not typical of modern back-arc basins.

VARISCAN COMPLEXES AND THEIR EVOLUTION

Stratigraphy of Variscan complexes.
There are two distinct types of Variscan complexes due to the coexistence of various regimes. Orogenic molasse and volcanic complexes were formed within the Western domain and the Boshekul-Chingiz system since the early Devonian [34] or even since Wenlockian [6] when flysch sedimentation took place in the Junggar-Balkhash system (Fig. 5A). Early Devonian orogenic volcanics reveal a decrease in total alkalinity toward the developing basin of the future Junggar-Balkhash system [29]. But the Middle Devonian is characterized by the coexistence of orogenic volcanism (Devonian), tectonic accretion (Telbessian event) and continuous formation of ophiolites and shale-clastic sedimentation in the southeastern part of the Junggar-Balkhash system (Fig. 5B). Since Famennian-Visean time the same environments existed, but migrated southeastward. Initial formation of the following units took place: Late Paleozoic volcanic belt (orogenic); rifting on Telbessian basement [30]; tectonic accretion originating in the North Balkhash suture accompanied by continuous shale-clastic sedimentation in the southeastermost part of the system (Fig. 5C). The basin evolution of the Junggar-Balkhash system was completed during the Middle Carboniferous (Sayakian) event after which the final granitization of almost the whole Central Kazakhstan took place (Fig. 5D). So, the Variscan evolution of Central Kazakhstan can be regarded as successive closing
of relict back-arc basins in specific association with orogenic volcanism as noted by [34, 39, 22]. The arc-like structure of Middle-Late Paleozoic orogenic belts provides a reason to discuss the former straight alignment of the belts that were deformed during the accretional events [39, 22]. This question may be resolved with paleomagnetic data. However, good paleomagnetic data on Devonian volcanics is not abundant and data on the Ordovician complexes within the Devonian arc do not support this idea.

Late-Early Cambrian history of the already rigid Kazakhstanian block was studied by [9], who considers the strike-slip disruption of the rigid block took place during gradual cooling of the new continental crust after granitization. The effect of such cooling is reflected in a 4-stage evolution of strike-slip faulting (Fig. 6) which became longer and longer during the first three stages according to the law of successive mechanical destruction of a rigid body. All these events took place in a northeast-oriented compressional stress field. The fourth stage occurred under an extensional field of the same orientation [9].

PALEOMAGNETIC DATA

Preliminary paleomagnetic data permit us to judge the former size of basins at the stage of their maximal opening (Fig. 7). The Caledonian evolution became clearer after the recent publication of paleomagnetic data by [28]. The data show, that the modern distribution of structures within the Eastern domain is almost the same as it was during the Ordovician (Fig. 8), so it is thought that the main displacement took place along a . North-South boundary which indicates a total of 1000 km northward displacement of the Western domain relative to the Eastern domain. Paleomagnetic directions show that the Kazakhstanian block rotated $160°$ counter clockwise since the Early Paleozoic when Kazakhstan was near the equator.

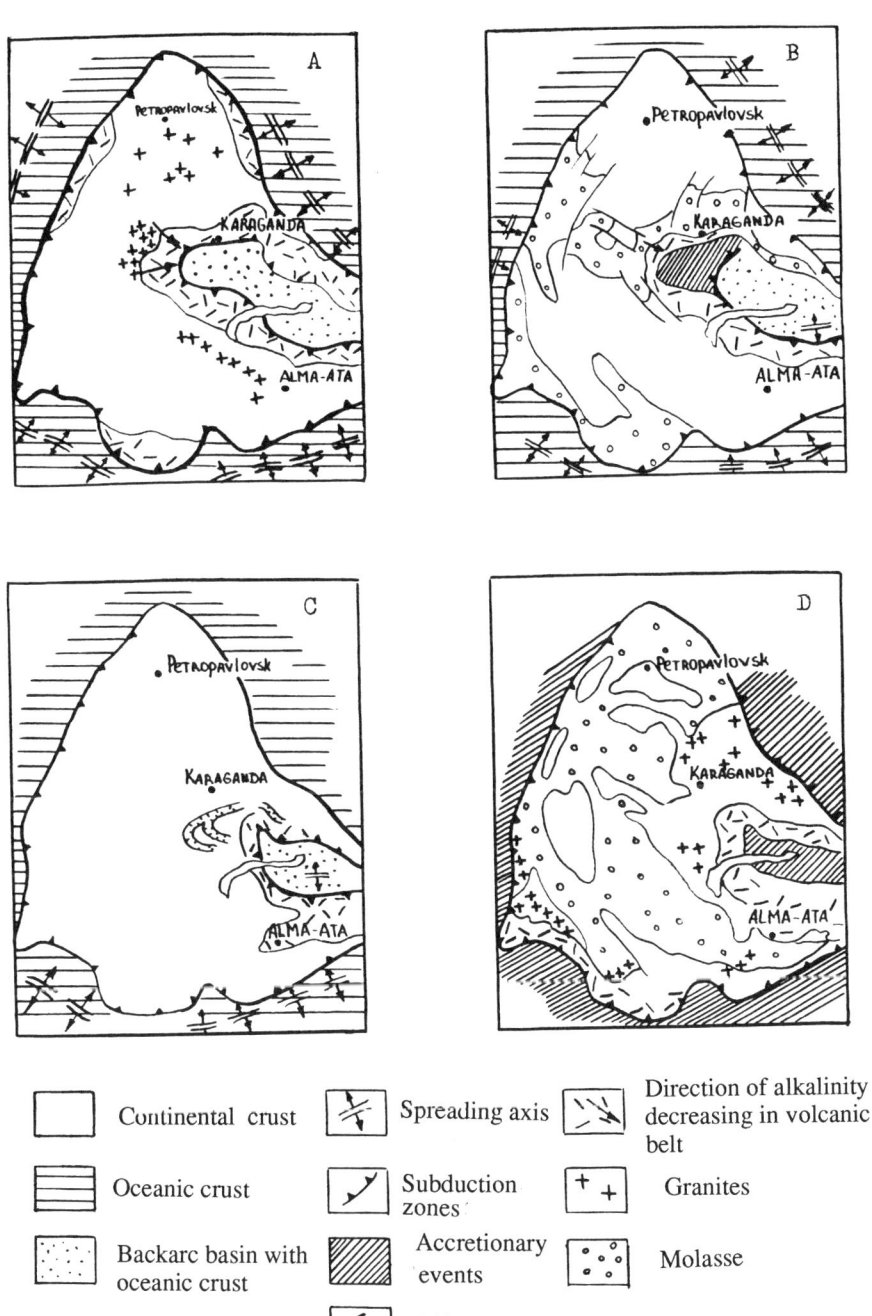

Figure 5 Schemes of Devonian-Permian evolution of Central Kazakhastan (modified after [36].). A - Early Devonian, B- Givetian-Fransian, C- Famennian, D- Middle-Late Carboniferous

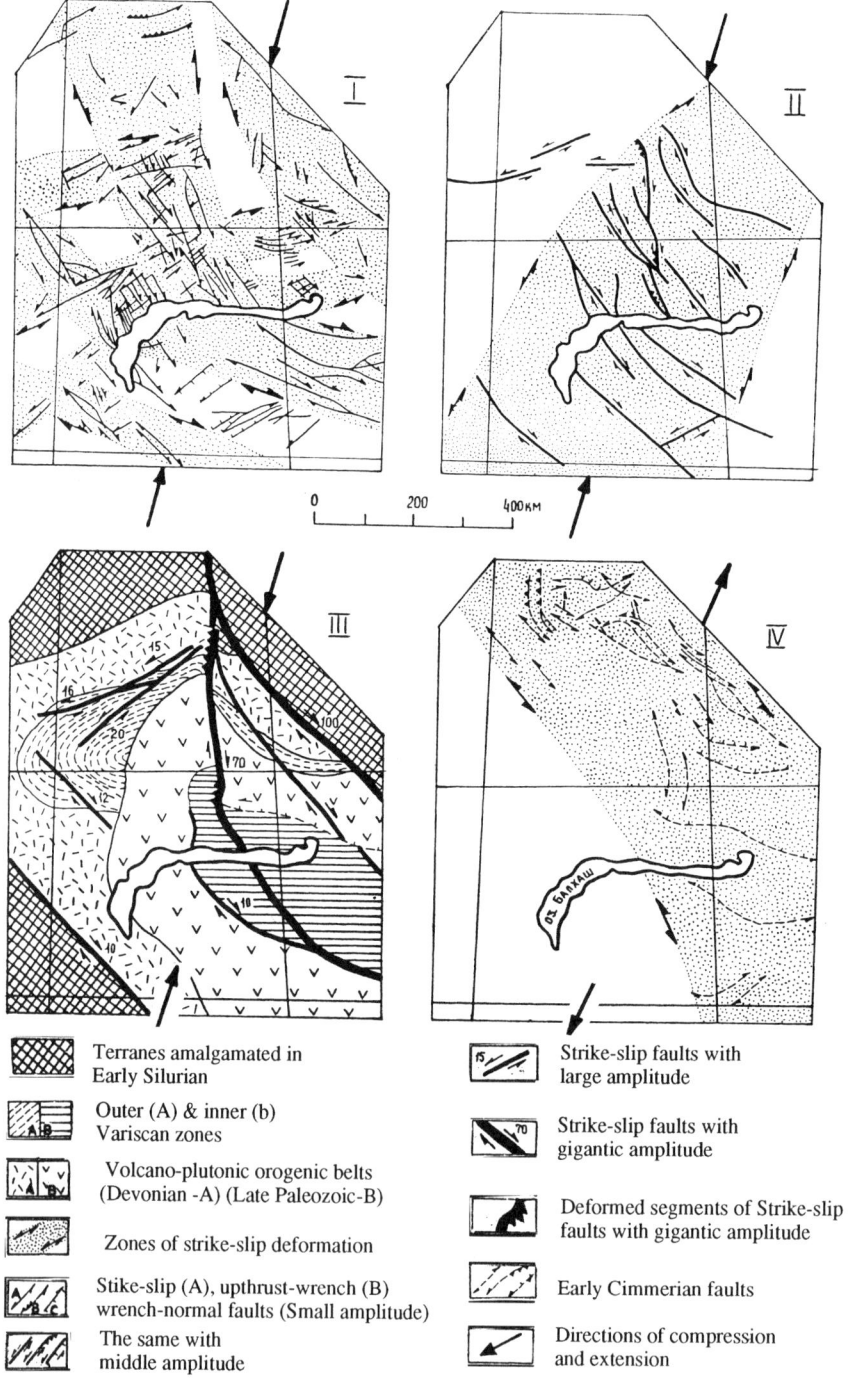

Figure 6 Tectonics of strike-slip faults of four different stages [9].

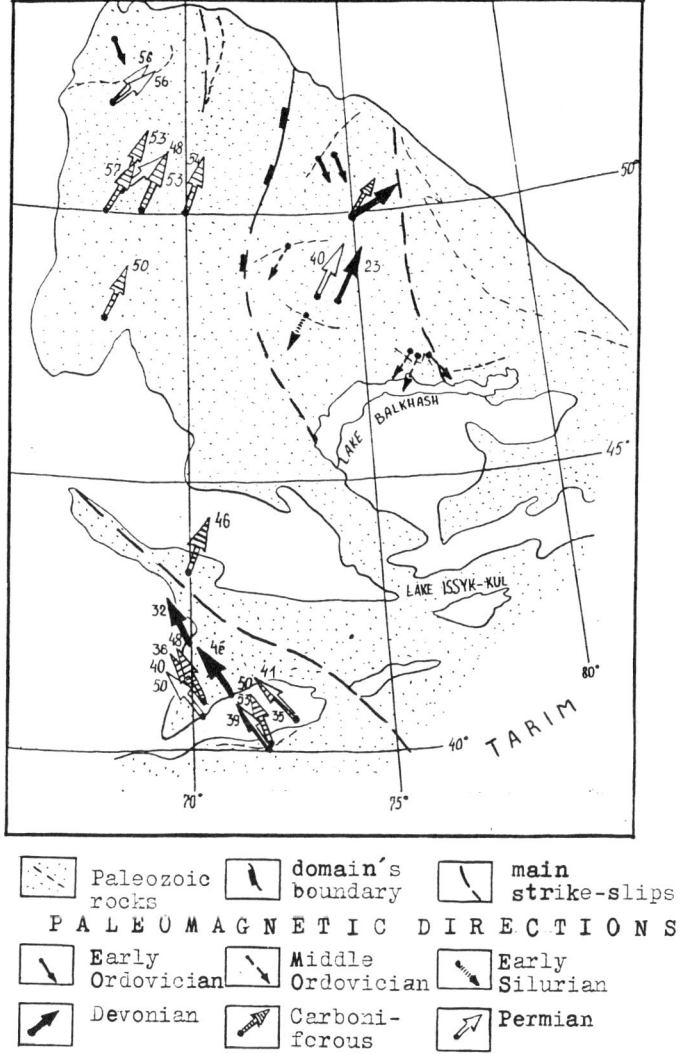

FIGURE 7 Paleomagnetic directions of Central Kazakhstan and adjacent areas [10, 28].

Variscan evolution is more simple during the movement of the larger blocks (Fig. 9) according to [10] but gradual clockwise rotation of Kazakhstania took place during this epoch. Since the Middle Carboniferous Kazakhstania became part of Eurasia welded together with the East European platform, Siberia and Tarim.

These schemes are supported by the uniform orientation of the Devonain-Permian paleomagnetic directions except in the southernmost part of the region where these directions have a different modern orientation as a result of Mesozoic strike-slip faulting. The accretionary nature of these terranes is made clear by the disparate orientations of the Early Paleozoic units.

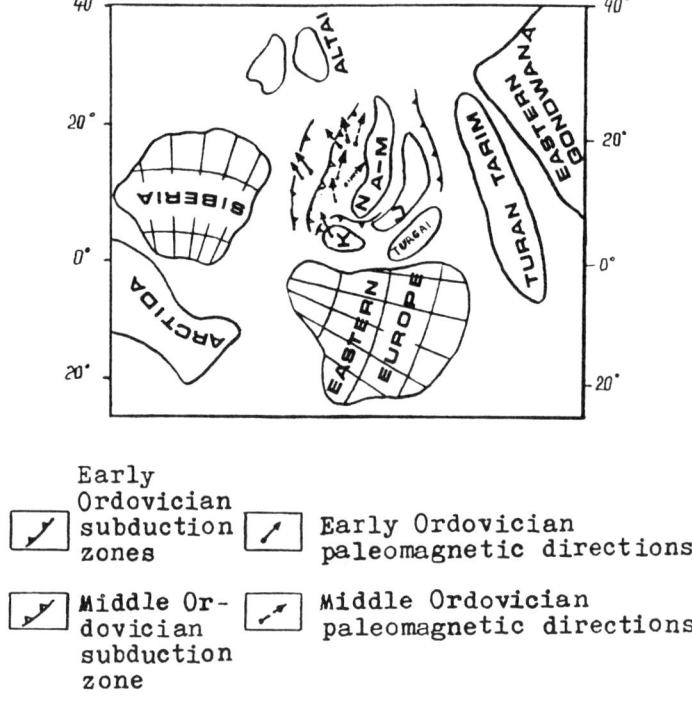

Figure 8 Palinspastic reconstruction of Early-Middle Ordovician [39, 28]. K- Kokchetav terrane, N - Niyaz terrane, A-M - Aktau-Mointy terrane, U- Ulutau terrane.

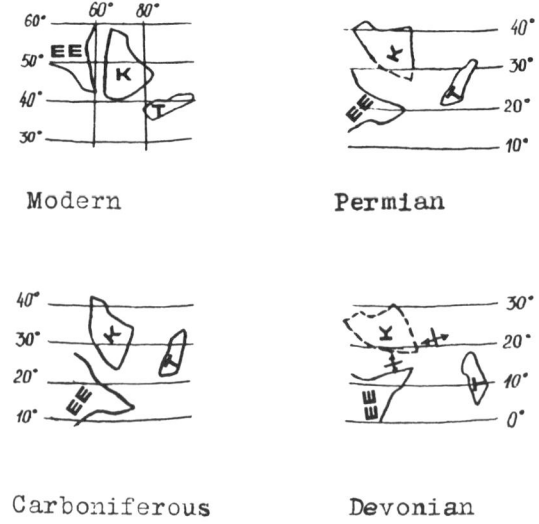

FIGURE 9 Palinspastic reconstruction of Devonian-Permian modified and simplified after [10]. EE - Eastern European platform, T - Tarim block, K - Kazakhstanian block.

DISCUSSION

Evidence on the evolution of Central Kazakhstan shows that the origin of new continental crust resulted from successive amalgamation of continental and arc terranes stitched by ophiolite sutures and accretionary complexes. Central Kazakhstan is not the result of reworking of continental crust with imbricated oceanic lithosphere followed by its granitization. Orogenic complexes lie unconformably on the accretionary structures.

The conjunction of the Indian, Eurasian and Pacific plates is regarded by [4] as an actualistic equivalent of Kazakhstania. Petrochemistry of magmatic rocks and sedimentary formations also show arc-back-arc environments in the Early Paleozoic [15].

Initial relationships among synchronous ophiolite sutures are still not completely clear. For example, Tekturmas and North Balkhash sutures contain synchronous and similar ophiolites, but they have been disrupted by modern structures [1]. If they had been a single stucture before disruption, a giant 200 km strike-slip displacement had to occur during the Middle Devonian, however it is possible that a transform-like junction between the paleobasins may have been present during the Middle Ordovician. The described relationships between the Western and Eastern domains may also have been the result of transform-like movement.

CONCLUSIONS

Central Kazakhstan is characterized by the presence of oceanic and backarc complexes formed in the Caledonian and Variscan epochs.

The Caledonian epoch is subdivided into three stages: (1)The Vendian-Early Cambrian stage with destruction of 1100 Ma continental crust and development of new oceanic crust; (2) Cambrian-Middle Ordovician stage of coexistence with subduction, arc magmatism and backarc spreading in three independent basins; (3) Caradocian-Llandoverian stage of predominately accretional and collisional events without ophiolite formation. Continental terranes existed as a back stop to the backarc basins during their evolution. All stages were reflected by sedimentary cover. It is important to note that the Eastern and Western domains of Central Kazakhstan developed relatively independent and were amalgamated only during the Early Silurian. The Caledonian epoch was completed in Central Kazakhstan by Middle Devonian, however, accretionary complexes of this stage are subordinate to the later Variscan structural style.

The Variscan epoch is characterized by coexistence of orogenic processes on the Caledonian basement within the already accreted Western domain and the Boshekul-Chinghiz system. Here synchronous magmatic accretion is present within the relict back-arc basin of the Junggar-Balkhash area. Additional tectonic accretion of this system stabilized this area during the Middle Carboniferous with successive wide spread Kazakhstanian granitization and synchronous collision with the Siberian and Eastern European cratons.

The Late Variscan-Early Cimmerian stage is characterized by strike-slip disruption of the cooling Kazakhstanian block.

Acknowledgements
We thank sincerely Professor R.G. Coleman, Stanford University for informal interest and stimulation of this papers creation. The first author thanks Professor Roger Laurent, University Laval for fruitful discussion on problems of Central Asian ophiolites during the IGC 1992 meeting in Kyoto.

REFERENCES

1. N.A. Afonichev, V.Y. Koshkin, A.E. Mikhailov, and N.A. Pupyshev. About age of the Urtynzhal series of Central Kazakhstan, *Izvestia Akademii Nauk SSSR, ser. Geol.* **7**, 90-93.(1976) (*in Russian*).

2. R.M. Antoniuk, Oceanic crust of eugeosyncline area of Central Kazakhstan, In: *Tectonics of the Ural-Mongolian fold belt.*, A.L. Yanshin (Ed.), pp. 67-74, Nauka, Moscow. (1974) (*In Russian*)

3. R.M. Antoniuk, G.F. Lyapichev, N.G. Markova, and O.M. Rozen, Oceanic crust of eugeosyncline area of Central Kazakhstan, In: *Tectonics of the Ural-Mongolian fold belt,* A.L. Yanshin (Ed.),. pp. 67-47, Nauka, Moscow. (1977) (*in Russian*).

4. A.V. Avdeev and A.A. Kovalev, *Ophiolites and evolution of the South-Western part of the Ural-Mongolian fold belt.* Moscow University Press, Moscow. (1989) (*in Russian*).

5. A.A. Bogdanov. Tectonic features of the Paleozoieds of Central Kazakhstan and Tienshan (In Russian), *Biul. Moskovskogo Obshchestva Ispytateley Pyrody, otd. geol.* **6**, 8-42.(1965) (*in Russian*).

6. V.I. Borisenok, A.V. Ryazantsev, K.E. Degtyarev, and A.S. Yakubchuk, Paleozoic geodynamics of Central Kazakhstan, In: *Tectonic research and middle-large scale geomapping,* Y.M. Pusharovskii (Ed.), pp. 81-95, Nauka, Moscow. (1989) (*in Russian*).

7. R.A. Borukaev,*The Pre-Paleozoic and Lower Paleozoic of northeastern Central Kazakhstan (Sary-Arka).* Nedra, Moscow. (1955) (*in Russian*).

8. A.G. Burdyniuk, E.S. Kichman, Y.A. Gabov, and V.N. Kokurkin, Basite-ultrabasite formation, In: *Geology of Chu-Ili region,* A.A. Abdulin and others (Ed.), pp. 174-186, Nauka, Moscow. (1980) (*in Russian*).

9. A.F. Chitalin. Late Variscan strike-slip tectonics of Central Kazakhstan, *Vestnik Moskovskogo Universiteta ser. 4, geologia.* **5**, 13-22.(1991) (*in Russian*).

10. A.N. Didenko and D.M. Pecherskii. Paleomagnetism of Middle Paleozoic rocks from ophiolitic complexes, the Alai Range, *Geotektonika.* **4**, 56-68 (1988).

11. N.A. Gerasimova, Stratigraphy of Ordovician of the Atasu anticlinorium In: *Geology of early geosynclinal complexes of Central Kazakhstan,* Y. Zaitsev (Ed.), pp. 53-96, Moscow University press, Moscow. (1985) (*in Russian*).

12. N.M. Gridina and T.V. Mashkova. Conodonts from siliceous-terrigenous suites of the Atasu anticlinorium., *Izvestia Akademii Nauk Kazakh SSR, Ser. Geol.* **6**, 48-55.(1977) (*in Russian*).

13. A. Ishiwatari, Time-space distribution and petrologic diversity of Japanese ophiolites, In: *Ophiolite genesis and evolution of the oceanic lithosphere,* T. Peters, A. Nicolas, and R.G. Coleman (Ed.), pp. 723-743, Kluwer Academic Publishers, Boston. (1991).

14. K.S. Ivanov, V.A. Sakharova, and others. New data on age of volcanic-sliceous units of the Kokchetav massif rim (Northern Kazakhstan), *Doklady Akademii Nauk SSSR.* **301**, 158-163.(1988) (*in Russian*).

15. T.N. Kheraskova, *Vendian-Cambrian formations of the Caledonides of Asia.* Nauka, Moscow. (1986) (*in Russian*).

16. T.N. Kheraskova, M.Z. Novikova, and N.I. Zardiashvili. Specification of composition of early geosynclinal volcanic formations of Central Kazakhstan, *Izvestia Akademii Hauk SSSR, Ser. Geol.* **6**, 47-61 (1979) (*in Russian*).

17. B.F. Khromykh. New data on Vendian-Paleozoic evolution and metallogenesis of the Boshekul ore region , *Izvestia Akademii Nauk Kazak SSR, ser. geol.* **6**, 20-34.(1986) (*in Russian*).

18. L.A. Kurkovskaya, Complexes of conodonts from siliceous and volcanic-siliceous units, In: *Geology of early geosynclinal complexes of Central Kazakhstan,* Y. Zaitsev (Ed.),. pp. 164-176, Moscow University Press, Moscow. (1985) (*in Russian*).

19. I.E. Kuznetsov, M.Z. Novikova, and A.S. Yakubchuk, Evolution of magmatism of ophiolite zones of Central Kazakhstan, In: *Geodynamic environment of origin, geochemical aspects of genesis of basites and ultrbasites*, O.M. Glazunov et al (Ed.), pp. 6-10, Irkutsk. (1990) (*in Russian*).

20. N.P. Mikhailov, V.N. Moskaleva, et al, *Petrography of Central Kazakhstan*. Vol. 2., Nedra, Moscow. (1971) (*in Russian*).

21. A.V. Peive and A.A. Mossakovskii, Tectonics of Kazakhstan. Explanatory note to the tectonic map of Eastern Kazakhstan (1:2,500,000). Nauka: Moscow. 1982) (*in Russian*).

22. A.M.C. Sengor. The Paleo-Tethyan suture: A line of demarcation between two fundamentally different architectural styles in the structure of Asia, *Island Arc*. 1, 78-91.(1992).

23. G. Serri and M. Saitta. Fractionation trends of the gabbroic complexes from high-Ti and low-Ti ophiolites and the crust of the major oceanic basins: a comparison, *Ophioliti*. 5, 241-264.(1980).

24. N.S. Shatskii and A.A. Bogdanov, Tectonic map of the USSR (1:5, 000, 000) . Moscow. 1954) (*in Russian*).

25. V.S. Shatsky, E. Jagoutz, O.A. Kozmenko, and N.V. Sobolev. Rare-earth and trace element variation in metamorphic ultra-high pressure rocks from Kokchetav massif (Northern Kazakhstan, USSR), *Abstracts of 29th IGC, Kyoto*. 1, 599.(1992).

26. E.M. Spiridonov, Geosynclinal basite complexes of northeast Central Kazakhstan, In: *Problems of geology of Central Kazakhstan*, Y. Zaitsev (Ed.), Vol. 1, pp. 102-121, Moscow University Press, Moscow. (1980) (*in Russian*).

27. E.M. Spiridonov, S.P. Sigachev, N.K. Ivshin, O.V. Minervin, L.V. Bulygo, and I.A. Poslavskaya. Specification of Tremadocian island arc complex of Northern Kazakhstan, *Doklady Akademii Nauk SSSR*. 301, 415-420.(1988) (*in Russian*).

28. T.L. Turmanidze, D.V. Grishin, D.M. Pecherskii, and V.G. Stepanets. Paleomagnetism of Ordovician ophiolites of allochthonous massifs of Central Kazakhstan, *Geotektonika*. 4, 54-68.(1991) (*in Russian*).

29. A.P. Uryvaeva and N.P. Chetverikova, Main features of Devonian volcanic series composition of southwestern branch of Kazakhstanian marginal volcanic belt, In: *The problems of geology of Central Kazakhatan*, Y. Zaitsev (Ed.), Vol. 2, pp. 37-58, Moscow University Publishers, Moscow. (1980) (*in Russian*).

30. A.B. Veimarin and E.E. Milanovskii. Famennian rifting in Central Kazakhstan and some other areas. Paper 1, *Biul. Moskovskgogo Obshcestva Ispytalelei Prirody, otd. geol*. 65(5), 55-68.(1990) (*in Russian*).

31. A.S. Yakubchuk. Tectonic setting of ophiolitic zones in the structure of Paleozoide of Central Kazakhstan, *Geotektonika*. 5, 55-68.(1990) (*in Russian*).

32. A.S. Yakubchuk, V.G. Stepanets, and L.L. German. Swarms of sheeted dykes, subparallel to the layering in ophiolitic massifs-evidence of spreading, *Doklady Akademii Nauk SSSR*. 298, 1193-1198.(1988) (*in Russian*).

33. A.S. Yakubchuk, V.G. Stepanets, M.Z. Novikova, N.A. Gerasimova, and L.A. Kurkovskaya. Specification of axial paleospreading zone in Ordovician ophiolites of Central Kazakhstan, *Doklady Akademii Nauk SSSR*. 307, 1198-1202.(1989) (*in Russian*).

34. Y. Zaitsev A.,*Evolution of geosynclines*. Nedra, Moscow. (1984) (*in Russian*).

35. Y.A. Zaitsev, Mantle protrusions-a specific type of geosynclinal deep structure in Paleozoic eugeosyncline of Central Kazakhstan, In: *Problems of geology of Central Kazakhstan*, Y.A. Zaitsev (Ed.), Vol. 1, pp. 140-182, Moscow University Press, Moscow. (1980) (*in Russian*).

36. Y.A. Zaitsev. Areageosynclines and their role in geotectognesis, *Acta Univ. Carolina Geologica*. 1, 55-73.(1990) (*in Russian*).

37. Y.A. Zaitsev and M.Z. Novikova, Correlation of early geosynclinal volcanic-sliceous and siliceous-terrigenous complexes of the Lower Paleozoic of Central Kazakhstan, In: *Geology of early geosynclinal*

complexes of Central Kazkhstan, Y.A. Zaitsev (Ed.), pp. 177-191, Moscow University Press, Moscow. (1985) (*in Russian*).

38. L.P. Zonenshain, *Geosynclines and their implication to Central-Asian fold belt*. Nedra, Moscow. (1972). (*in Russian*).

39. L.P. Zonenshain, M.I. Kuzmin, and L.M. Natapov, *Plate tectonics of the territory of USSR*. Vol. 1., Nedra, Moscow. (1990) (*in Russian*).

Tectonic Evolution of Northern Xinjiang, N.W. China : an introduction to the tectonics of the southern part of the Paleo-Asian Ocean.

XIAO XUCHANG, TANG YAOQING, ZHAO MIN, and WANG JING

Institute of Geology, Chinese Academy of Geological Sciences, Beijing, China

Abstract - The fold belt between the Siberian and Sino-Korean-Karakum platform is one of the typical Paleozoic fold belts converted from Paleo-Asia oceanic basins, which is more recently called the Paleo-Asian composite megasuture. The northern Xinjiang is situated within the southern margin of this Paleo-Asia composite megasuture. Our recent studies have focused on the ophiolites, high pressure metamorphic zones and tectonics of this area. The authors have divided the tectonic evolution of this Paleozoic fold belt into 7 stages and do not believe that once there existed vast oceans like the Tethys in Xinjiang and its neighboring region during the Paleozoic period.

Key Words - Xinjiang, tectonics, ophiolite, blueschist, amphibole

INTRODUCTION

Based on recent investigations of the Paleozoic fold belt, ophiolites and blueschists of northern Xinjiang, and on integration of previous geological and geophysical data, the tectonic evolution of this area since the Late Proterozoic is systematically discussed. Three major stages of the tectonic evolution are recognized, i.e., the formation stage ($Z-\mathcal{E}_1$) of the "Xinjiang Paleocraton", the stage ($C_2-\mathcal{E}$) of the "limited oceanic basin" and the stage (P-Kz) of the development of the intraplate orogens. The authors do not believe that there were vast oceans like "Tethys" in this area during the Paleozoic. The Paleozoic tectonic framework of this area exhibited "limited oceanic basins", alternating with microplates (blocks) and island arcs. The concept of "limited oceanic basin" and "residual oceanic basin", using the tectonic evolution of northern Xinjiang as an example, are illustrated and defined.

Since the realization that ophiolites represent fragments of oceanic lithosphere, the correlation studies between the ophiolite and oceanic lithosphere and the classification of ophiolites are very important for us to understand the tectonic history of oceanic basins and to reconstruct past tectonic environments, therefore we first focus our attention on the ophiolite study of this area.

MAIN CHARACTERISTICS OF NORTH XINJIANG OPHIOLITES

All of the ophiolites in northern Xinjiang belong to the Paleozoic period, and most of them are tectonized and dismembered, occuring as ophiolite melange; the lower part of these ophiolites consist mainly of harzburgite, dunite and lherzolite. The upper part usually with

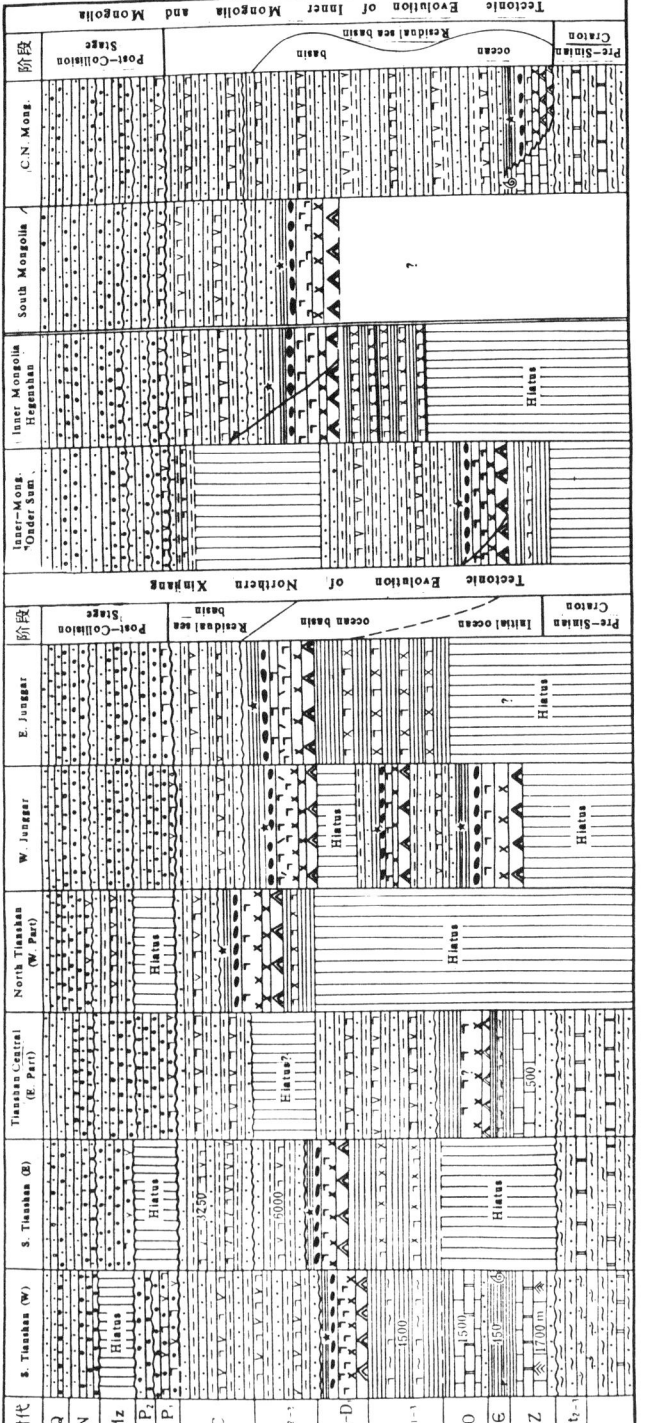

Figure 1 Ophiolites and stratigraphic correlation of the southern margin of the Paleo-Asian composite megasuture. 1. Intermediate-basic volcanics and tuff with marine sediments (formation of the Initial or Embryo oceanic basin). 2. Intermediate-basic volcanics and schists (formation of the Initial or Embryo oceanic basin). 3. Ophiolites (Star = radiolarians). 4. Intermediate-acidic volcanics and tuff with Epi-continental sediments (formation of the Residual sea basin) 5. Sandstone, conglomeratic sandstone, conglomerate, intermediate-acidic volcanics and tuff. 6. Mainly schists intercalated with marble and gneiss.. 7. Glaucophane Schist, 8. Limestone, marble, dolomite and sandy shale. 9. Unconformity. 10. Fossils, phosphate-bearing rocks/ stromatolite.

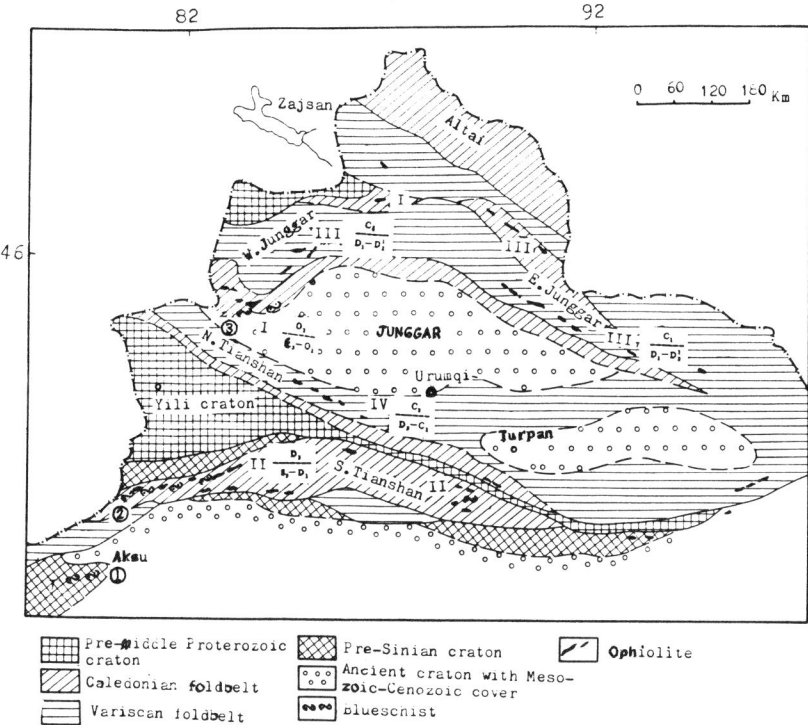

Figure 2. Distribution of ophiolites and blueschists, major fold belts and the Precambrian craton of Northern Xianjiang.

thin cumulate mafic-ultramafic rocks (indicative of small magma chambers) and minor mafic dikes and the top of the section consists of basic lava with pillow structure interlayered with radiolarian Chert. The total thickness of these ophiolites is usually from 2500-3500m. (Fig. 1 & 2)

On the basis of the radiolaria and isotopic determination form the leucogabbro or plagioclase-granite and of the tectonic setting, the ophiolites in north Xinjiang can be chiefly divided into four groups:

I. Middle Caledonian (ϵ_3 - O) Ophiolite.
1. Tangbale ophiolite near the north-western margin of the Yili--central Tianshan microplate (craton).
2. The Honggulelan ophiolite near the south margin of the "Altai continental crust (microplate)".

II. Late Caledonian (S - D_1) Ophiolite.
1. Northern branch ophiolite of the south Tianshan.
2. Southern branch ophiolite of the south Tianshan.

III. Early Variscan Ophiolite (D_1 - D_2)
1. Darbut-Karamaili ophiolite zone in west Junggar.
2. Armantai ophiolite.

IV. Late Variscan Ophiolite
1. North Tianshan (Baiyingou) ophiolite.

Judging from the geochemical data, the associated sediments and the rock assemblages of the ophiolites in the N. Xinjiang, especially from the thin cumulates (small magma chamber) resulting from a slow spreading rate [2, 6], the ophiolites in the north Xinjiang are relics of oceanic crust-upper mantle but formed in a small oceanic basin such as a back-arc basin, marginal sea or intracontinental sea. To support this idea, we have roughly estimated the spreading rate of the paleo-oceanic basin in N. Xinjiang and use the complete ophiolite at Tangbale as our example, because it has an accurately determined age of 508Ma using the lead isotope results from the leucogabbro [3] and it also contains the fossils (radiolarian) of Early to Middle Ordovician (taking 480Ma as the average age) in the topmost interlayered chert beds with pillow lava. We have not yet found any ophiolites older than Cambrian in N. Xinjiang, so we suppose that the Tangbale ophiolite formed from 570Ma (Early Cambrian) to 480 Ma and the 2 cm/yr (half spreading rate) as an average. Thus the Paleozoic Tangbale ophiolite in the W. Junggar Mts. could be representative of an ocean basin of about 3600 km in width (570-480 Ma) x 4 cm/yr = 3600 km). From this rough estimate combined with the very thin cumulates in the ophiolites in N. Xinjiang we consider that there was no vast oceanic basin like "Tethys" in this area during the Paleozoic.

HIGH-PRESSURE METAMORPHIC BELTS IN NORTH XINJIANG

The high-pressure metamorphic belts occurring in the northern Xinjiang are the best example in China, which can be divided from north to south into three zones as follows:
1. Pre-Sinian blueschist belt, the Aksu blueschist, in northwestern margin of
   Tarim Craton (Fig. 2, No. 1).
2. Paleozoic blueschist belt along the northern margin of south-Tianshan (to the south
   of the Yili microcraton) (Fig. 2, No. 2).
3. Paleozoic blueschist belt in southwestern Junggar Mts. (to the north of the Yili
   microcraton) (Fig. 2, No. 3).

It is emphasized that these high-pressure-low-temperature metamorphic belts are often associated with ophiolites and olistrostromes and we call this association a "Trinity". Only in this "Trinity" case we are confident that once there existed an oceanic basin which then disappeared through subduction. The Aksu blueschist terrane is exposed in the NW part of the Tarim plate (craton) and has a structural thickness of about 2.5 km and about 40 km in length. Rocks of the Aksu blueschist terrane were subjected to multi-stage deformation and transitional blueschist-greenschist facies metamorphism. The blueschist mainly consists of crossite, epidote, chlorite, stilpnomelane, actinolite, phengite and minor winchite [4]. It is unconformably overlain by upper Proterozoic Sinian formation along its southern boundary. The Sinian formation mainly consists of siliciclastic and carbonate rocks with m inor basic volcanics. The carbonate rocks contain many <u>Stromatolites, Panicolenia, Collenia, Jurusania, Conophyton</u>, which typify upper Proterozic strata in east China and Russia. The basal conglomerate of the Sinian formation is composed of subrounded to rounded clasts of the underlying blueschist and greenschist, clearly demonstrating the Pre-Sinian age of the blueschist facies metamorphism at this locality [8]. The isotopic ages from the Aksu blueschist using Rb-Sr method are 711 and 943 Ma (Fig. 4, 5, & 6). Rb-Sr isotopic dating for phengite of the blueschist is 710-720 Ma (Dr. Wang Xiaomin and Prof. Shige Maruyama, pers. com., 1988). This geochronological data along with the geologic relationships indicate that the Aksu blueschist is one of the best authenticated examples of Pre-Cambrian high pressure metamorphism

The Paleozoic blueschist belt along the northern margin of the southern Tianshan extends east-west about 300-400 km and is about 20-20 km wide within China. In South Kazakhstan this same belt probably connects with the blueschist belts of Maykisu, Atabasi

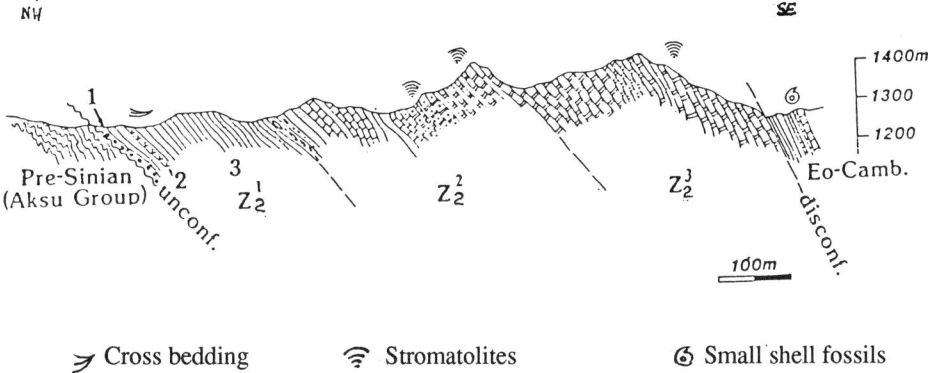

Figure 3. Cross section showing the unconformity between the Sinian system and Aksu group (Green-blueschist).

Z (1-2) Lower part of the Upper Sinian
1. Basal conglomerate, 2. Basaltic lava (dolerite) conformable contact with sandstone. 3. Brick red sandstone including calcareous, ferriferous and glauconite bearing sandstone.

Z (2-2) Middle part of the Upper Sinian
1. "Wormkalk limestone". 2. Dolomite intercalated with dolomite feldspathic sandstone and limestone containing Stromatolite (Conophyton, Collena)

Z (3-2) Upper part of the Upper Sinian
Mainly consisting of siliceous rock, phosphorus rocks and limestone containing "small shell fossils".

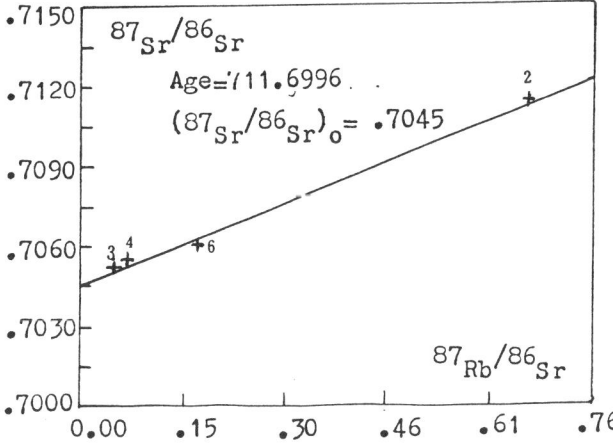

Figure 4. The whole rock Rb-Sr isochron diagram of blueschist facies rocks of the Aksu Group [5].

and Kansk. The mineral assemblage of the Chinese blueschists consists mainly of epidote, zoisite, chlorite, albite, phengite and glaucophane. Microprobe analyses of these glaucophanes indicate that their compositions are similar to crossite (Fig. 7 and 8).

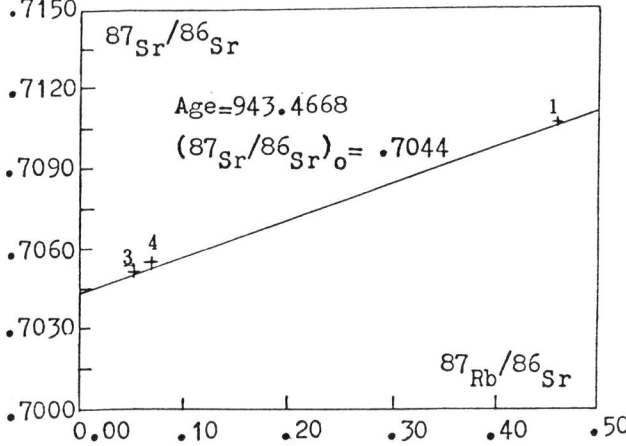

Figure 5. The Rb-Sr isochron diagram of blueschist facies of the Aksu Group [5].

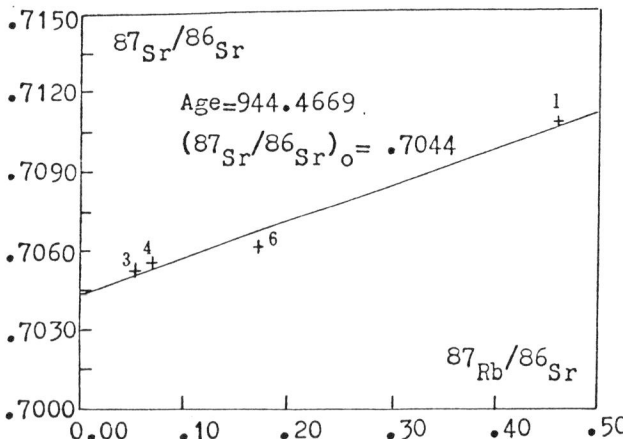

Figure 6. The whole rock Rb-Sr isochron diagram of blueschist facies rocks of the Aksu Group [5].

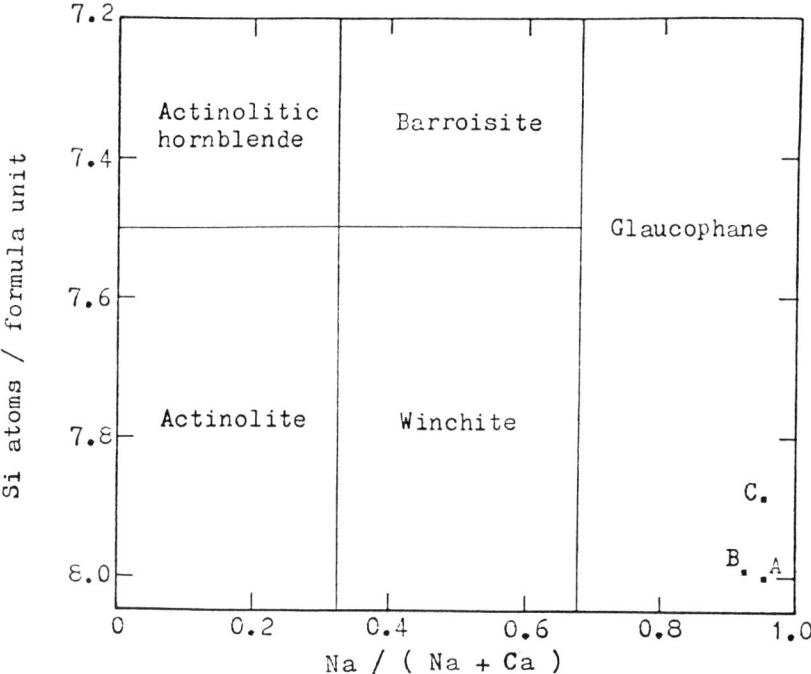

Figure 7. Compositional plot of Si versus Na/ (Na + Ca) for amphiboles from the extreme west of South Tienshan, Xinjiang (Classification scheme after Leake, Data from Wu Hanquan)

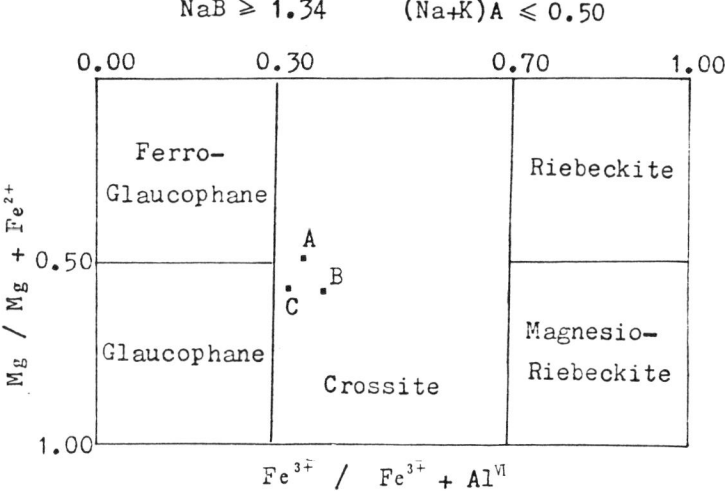

Figure 8. The composition of three of the alkali amphiboles (A, B, C) from the extreme west of South Tianshan, Xinjiang province.

The plateau and isochron age derived from isotopic dating for phengite of the blueschist, using the Ar^{40}-Ar^{39} method are consistent with each other yielding ages of 415-419 Ma (Fig. 9 & 10)

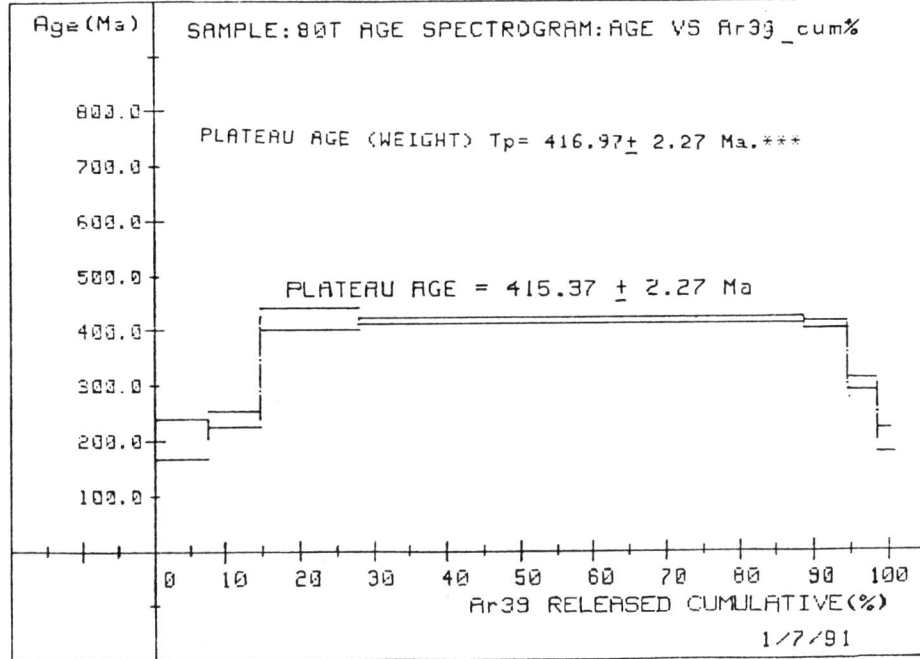

Figure 9. Isotopic dating for phengite of the blueschist from South Tianshan [7].

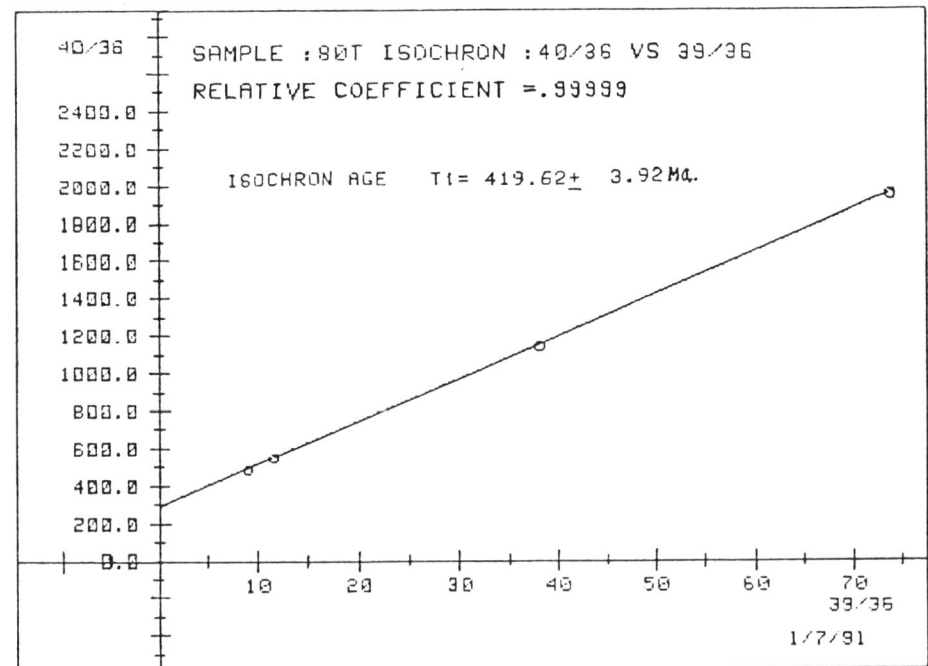

Figure 10. Isotopic dating for phengite of the blueschist from South Tienshan [7].

The Paleozoic blueschists of S. Tianshan are associated with dismembered ophiolite and olistostrome, the "Trinity", the indicator of ancient subduction zone, which represents the remnant slab of the Paleo-south Tianshan oceanic basin and a conspicuous fracture zone which may connect to the west with the well known "Nicolaea fracture zone". To the north of this fracture zone, there exist amphibolite, migmatite and gneissic granite probably altered by the acidic magmatism that forms a low pressure-high temperature belt. This low pressure-high temperature belt is parallel to the south Tianshan high P/T blueschist and they could be considered to make up a paired metamorphic belt.

The Paleozoic blueschist in the W. Junggar is associated with the Tangbale ophiolite. Its mineral assemblage is mainly epidote, actinolite, chlorite, albite and glaucophane. The microprobe analyses for the glaucophanes show also that they are mostly crossite as shown in Figures 11, 12, & 13.

Since the glaucophane has been found as a detrtal mineral in the coarse sandstone of Silurian age overlying the Tangbale ophiolites and blueschists [1] it is possible to suggest that the age of the Tangbale blueschist is late Cambrian to early Ordovician. The Tangbale blueschist and the dismembered ophiolite are the products of subduction resulting in the consumption of the south Junggar paleo-oceanic basin. This oceanic crust might have formed to the north of the Yili microcraton but somewhat earlier than the south Tianshan oceanic basin.

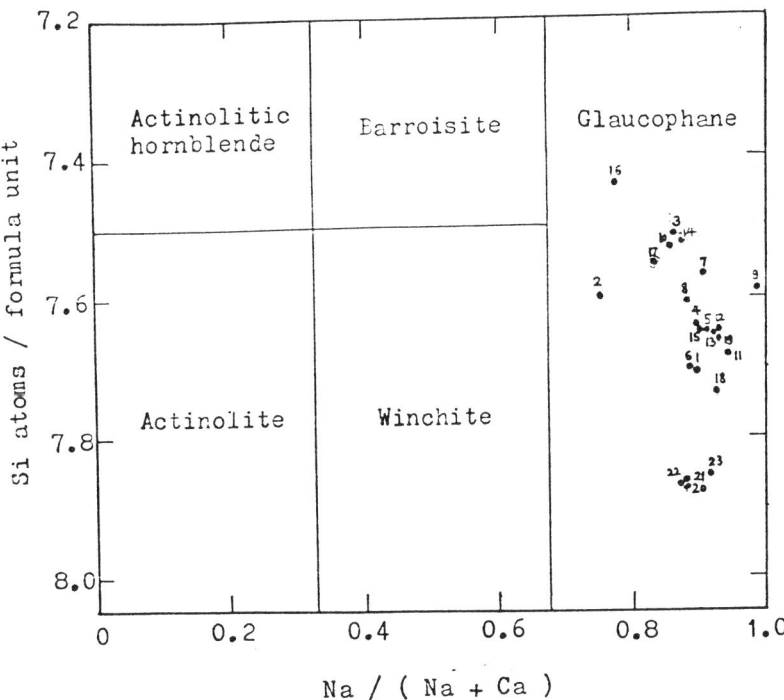

Figure 11. Compositional plot of Si versus Na / (Na + Ca) for amphiboles from Tangbale blueschist, West Junggar, Xinjiang (Classification scheme after Leake; Data from Guo Yihua)

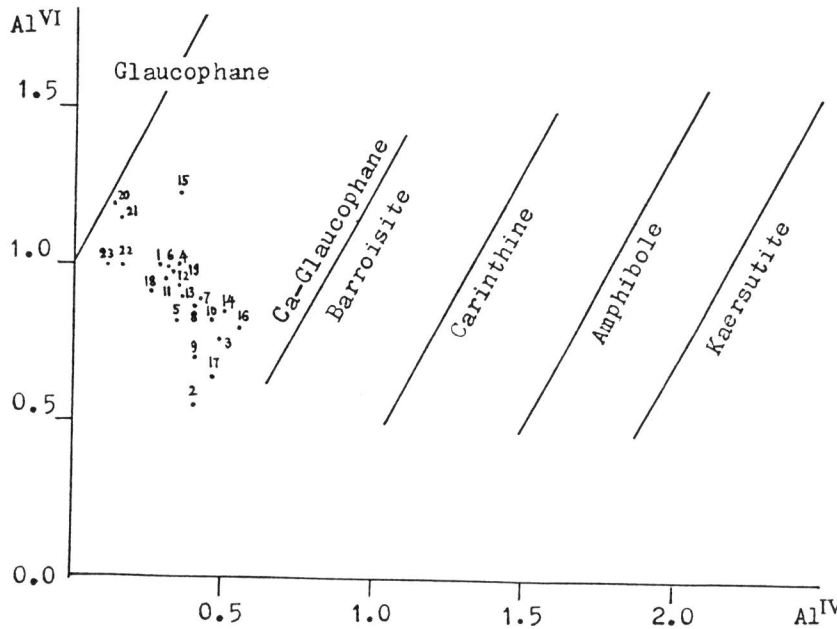

Figure 12. Diagram of Al^{IV} to Al^{VI} of glaucophane from Tangbale blueschist, West Junggar, Xinjiang (Data from Guo Yihua)

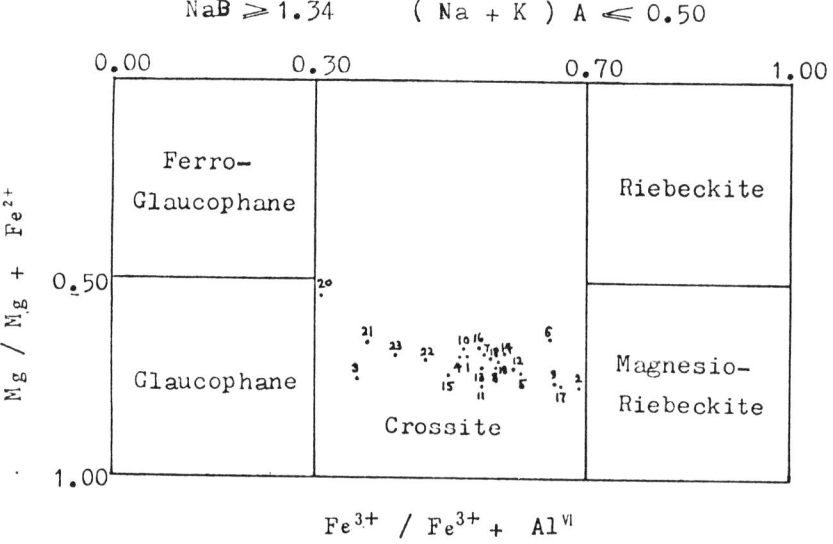

Figure 13. The compositional nomenclature of the alkali amphiboles from Tangbale blueschist West Junggar, Xinjiang. Data from Guo Yihua.

TECTONIC HISTORY AND EVOLUTION OF NORTH XINJIANG

Studies of the ophiolites, the sedimentary facies and the magmatism of the northern Xinjiang and its neighboring region indicate that the Late Proterozoic (Vendian-Sinian) oceanic basin occurred only in western Sayan, Russia and southwestern Mongolia. However, during that time most of the northern Xinjiang was consolidated, forming the so-called "Xinjiang Paleocraton". The Sinian-Cambrian formations are mainly platform sediments including carbonate facies, clastic deposits and locally glacial deposits. In the Early Paleozoic, the "Xinjiang Paleocraton" was dismembered and several ocean basins including the southwestern Junggar oceanic basin in the north and the south Tianshan oceanic basin to the south were formed (Fig. 14). The Early Devonian (D_1-D_2) was the most pronounced phase of extension in northern Xinjiang, the Darbut-Karamaili ophiolite representing the upper mantle-oceanic crust was originated in this period, constituting the northern Junggar oceanic basin. The Early Carboniferous oceanic basin was formed only in the north Tianshan represented by the Baiyingou ophiolite.

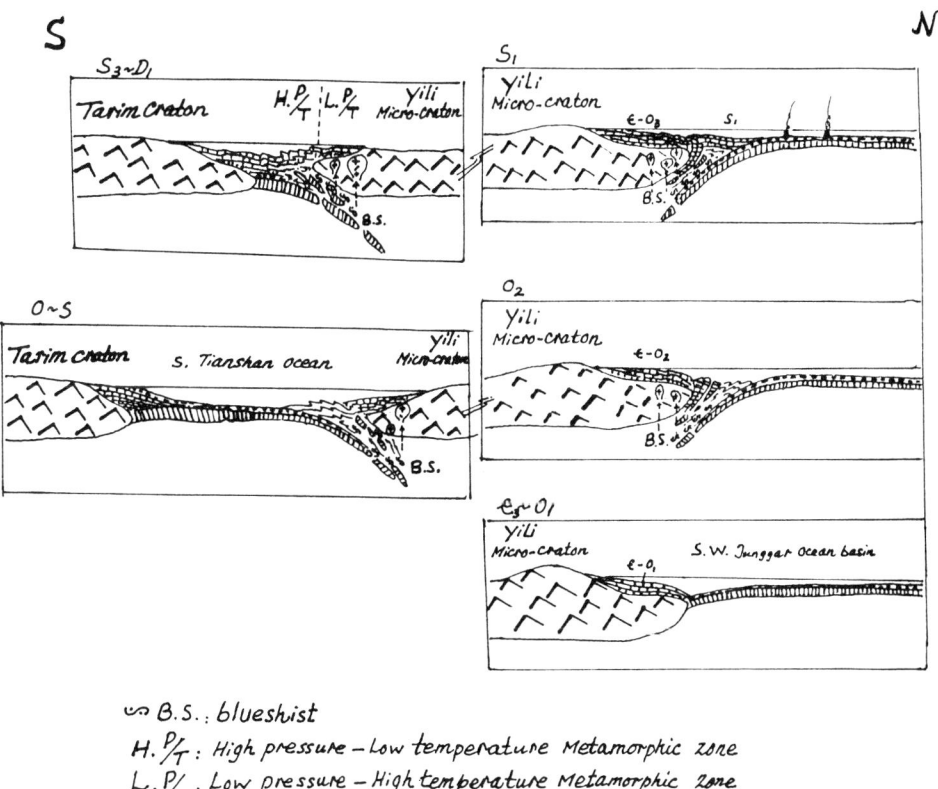

Figure 14. Teconic evolution of S. Tianshan and N. Tianshan (S.W. Junggar) in the Early Paleozoic.

Table 1. Simplified table showing evolution of seven stages of the Tianshan orogenic belt

		Major stages of tectonic development in orogenic belts	Associated sedimentary rocks	Magmatic activity
T E N S I O N	I	RIFTING	Clastic sediments, mudstone, ooze sediment, carbonate rock evaporite & anhydrite	Mainly alkaline basalt, and layered mafic-ultramafic rocks. (sometimes subophiolite)
	II	INITIAL OCEANIC BASIN	Ooze sediment, turbidite siliceous sediments including radiolarian chert.	Calc-alkaline (Na-rich to mafic volcanics and intrusions, mafic-ultramafic rocks, incomplete ophiolite
	III	OCEANIC BASIN 1. Fast spreading	1. Thick sequence of submarine sediments, including pelagic deposits, radiolarian siliceous, ooze sediment, turbidite and Fe-Mn bearing sediments.	Mainly ophiolite type I
		2. Slow spreading	2. Thin and discontinuous sequence of submarine sediments.	Mainly ophiolite type II
C O M P R E S S I O N	IV	CONVERGENCE OF OCEANIC BASIN 1. Subduction of interoceanic plate. 2. Formation of oceanic island arc & volcano-magmatic	1. "Submarine olistrostromic melange", sediments of fore-arc basin. 2. Mainly basic to intermediate-acidic tuffs and sediments of back-arc basins.	1. Mainly mafic-intermediate basic (mafic) magmatism. 2. Mainly mafic to calc-alkaline volcanics and intrusions.
	V	RESIDUAL SEA BASIN	Clastic sediments, neritic deposits, tuffaceous sandstone and sub-flysch sediments	Mainly calc-alkalic (K-rich) volcanics and intrusions
	VI	COLLISONAL (OROGENIC)	Clastic sediments, subflysch sediments and olistostromic melange.	Mainly acidic-calc-alkaline magmatic activity
	VII	POST COLLISONAL (Post Orogeny)	Continental sediments including lacustrine deposits, and sediments of rift and fault trough.	Acidic to alkaline magmatic activity.

The features of ophiolites and the tectonic history of northern Xinjiang mentioned above reveal that there was no large ocean like the Tethys, Atlantic, or the Pacific existing in N. Xinjiang during the Phanerozoic. Therefore we suggest that only " limited ocean basins" existed during that time, when the tectonic framework was characterized by island arcs, fore-arc and back-arc basins, marginal basins (seas) and even an intra-continental sea. The Paleozoic orogenic belts derived from the paleo-oceanic basin have undergone the following tectonic processes (Table 1):

    I.   Stage of rifting of Pre-Cambrian basement.
    II.   Stage of initial (embryonic) ocean basin.
    III.   Stage of spreading of the "limited oceanic basin".
    IV.   Stage of the convergence and consumption of the "limited oceanic basin".
    V.   Stage of the "residual oceanic (sea) basin"
    VI.   Stage of the collision (orogenic period).
    VII.   Stage of post-collsion.

Acknowledgements

This study was supported by the National Natural Science Foundation of China (NNSFC), the Chinese Foundation for Development of Geological Sciences and Techniques (CFDG) and the IGCP-Project 283. We greatly appreciate Profs. Lu Songnian, Gao Zhenjia, Wang Zouxun, Wu Hanquan and others for providing the isotopic ages and microprobe analyses. We also greatly acknowledge Prof. R.G. Coleman for his detailed revision of the manuscript.

REFERENCES

1. Y. Guo, The glaucophane schist from Tangbale district, Xinjiang, In: *Contributions for the Project of Plate Tectonics in North China, No. 1 (In Chinese with English Abastract)*(Ed.), Vol. 1, pp. 89-104, CAGS, Beijing. (1983).

2. N.J. Kuznir. Thermal evolution of the oceanic crust: Its dependence on spreading rate and effect and crustal structure, *Geophys. J.R. Abstr. Soc.* **61**, 167-181.(1980).

3. S.T. Kwon, G.R. Tilton, R.G. Coleman, and Y. Feng. Isotopic studies bearing on the tectonics of the west Junggar region, Xingiang, China, *Tectonics.* **8**, 753-757.(1989).

4. J.G. Liou, S.A. Graham, S. Maruyama, X. Wang, and X. Xiao. Proterozoic blueschist belt in W. China: best documented Precambrian blueshict in the world., *Geology.* **17**, 1127-1131.(1989).

5. S. Lu and Z. Guo, Discussion on geochronology of Precambrian system of N. Xinjiang (in Chinese). Research Report of Project 305, CAGS: (1989)

6. I. Reid and H.R. Jackson. Oceanic spreading rate and crustal thickness, *Marine Geophys. Res.* **5**, 165-172.(1981).

7. X. Xiao et al,. A synthesis of ophiolites and tectonic evolution of N. Xinjiang, NW China, Project 283, *IGCP Reports*.(1990).

8. J. Xiong and W. Wang. Preliminary research on Aksu Group of the Pre-Sinian, *Xinjiang Geology.* **4**, 33-46.(1986). (*in Chinese*).

Evidences for the West China Craton and its Evolution

YUAN XUECHENG [1]; ZUO YU [2]; ZHANG CHAOWEN [3].

1. Chinese Academy of Geoexploration, Ministry of Geology and Mineral Resources, Xisi, Beijing, 100812. 2. Center of Regional Gravity, Ministry of Geology and Mineral Resources, 5, Longshou Beilu, Xian, Shanxi, 710014. 3. Chengdu College of Geology, Chengdu, Sichuan, 610050.

Abstract

Geophysical evidences show that the Junggar basin in northwest China is a continental massive comprised Archean core. The Junggar basin, together with the Tarim basin, and the Kazakhstan, formed a craton disconnecting with the Sino—Korean Craton, we named it as West China Craton. The complicated evolution of the West China Craton in Paleozoic period is owing to the sandwich model of crust up to the Paleozoic have well—developed, brittle upper—middle crust splitted and formed terranes of different size, moved above ductile lower crust, and their divergence and convergence led to the opening and closing of sea basins.

1. INTRODUCTION.

The Xinjiang Autonomous Region in northwestern China was the southwestern margin of Paleo—Asian Ocean. It can be divided into three parts according to the geological and geomorphological features, the Altay area in the north, the Junggar basin in the center, and the Tarim basin in the south. Between the Junggar and the Tarim basin stands the Tianshan Mountain, and stretches westward into territory of Kazakhstan. The Turpan basin is an intermountain basin lying within the Tianshan.

Most of the published literatures put the Altay area under the Siberian Plate, the Tarim massive and Sino—Korean massive were referred to an unified Tarim—Sino—Korean Plate, and the Junggar and Kazakhstan were regarded as one plate made up small median massifs and Paleozoic geosynclinal foldbelts without PreCambrian. The Krameili—Ertix fault was taken for boundary between Junggar—Kazakhstan Plate and Siberian Plate, and the South Tianshan fault was the boundary between the Junggar—Kazakhstan Plate and the Sino—Korean Plate [10].

Xiao Xuchang et. al. [14] perceived the northern Xinjiang as micro continental massifs mosaiced by foldbelts of different period, and argued that the Xinjiang Craton had splitted by limited extension with limited oceans, and broad, deep, and vast ocean never occurred.

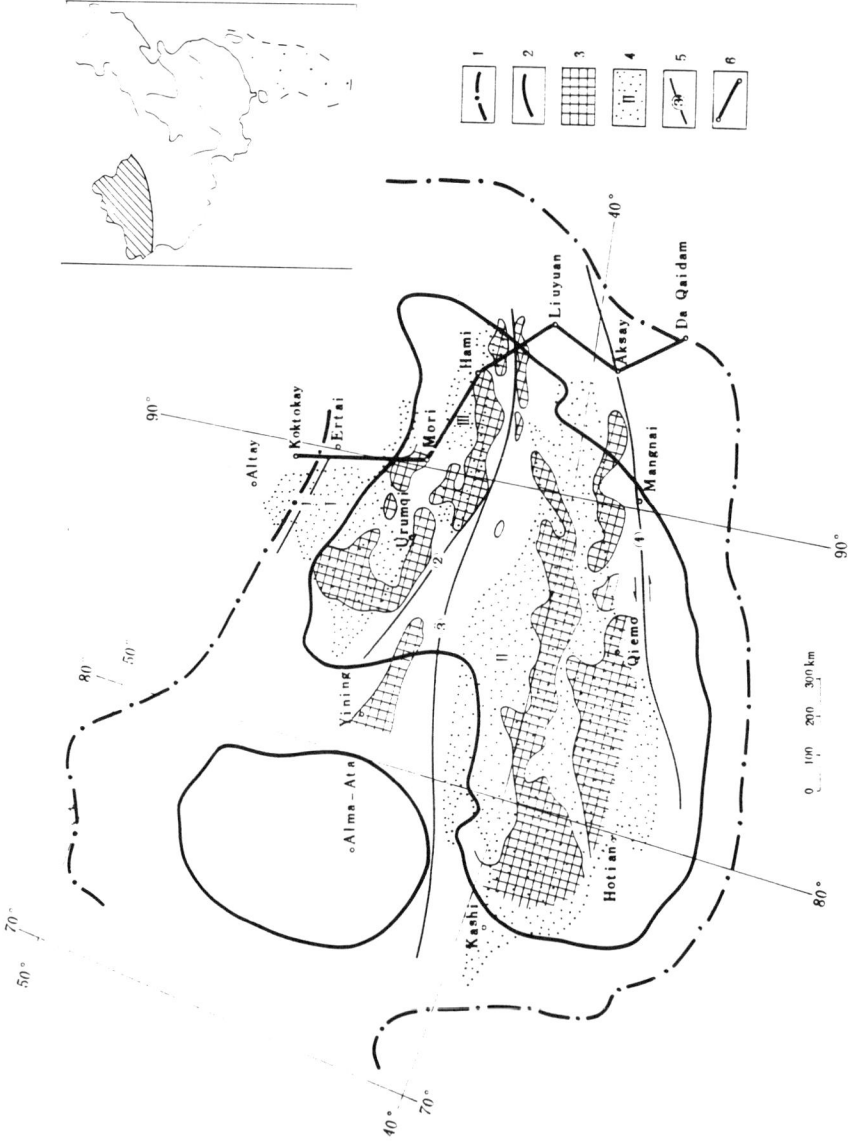

Fig. 1 The West China Craton. 1—Contour zero in MAGSAT, 2—Contour 3nt in MAGSAT, 3—Archean core, 4—Accreted Proterozoic basement: I, Junggar basin, II, Jurpan basin, III, Tarim basin. 5—Faults: (1) Almantai, (2) Bolukulu, (3) South Tianshan, (4) Altun —Layout of transect.

On the basis of geophysical studies, we believe that the Junggar is a continental massive with Archean core, it gethered with Tarim, Kazakhstan, made up an unified craton, We named it as the West China Craton, which is disconnected with Sino—Korean Craton, and is an isolated craton in Mid—Asia. In addition, a new explanation is given to the evolution of Paleozoic West China Craton, its frequent changes from land to ocean and ocean to land is difficult to understand by Wilson cycle.

2. THE NATURE OF CRUST IN XINJIANG

There is no doubt that the Tarim Plate is continental. But there is great controversy regarding the nature of the Junggar Plate. In the seventies, most geologists perceived the Junggar Plate as relics of Mid—Asian Mongolian oceanic crust, denied that there are PreCambrian blocks inside. Aeromagnetic survey accomplishing in 1982 revealed a strong, lumpish anomaly in the hinterland of the Junggar basin, the apparent magnetic susceptibility of anomaly source reached $12570--25140(10)^{-6}$ SI. Two interpretations arised, one interpreted it as ridge basalt [8], the other interpreted it as PreCambrian basement [6].

In 1989, a geoscience transect sponsored by national 305 project and Ministry of Geology and Mineral Resources extending for 1170 km including seismic sounding, MT, geomagnetic differential sounding, heat flow, gravity, and mgnetic measuements was completed (fig.1), it deepened the knowledge of the crustal structure and nature of Xinjiang region. Upon geophysical investigation, we believe that the crust of the Junggar basin and the East Junggar are continental.

The velocity structure of a continental region can be divided into four layers: the uppermost is loose sediment, its velocity is less than 5.5 km/s, the velocity of upper crust is 5.8—6.3 km/s, middle crust 6.4—6.7 km/s, lower crust is 6.8—7.6 km/s, the velocity of the top of upper mantle is $V_n \cong 8.1$ km/s. The uppermost layer of a oceanic crust is incoherent sediments ($V<2.7$ km/s), underlying are volcanic sediments ($V \cong 5$km/s), and basic or ultrabasic rocks ($V=6.7-6.9$ km/s). If Mesozoic and Cenozoic continental deposits rest directly on the Paleozoic oceanic crust, the velocity should jump from about 6.0 km/s to 8.1 km/s at a depth about ten kilometers.

Fig.2 shows the velocity profiles of 12 shotpoints along the transect. All of the velocity profiles show typical continental structure, three of them have high velocity intercalation. The first is north of Erdai, the high velocity layer is up to 9 km thick, at 17 km deep, the velocity is 7.31 km/s, the second situates between Kubusu and Jiangjunmiao, at 21 km depth, 15km thick, velocity ranges 7.30 km/s, the third locates between Daquanwan and Kushui, a high velocity layer of 6.99km/s intercalates at a depth 13—24km.

The thickness of Archean crust by world statistics is $\leqslant 36$ km, average crustal velocity is about 6.2 km/s, the thickness of Proterozoic crust is about 45—55 km, average velocity is approximately 6.6km/s, the thickness of Phanerozoic crust varies from 22 km to 66 km according to the geological setting. The thickness of oceanic crust is only 7 ± 3 km, much thinner than continental crust.

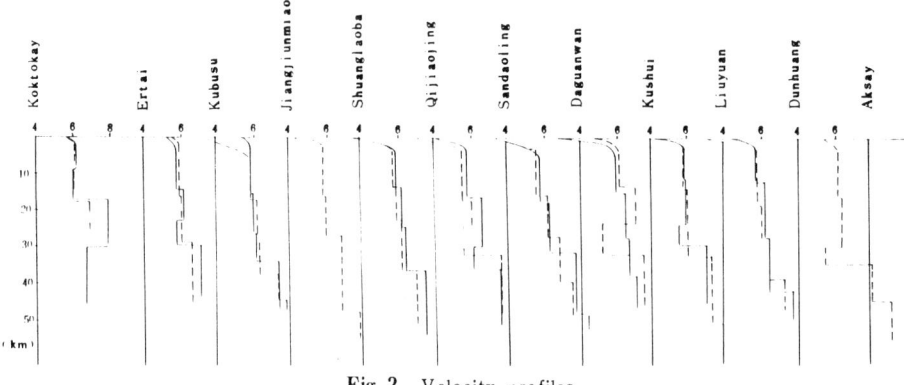

Fig 2. Velocity profiles

Solid lines show records south of shotpoint, dash lines show records north of shotpoint.

The average crustal thickness in East Xinjiang is 44—45 km, can correlate with Proterozoic crust, and its average crustal velocity/is about 6.39—6.47 km/s, a little smaller than Proterozoic crust.

The heat flow of a continental crust is correlated with geological age. The heat flow of PreCambrian shield is about $38 mwm^{-2}$, PreCambrian platform about $44 mwm^{-2}$, Calidonian orogen about $46 mwm^{-2}$, Hercynian orogen about $52 mwm^{-2}$, Mesozoic orogen about $59 mwm^{-2}$, Tertiary orogen about $73 mwm^{-2}$, Tertiary volcanic region about $92 mwm^{-2}$ [12]. Along the transect 9 heat flow measurements were performed, the values are shown in circles in fig. 4—6. The highest heat flow values found in Xinxinxia and Anxi (fig.6), coincident with Beishan rift zone. The heat flow in other places is similiar to the heat flow value of PreCambrian shield.

Another problem merits attention is the asthenosphere. The asthenosphere is 150—200 km deep in continental and about 80 km in oceanic region. The viscosity of the continental asthenosphere is higher than the oceanic asthenosphere and can be preserved for several hundred to several thousands million years [9]. If the basement in Xinjiang is a Paleozoic oceanic crust, we can find it at a depth of ten kilometers. But no conductive layer down to 200 km were found by MT, except a thin layer has a thickness only several kilometers (fig. 3).

Therefore, based on geophysical investigation, both the Junggar basin and the East Junggar are continental crust.

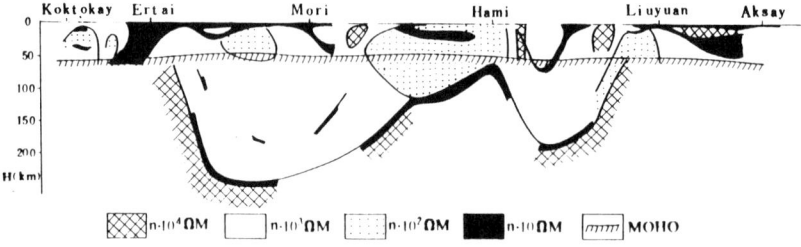

Fig.3 The resistivity section

3. THE WEST CHINA CRATON

One of the important contribution of MAGSAT is the discovery of strong anomalies in continent which coincide with shields.

There are two strong MAGSAT anomalies in China, one situated in the east, enclosing the North China and the Sichuan basin, with a north east axis. We believe this is an old craton, called it as East Asia Craton [15]. The other includes Tarim, Junggar, and Kazakhstan, we named it as the West China Craton. The area surrounded by zero contour of MAGSAT anomaly may consider roughly as the extent of the West China Craton. The 3nT contour defined two local anomalies, the Kazakhstan and the Tarim—Junggar basin. (fig. 1)

It is generally accepted that the Tarim Craton consolidated in late Proterozoic [3]. The upper Proterozoic system in Tarim region is neritic carbonate, middle Proterozoic Jixian system is marble, underlying Changcheng system is neritic clastic rocks, lower Proterozoic system is consisted of meta—sandstone, quartzite, and marble, late Archean complex is felsic volcanics. All of these rocks are non—magnetic or weak—magnetic, they can't produce such a strong MAGSAT anomaly. Therefore, a hypometamorphic median—basic complex of Archean eon must exist, which havn't found out at the surface, and should be the source of anomaly. Archean cores can be defined precisely by aeromagnetic anomaly (fig. 1). There are two strips of aeromagnetic anomalies which are produced by Archean core in Tarim basin, one situated at the center of the basin along $40°$ north latitude, coincident with the central upwelling of Tarim, another extended along the southern margin, running through Hotian, Qiemo etc. In the north of Tarim basin, a strip of negative magnetic anomaly corresponds the exposures of Proterozoic marble, schist, and sandstone.

It is reported that 90% of recent continents of the world were consolidated during late Archean to early Proterozoic [11][1][13][4]. The continents formed in early Proterozoic pieced together to get a supercontinent, vast continental crust turned to be stable and thick platformal sediments deposited on them.

The Junggar basin is within 3 nT contour too, most possibly, the Junggar basin have Archean—Proterozoic basement as well. The interpreted magnetic basement by aeromagnetic data in Junggar situated at the western margin of the basin, and extended through Urumqi to the south. All of the magnetic Archean core of Tarim, Junggar and Turpan located in the southern part of basins and the vast area of negative magnetic anomaly produced by Proterozoic system located in the north, these led to the assumption that the geodynamic settings of these basins in the Archean—Proterozoic times are similiar.

4. THE CRUSTAL STRUCTURE OF EAST JUNGGAR

We'll clarify further the crustal structure of Junggar along the geoscience transect.
The transect may be divided into three parts (fig. 1), the first part extends north—south, locating in the Altay and East Junggar (fig. 4), the second part turns to southeast direction crossing the Tianshan (fig. 5), the third part have also in a north—south direction, sits on the east of Tarim or west of Heixi corridor (fig. 6).

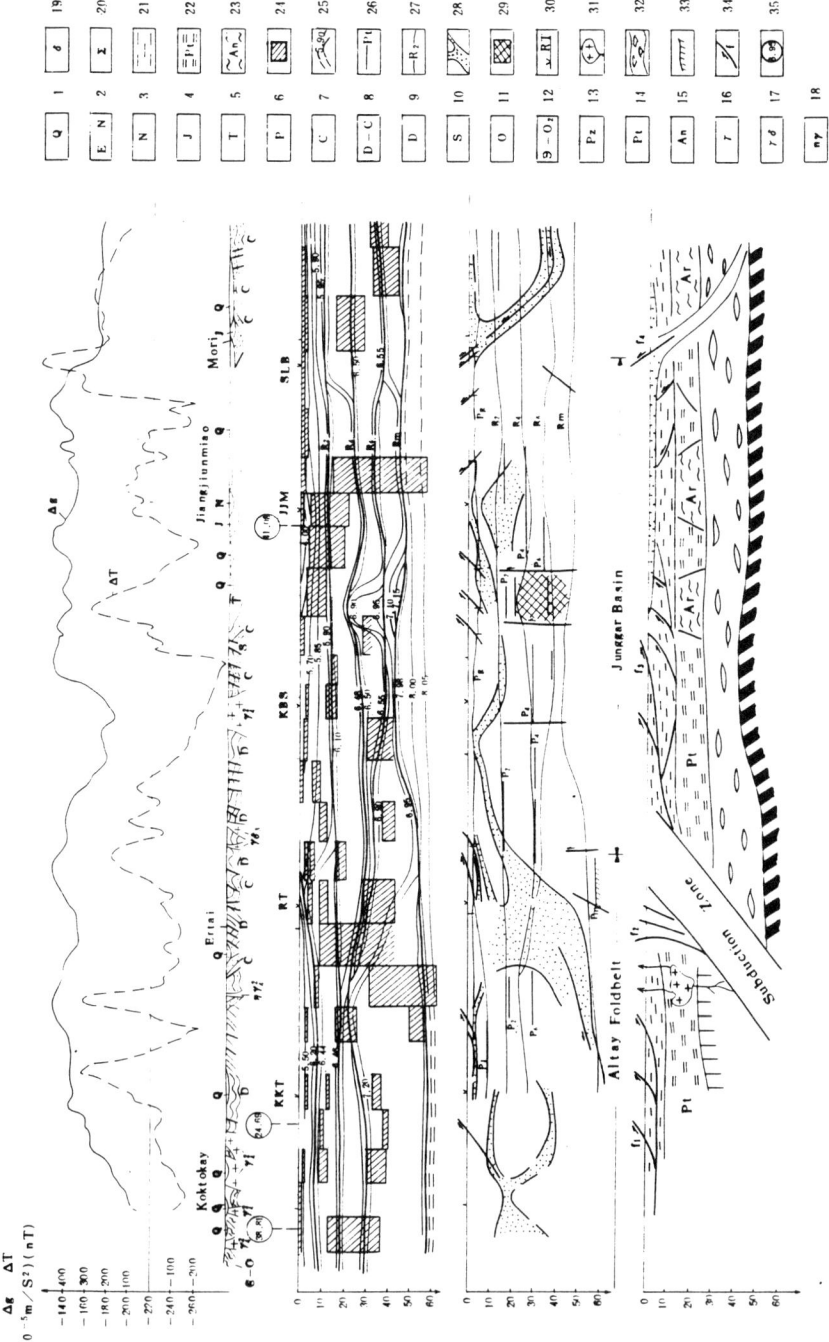

Fig. 4. The northern part of Mid-Asia transect, annotation of symbols see next page

Symbols in fig. 4:

1—Quaternary; 2—Neogene—Paleogene; 3—Neogene; 4—Jurassic; 5—Triassic; 6—Permian; 7—Carboniferous; 8—Devonian—Carboniferous; 9—Devonian; 10—Silurian; 11—Ordovician; 12—Cambrian—middle Ordovician; 13—Paleozoic; 14—Proterozoic; 15—Archeozoic; 16—granite; 17—granodiorite; 18—monzogranite; 19—diorite; 20—ultrabasic rocks; 21—Paleozoic; 22—Proterozoic; 23—Archean; 24—Low—resistivity layer; 25—isoline of velocity; 26—reflectors; 27—boundary of seismic layering; 28—interpreted conductive zone; 29—high velocity blocks; 30—shotpoint; 31—magmatic intrusion; 32—lower crust and mafic blocks; 33—Moho; 34—fault; 35—heat flow value;

(1) The Altay and the East Junggar

The northern part of transect include two tectonic unit, the Altay and the East Junggar. The Altay is a foldbelt of Siberian Plate. There is intense debate as to where is the suture between the Altay and the Junggar Plate. Chen Zhefu et. al. [2] considered that Ertix fault (fig. 4, f_1) is the boundary between Siberian Plate and Junggar Plate, Li Chunyu et. al. [10] convinced that Klameili fault is the Paleo—suture. Xiao Xuchang et. al [4] also championed the Klameili fault is the suture between the Siberian Plate and Juuggar Plate.

According to geophysical results, we believe that the Almantai fault (fig. 4, f_2) is the right place of this suture. Hercynian ultrabasic rocks and alkali basalt distributed along Almantai fault. A wide northerly dipping conductive zone extends from surface to the Moho. This zone is devoid of reflections, and coincides with gravity low and magnetic low. The crustal velocity to the north of Almantai fault is rather high than to the south. In the north, phase P_1 from a depth about 7—8 km can be defined, but it can't be found in the south.

Differed from Almantai fault, the crust under Klemeili fault is reflective, better layering, no clear indiction for subduction. In the upper—middle crust, a continuous conductive zone, extending from Almantai fault to the south, which might be a detachment surface. Below the nappe, from Almantai suture to Urumqi fault (fig. 4, f_4) there might be Proterozoic stable continental massive.

The Altay foldbelt can be divided into two different region. To the north of Ertix fault (fig. 4 f_1) PreCambrian crystalline rocks exposed, both gravity and magnetic field is low. There havn't oceanic sediments of Vendian to early Cambrian period. The rocks from Ordovician to Silurian are thick neritic clastic flysch derived from northern newly born folded mountains. To the south of Ertix fault, the exposed oldest rocks are shelf deposit of middle to upper Ordovician. After Devonian, it turned to be an island—arc which was strong magnetized and produced positive and negative anomaly alternately.

Therefore, Almantai fault is a suture of New Guinea type collision between Junggar continental plate and Altay arc.

(2) The Tianshan

Due to the transect make an acute angle with the strike of Tianshan, the length of this section was lengthened.

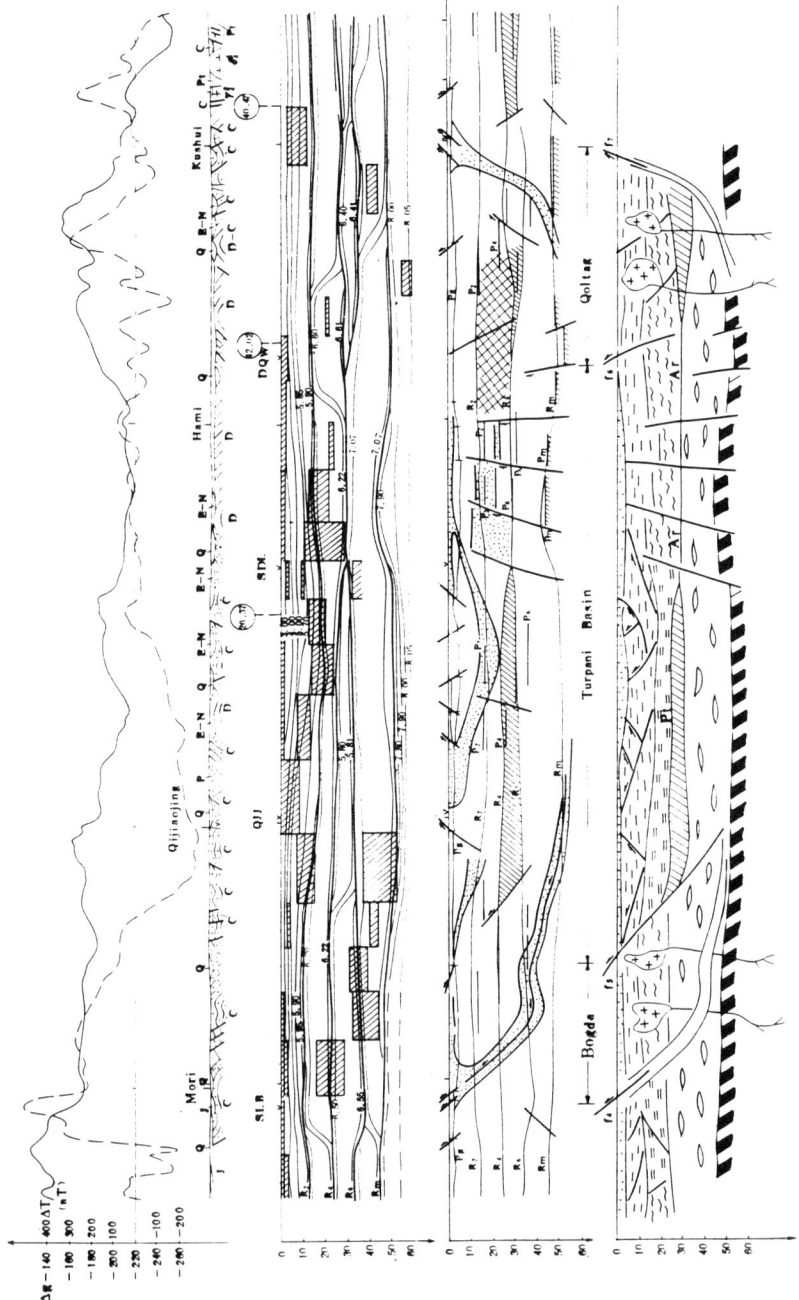

Fig. 5 The central part of Mid-Asia transect, symbols as fig. 4

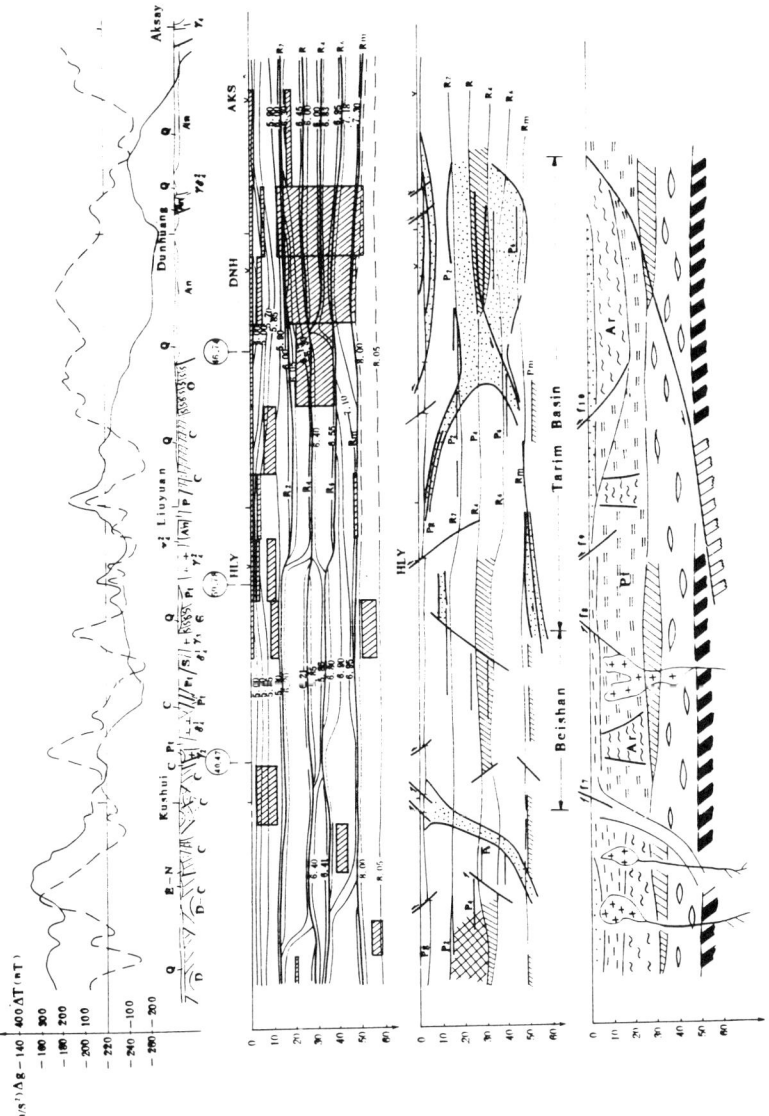

Fig. 6 The southern part of Mid-Asia transect, symbols as fig. 4

The East Tianshan where transect crossed, can be divided into three parts, the northern part is the Bogda Mountain, the southern part is the Qoltag Mountain, and the Turpan basin located in the middle.

Urumqi fault (fig. 5, f_4) dipping southerly is the northern boundary of Bogda Mountain, its dipping angle is about $50°-70°$. The upper Paleozoic system to the north of the fault is neritic—littoral clastic—carbonate formation, and to the south is volcanic, pyroclastic formation. A low resistivity zone with a width of about ten kilometers occurs at the fault, flattens downwards, and disappears in the lower crust. The southern boundary of the Bogda Mountain is the northern marginal thrust of the Turpan basin (fig. 5, f_5).

Recent geological investigations suggest the Bogda region is an intercontinental rift. Rifting began in early middle Carboniferous with the deposition of neritic, littoral pyroclastic rocks, in late middle Carboniferous volcanic—carbonatite—clastic mixed formation developed, and in late lower Permian, the rift culminated. The volcanics is alkaline, devoids of metamorphic rocks and acid intrusion. No indications of ophiolite, or melange were found [5]. This region occured strong magnetic anomaly, the Moho reflector is not clear, the velocity of the top of upper mantle is relatively low, yields 7.8 km/s. All of these give evidences that the Bogda region is a zone of mobility, its crust is reformed by mantle materials.

The Qoltag mountain situated between the southern marginal thrust (fig. 5, f_6) and Kushui fault (fig. 5 f_7). It is also a rift during late Paleozoic, well—developed thick intermediate to basic and basic to acid volcanics, pyroclastics rocks, and gabbro, diabase, and pyroxene hornblende peridotite, lherzolite ……ultabasic complex rocks occured [14]. The geophysical features of the Qoltag mountain is also similar to the Bogda mountain whose crust is also resistive, reflective, P_m events not very clear, coincident with gravity high and magnetic high.

The crust of Turpan basin can be divided into two parts by shotpoint Sandaoling. The south part responsed magnetic high, is explained as Archean basement. A high velocity layer was defined in the middle crust, high angle brittle faults were found in the middle—lower crust, the interpreted Moho is about 44 km, the shollowest site along the transect. The north part revealed magnetic low, was explained as Proterozoic basement. A huge low velocity layer occurs in the middle crust and a huge low resistivity layer occurs in the upper crust, the Moho despressed. The P_m phases are clear, the velocity structure near surface show obviously a sedimentary basin.

(3) Beishan—Tarim basin

The magnetic high coincides with Xinxinxia paleouplift between Kushui fault (fig. 6, f_7) and Yushishan fault (fig. 6, f_8), is produced by Archean basement, which located at the strike extension of the central uplift of Tarim basin and is inferred as the extension of the central Tarim Archean core (fig. 1).

The magnetic high at the south of Shulehe fault (fig. 6, f_{10}) coincides with the outcrops of lower Proterozoic Dunhuang group made of mesometamorphic gneiss, amphibolite, migmatite, greenschist, and marble. The magnetic anomaly is isolated, oval—shaped, situated along the strike of the south Tarim Archean core (fig. 1), They must have been connected in the history. The middle—lower crust of the West China Craton, which is a huge low—velocity low—resistivity zone, wedged into the crust of Qinghai—Tibet Plate (fig. 6, f9).

The region between Yushishan fault (fig. 6, f8) and Huaniushan fault (fig. 6, f9) is Proterozoic Beishan median massive. Both Yushishan fault and Huaniushan fault are high—angle fault, they might be formed in early Paleozoic, at the same time as the Beishan median massive uplifting. The crustal resistivity is normal of several hundred ohmmeter in this region, and the crust is devoid of reflective events.

Between Huaniushan fault and Shulehe fault is Hongliuyuan rift, which opened in Ordovician and closed in late Permian.

5. THE EVOLUTION OF WEST CHINA CRATON

Does the Archean core of the West China Craton joined into a whole or scattered as pieces at the very beginning? nothing but conjection. But the similar shape between the boundary lines of basins, and the identical accretion model of basins (that is, Archean core always locats in the southern part and early Proterozoic accreted to the north,) can help us to infer that very probably the original continent was an unified massive. Fig. 7 shows one possibility of this original continent.

In late Archean, continent splitted into three pieces, drifted northward, Proterozoic accreted at the north margin. Zhang Liangcheng et. al. (16) argued that the stratigraphic sequence of Yili, Bolukulu, Junggar, Wulutao, Qiatkar is very similar. The tillites of early Sinian in Kuluklitag, Kurguqinshan, Zhetemtao are also very alike. The phosphate—bearing silicolite formation of lower Cambrian can be correlated each other. Therefore, the basement of the West China Craton had been formed prior to middle Proterozoic, and up to the end of Proterozoic, when the first supercontinent Pangea I appeared.

In the early Paleozoic Pangea I broken, rifts and sea basin developed in Xinjiang. In the late Paleozoic, sea basins closed, PaleoAsian Ocean consumed, and Pangea II appeared.

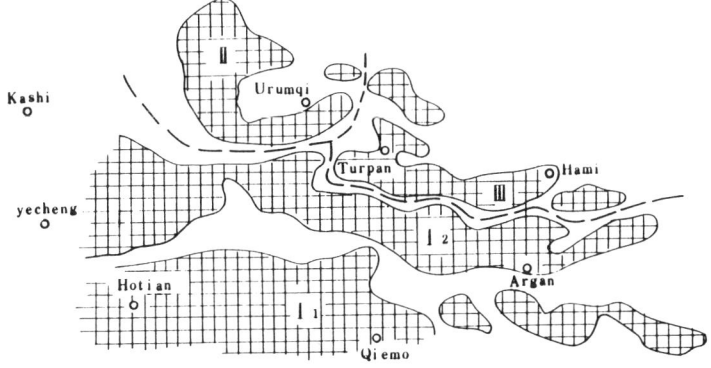

Fig. 7 One possibility of the shape of the original Archean West China Craton I1—Southern Tarim core. I2—Central Tarim core. II Turpan core. III—Junggar core

According to plate concept, the openning of oceanic basin is due to the split and divergence of lithosphere, and the closing of oceanic basin is due to the subduction and convergence of plates.

Surface geological studies show that oceans had been opened and closed between the Siberian Plate and Junggar Plate for four times, i.e. Proterozoic Asian Ocean, early Paleozoic Asian Ocean, Devonian Asian Ocean, and Carboniferous Asian Ocean. Three oceans had been opened and closed between the Tarim and Junggar, i.e. Proterozoic South Tianshan Ocean, early Paleozoic Middle Tianshan Ocean, and early Paleozoic South Tianshan Ocean[7]. But geophysical data revealed that the lithosphere is resistive down to depth 200 km (fig.3). Perhaps the asthenosphere is not existing everywhere in Xinjiang, even if it exists, it is very deep. It seems not reasonable that such a huge lithosphere splitted and merged frequently in a not too long geological duration.

One possible explanation is that there arn't oceanic basin at all since the Proterozoic, but only the continental sea made up by the splitting and merging of upper—middle crust. Brittle middle—upper crust, ductile lower crust and rigid upper mantle comprised the so—called sandwich model of crust, perhaps have formed up to the Phanerozoic. If the splitting of crust only took place in the middle—upper crust, then terranes were formed, which can moved above the soft, ductile lower crust, subducted one terrane under the other when they collided. In fig 3 we can see that collision zones which have defined by surface geology are low resistivity zone and all of them flatten downward ended in the lower crust or the top of upper mantle.

The tectonic features changed since Mesozoic. Tarim, Junggar, Turpan which were formerly upwelling area were subsided and the original rifts upwarped to be folded mountain with many thrusts. A possible geodynamic mechanism is that the lower crust was consolidated by the tectonic activity during the Paleozoic. When a force expert on craton, the consolided hard region, such as Tarim, Junggar Turpan will not be deformed and the soft region will be folded to be a mountain.

Although the Wilson cycle is truth in the explanation of the evolution of oceanic crust, perhaps it can't be applicable directly to the continental crust, In the continent, ductle lower crust may play a role as asthenosphere, the behavior of lower crust in different period give constraint to the features of evolution.

REFERENCES

1. G. C. Brown and A. E. Mussett. *The Inacecessible Earth* (1981).

2. Chen Zhefu. The Framework of Tianshan. *Xinjiang Geology*, **3**, (1981) (in Chinese)

3. Cheng Shoude, Wang Guangrui, Yang Shude, Jin Jinshui, and Zhu Jieshui. The Paleo Plate Tectonic of Xinjiang, *Xinjiang Geology*, **4**, 2, 1—26 (1986) (in Chinese)

4. J. F. Dewey and B. F. Windley. Growth and Differentiation of the Continental Crust, *Phil. Trans. R. Soc. Lond.* A, **301**, 189—206 (1981)

5. Fang Guoqing. Initial Studies about Bogda Mt, Late Paleozoic Aulacogen, *Xinjiang Geology*, **8**, 2, 183—141, (1990). (in Chinese)

6. Fei Ding and Zhang Xinsheng. Interpretation of the Magnetic Field in the Junggar Area and its Regional Structural Features, *Acta Geophysica Sinica*, **3**, 5, 459—468 (1987) (in

Chinese)

7. Huang Jiqing, Jiang Chunfa, and Wang Zuoxun. On the Opening—Closing Tectonics and Accordion Movement of Plate in Xinjiang and Adjacent Regions, in: *Geoscience of Xinjiang* No. 1, 3—16 (1990).

8. Jiang Yuanda A Preliminary Approach to the Basement of Junggar District, *Xinjiang Geology*, **2**, 1/1, 11—16 (1984). (in Chinese).

9. T. H. Jordan The Deep Structure of the Continents, *Scientific American*, **1**, 92—107 (1979).

10. Li Chunyu and Wang Quan The Paleo—Plate Tectonics of Northern China and Adjacent Regions and the Origin of Eurasia, in: *Contributions for the Project of Plate Tectonics in Northern China*, No. 1, 3—16 (1983) (in Chinese).

11. J. D. A. Piper Movements of the Continental Crust and Lithosphere—Asthenosphere systems in PreCambrian Times, in: *Tidal Friction and the Earths Rotation*, 253—321 (1982).

12. B. G. Polyak and Ya B. Smirnov Heat Flow on Continents, *Doklady of Acad. Sci. USSR, Earth. Sci. Sect.* **168**, (Engl. Transl.) 26—29 (1966)

13. D. H. Taring Plate Tectonics: Present and Past, in: *Evolution of the Crust*, 239—254 (1978)

14. Xiao Xuchang, Tang Yaoqing, Li Jingyi, Zhao Min, Feng Yimin and Zhu Baoqing On the Tectonic Evolution of the Northern Xinjiang, Northwest China, in: *Geoscience of Xinjiang*, No. 1, 47—68 (1990) (in Chinese)

15. Yuan Xuecheng Deep Structure and Tectonic Evolution of Qinling Orogenic Zone, in: *A Selection of Paper Presented at the Conference on the Qingling Orogenic Belt*, 174—184. (1991) (in Chinese)

16. Zhang Liangcheng, and Wu Naiyuan The Geotectonic and its Evolution of Tianshan, *Xinjiang Geology*, **3**, 1—14 (1985) (in Chinese)

Geodynamic evolution of Southern Siberia in Late Precambrian - Early Paleozoic time

NIKOLAI A. BERZIN and NIKOLAI L. DOBRETSOV

United Institute of Geology, Geophysics and Mineralogy, Russian Academy of Sciences, Novosibirsk, 630090 Russia

Abstract. Southern Siberia is an important part of the Central Asian fold belts which was formed due to the evolution of the Paleoasian ocean. In the Late Precambrian stage (800-600 Ma) many microcontinents were separated from the Siberian Craton by the active spreading oceanic basin located near the modern Baikal corner of Siberian Craton. During Pre-Vendian and Vendian collision of these continents with Siberian Craton a new spreading basin was active (600-500 Ma) outboard of the former micro continent. This Cambrian ocean was separated from the continental block by two systems of island arcs (Kumetsk-Tuva and Dzhida) with subduction zone inclination towards the continent. In the Middle Cambrian (530 Ma) several seamounts had choked the subduction zone resulting in migration of the subduction zone oceanwards. A new system of Island arcs (Salair - West Ozernaya) were been formed here. Cambrian ocean closing was connected with Late Cambrian - Early Ordovician (505-480 Ma) collision of Altai-Mongolian micro continent (AMMC) which probably has been transported away from Gondwana. The new Ordovician-Silurian ocean was formed southwest of the Altai-Mongolian continental mass. At that time (400-480 Ma), South Siberia was a passive margin to a Paleoasian ocean with several deep-water troughs. An alternative model suggests that the Dzhida terrane in South-East Sayan and some of the West Sayan terranes and troughs were components of a large oceanic bay. Devonian closing of these bays and troughs and the spreading of the Devonian-Carboniferous ocean resulted in formation of the active continental margin at the periphery of South Siberia. The final collision stage and intensive Permian movement are referred to the creation of Mongol-Okhotsk and Paleotethys oceans.

Key Words - Paleoasian ocean, ophiolite, island arc, paleogeography

INTRODUCTION

Eurasia is a composite continent formed by amalgamation of several continental and micro-continental blocks [17,28,42]. Three types of blocks can be distinguished here (Fig.1):
1. Laurasia (Russian and Siberian platforms, possibly North China block and several micro-continents);
2. Gondwana type (South China, Indo-China, India and Tarim blocks and several microcontinents such as Altai-Mongolian);
3. Pacific type and undetermined blocks (Omolon, Okhotsk, Bureya, and others.). The main suture zones between them are marked by ophiolite and blueschist belts.

Central Asian fold belts between Russian, Siberian, Tarim and North China blocks (Fig.1) have being formed during Paleoasian and Uralian ocean evolution. South Siberia belongs to this folded area. Accretional and thrust-folded structures of South Siberia are composed of terranes originated in different geodynamic situations during the Paleoasian ocean evolution (Fig.2).

New paleomagnetic studies [33,34,43] and geological paleo-reconstructions [42] allow us to show relative position of thses large blocks during Paleozoic time. But many details of their related structures; such as the distribution of island arcs, back arc basins, marginal seas,

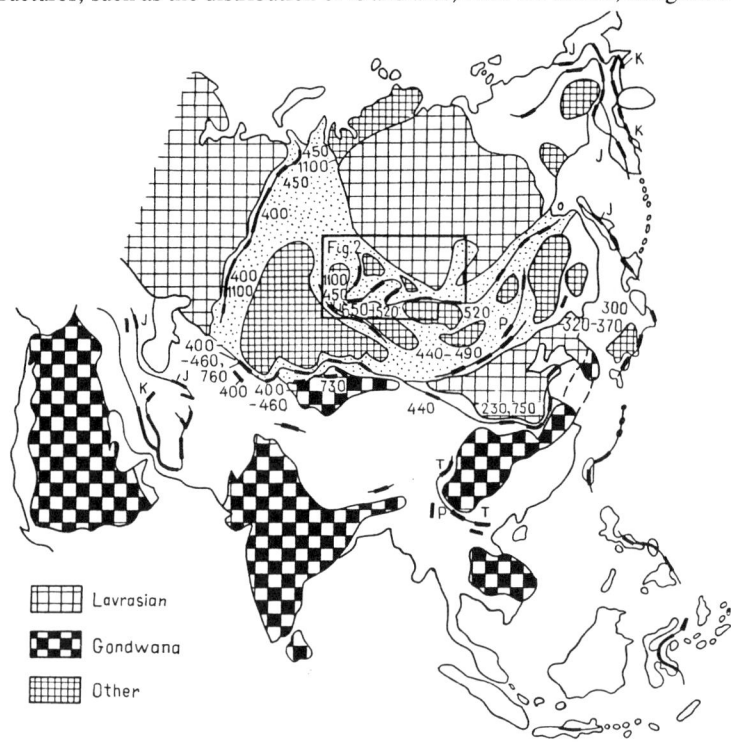

Figure 1. . Distribution of continental (micro-continental) blocks and blueschist belts between them [17]. Continental blocks are subdivided into three types - Laurasian, Gondwana and other (Pacific or undetermined); Precambrian and Paleozoic ages of blueschists are shown by numbers, and younger ones as P= Permian, T= Triassic, J= Jurassic, K- Cretaceous. Dotted areas correspond to structures originated during evolution of Paleoasian and Uralian oceans.

passive margins of continents could not be definitely reconstructed. For many areas such reconstructions have only been recently started resulting in different views of tectonic history (see below).

GEODYNAMIC EVOLUTION

We would like to reconstruct the geodynamic evolution of South Siberia folded areas based on authors' works [5,6,10,16,18,19,20,22] and on the preliminary version of Geodynamic Map of Southern Siberia, scale 1:2,500,000, which was compiled under the leadership of the authors during the IGCP Project 283 activity "Geodynamic evolution of Paleoasian ocean". The Main topics of our compilation and discussion will include:
 1. Early stages of evolution of Paleoasian ocean (Late Precambrian history).
 2. Vendian - Cambrian stage of the ocean active margins.
 3. Collisional stages and the closing of South Siberia segment of Paleoasian ocean
    (Late Cambrian - Early Ordovician time).
 4. Middle-Late Paleozoic evolution and deformations.

Early stages of evolution of Paleoasian ocean
Russian geologists believe that the Siberian craton was separated from its surrounding structures as early as Archean [5,29]. The further investigations of Precambrian tectonics and

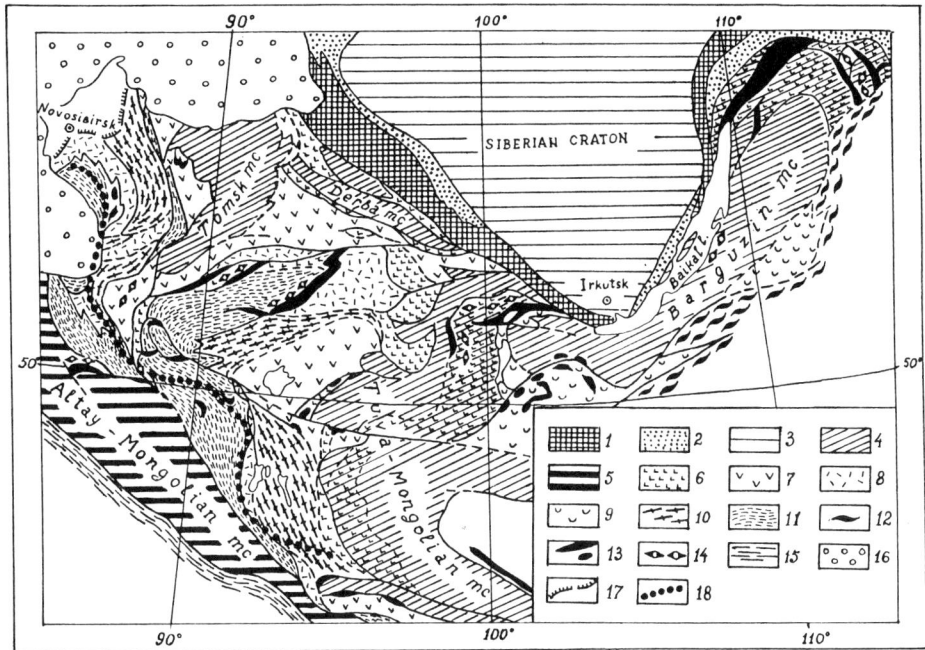

Figure 2. Geodynamic complexes of South Siberia and West Mongolia: 1- Basement of Siberian craton (**AR-PR$_2$**); 2 - Passive margin complexes and intracontinental rifts in the craton; 3 - Cover of the craton (**V-MZ**); 4 - Microcontinents of the Laurasian group; 5 - Altai Mongolian microcontinent (part of the Gondwana passive margin -(**PR$_3$-Є**); 6-9 - Island arc terranes: 6- **PR$_3$**, 7- **V-Є$_1$**; 8- **Є$_{1-2}$**, 9- **Є-O**; 10 - Seamount and arc-related terranes (**V - Є$_1$**); 11- Accretional collsional complexes (**Є- O**); 12 - Collisional complexes of the East Baikal type with granite metamorphic domes; 13 - Ophiolites including ultrabasites and serpentinite melange; 14 - Belts of high-pressure rocks (blueschists, eclogites, and others); 15 - Hercynian Paleooceanic zone (**O-S-D**); 16 - **MZ-KZ** cover; 17 Hercynian thrust; 18 - Boundary of the Hercynian active margin.

other ancient platforms and their margins allowed us to conclude that Siberian craton at that time was one of the largest sialic blocks of the Earth appearing only in the Riphean. Disintegration of this unknown giant super continent or Riphean Pangaea began by spreading and formation of new ocean crust [10].

Destruction of Riphean Pangaea and divergence of sialic blocks resulted in the opening of the Paleoasian ocean. This initial break-up was followed by riftogenic processes inside the large detached continental blocks and within the Siberian craton. Simultaneously, the systems of Riphean aulacogens and peri-craton depressions have been formed. within these paleocontinents. They were produced by convergence of three-axial rifts whose several branches were preserved inside the cratons as aulacogens, and in other cases were transformed into the passive continental margins. This model is suitable for Riphean rocks of Enisey Ridge and for Baikal-Patom mountainous country [42].

We have no clear signs of Pangaea destruction.earlier than Riphean. Some geoscientists pay attention to the marginal position of Akitkan volcanic-plutonic belt in relation to the Siberian Craton. This belt is located within northwest Trans-Baikalia and has the features of an active continental margin [9,42]. But the absolute geochronological data confirm its Pre-Riphean or Pre-Late Proterozoic age (1.9-1.65 Ga) [11]. Moreover, there is no evidence of wide-spread oceanic complexes outside the craton that could be part of the Akitkan belt formation.

According to the drilling and geophysical data, the Akitkan type discordant sequences occur beneath the platform cover large distances away from craton margin. Therefore, the marginal position of Akitkan belt may be considered atypical caused by peculiarities of Riphean Pangea destruction. Its geodynamic nature depends not on the initial stages of Paleoasian ocean evolution, but on the later events in the Middle Proterozoic along the boundaries of these giant sialic blocks during the final stages of Proterozoic Pangaea formation.

The first large oceanic basin has contacts with the Siberian craton only in the Late Proterozoic - Early Riphean. Their interaction produced Riphean island arcs, accretionary prisms and active and passive continental margins. The most ancient oceanic complexes related to Paleoasian ocean formation are ophiolites and island arc assemblages of Baikal-Muya belt [19,25]. They occur within the north and north-west margin of Bargusin microcontinent (Fig.2).

We have to note that Zonenshain's paleomagnetic and geological reconstructions [42] of the Siberian continent position are rotated 180° relative to the modern situation. In the Late Riphean the modern Baikal corner of the craton was situated near the equator. The age of Baikal Muya ophiolites should be Early Riphean [19,37]. However, more ancient rocks within such ophiolites are possible, especially those rocks associated with komatiite lavas and dikes. Their age is close to Akitkan series age [21].

Riphean ophiolites of Baikal-Muya belt include dunites, harzburgites and basaltic pillow-lavas. Stratigraphically upwards there are island arc volcanics and tuffogenous rocks of basalt-andesite-dacite Kelyansk series with which the gabbro-granite calc-alkaline association of Muya complex is connected. The age of Kelyansk series volcanics is 1100-1090 Ma [21]. Kelyansk series include metamorphosed olistostromes [15,23] confirming the first accretionary and thrust processes within the marginal part of Baikal-Muya paleotectonic basin. The major stage of accretion and nappe-formation is fixed by Riphean olistostrome resting on the upper sections of Olokit series [19]. This stage is marked by zonal metamorphism of 1.0-1.1 Ga. All these events took place before the accumulation of molassic volcanic-terrigenous Padrinsk series and intrusion of layered gabbro of Dovyren complex (720-750 Ma) which are related to an active continental margin environment. The Bargusin microcontinent collision with Siberian craton likely occurred in Pre-Vendian time and is fixed by the second stage of zonal metamorphism of 650-620 Ma. These complexes are overlain by shallow water Vendian-Cambrian terrigenous-carbonate platform-type deposits [19].

Late Proterozoic complexes also form big areas within the Tuva-Mongolian microcontinent. This microcontinent is divided into two sialic blocks: Sangilen and Central-Mongolian. The first block is characterized by Late Proterozoic predominantly carbonate sedimentation. The Central Mongolian block is mainly composed of metamorphic gneiss-schist sequences with moderate development of Pre-Vendian carbonate rocks. Among these blocks, a zone with ophiolites, complexes of active margins, accretionary and collisional complexes and Vendian-Cambrian post-collisional cover extends on to the adjacent sialic blocks. The complexes of this zone correspond to those of the Baikal-Muya zone complexes.

Precambrian ophiolites of Tuva-Mongolian massif are localized in several regions where they form belts of various length. The Ilchir belt, the largest, is traced from the marginal suture of Siberian craton far westwards into the northern areas of Mongolia. In the eastern part of these belts allochtonous ophiolite plates are thrust over Late Proterozoic schist-carbonate cover of the Gargan block [16,18]. To the west the ophiolites gradually disappear being overlain by younger sequences or overthrust tectonic sheets.

The ophiolites from south-east part of the East Sayan associate with tectonic sheets and lenses composed of schist-volcanic rocks with boninites, turbidites and olistostromes possibly reflecting the environments of primitive island arcs and subduction-accretion prisms.

A restricted belt of Precambrian ophiolites stretches along the eastern margin of Sangilen block. It relates to nappe-folded structure located at the boundary between continental block and paleooceanic zone. The fragments of Precambrian ophiolites are present in the western margin of Tuva-Mongolian micro-continent and in the Daribi ridge in Mongolia [41]. According to Tomurtogoo, an accretionary complex was distinguished in the western part of Tuva-Mongolian micro continent. The formation of these accretionary complexes can be related to subduction of the Late Precambrian oceanic crust beneath the Central-Mongolian block with local formation of the primitive island arcs.

Later, in place of these island arc - accretionary zones an active continental margin with volcanic-plutonic belts was formed. Here the marginal basins were filled by debris molasse-like sequences. Such sequences present by Sarkhoi and Darkhat suites on the south-east of East Sayan and in Trans-Khubsugul area, and by Dzabkhan suite on the west from Tuva-Mongolian micro continent. The first two suites include shallow-water sediments and gradually differentiated volcanics being referred to the complexes of active continental margins [31] or to island arcs [24]. The Dzabkhan suite is composed of acid effusives and sub-volcanic intrusive bodies formed within the active continental margin [41].

Radiometric ages of continental margin complexes of the zone are considered close. Sarkhoi suite effusives were dated in 718±30 Ma by Rb-Sr isochron [12], and Dzabakh suite rocks have the values of 725 Ma.

The problem on Sarkhoi suite interrelations with ophiolites and rocks of the primitive island arcs is now under discussion , although there are data showing the occurrence of ophiolite fragments in the clastic debris of the Sarkhoi suite. The Dzabkhan suite in some places occurs just above the accretionary complex rocks [41]. It proves once more the more ancient age of ophiolites and island arc rocks compared with the active continental margin complexes.

Collision of two continental blocks, which added to the Tuva-Mongolian microcontinent formation, occurred at the end of Late Proterozoic. These important events were followed by glaucophane metamorphism which relates to this time period [18]. A Rb-Sr isochron on several samples of glaucophane schists is 640 ± 20 Ma. Now, this belt of high pressure rock is traced for 250 km within the modern structure on the north and west it and outlines areas of active continental margin complex. This high-pressure belt is overlain by Vendian - Cambrian carbonate cover [21].

Do ophiolites belonging to Bargusin and Tuva-Mongolian microcontinent relate to a single oceanic basin or do they have various ages and represent separate oceanic basins? The question is considered as follows: The development of these paleo-oceanic zones is considered to have a common tendency during the final stages when the Vendian-Cambrian carbonate cover was formed. The Tuva-Mongolian ophiolite complexes formed before the development of the active continental margin complexes. The Padrin sequence of Baikal-Vitim zone corresponds to these complexes, but may have a more ancient age. The layered basic-ultrabasic bodies with dates of 750 Ma, are close to the age of Sarkhoi and Dzabkhan effusives and are comparable to such complexes. Berzin believes that ancient ophiolites and island arc rocks within the Bargusin and Tuva-Mongolian microcontinents are of the same age. Dobretsov considers that the ophiolites, olistostromes and zonal metamorphic complexes of Baikal-Muya complex are ancient (1500-1000 Ma), but the younger volcanics, differentiated intrusive bodies and metamorphic rocks of 750-600 Ma age are related to the history of the surrounding Paleoasian ocean .

Small fragments of ophiolites and island arc series of Late Proterozoic formed in a suture zone and were preserved in the central part of East Sayan. They are located between the Derba microcontinental block, resembling the Sangilen block and the ancient border of Siberian craton. Up to recent times, the geodynamic position of such rocks were not studied in detail. The radiometric data are absent, as well. Here, the alpinotype ultrabasics and greenstone altered basalts with subordinate amount of acid effusives, limestones and debris are known [5,39]. This suture includes thick terranes composed of Vendian-Lower Cambrian carbonate deposits of the Mirichun suite. The limited areas of such rocks within the marginal suture, incomplete information about the sections and absence of radiometric data do not permit us to make a reliable correlation with complexes previously characterized from the Bargusin and Tuva-Mongolian micro-continents, however the general geological situation and structural position make this correlation possible.

Late Precambrian ophiolites and their associated rocks could extend further to the north-west and north into Mansk region of East Sayan and the western slope of Enisey ridge.

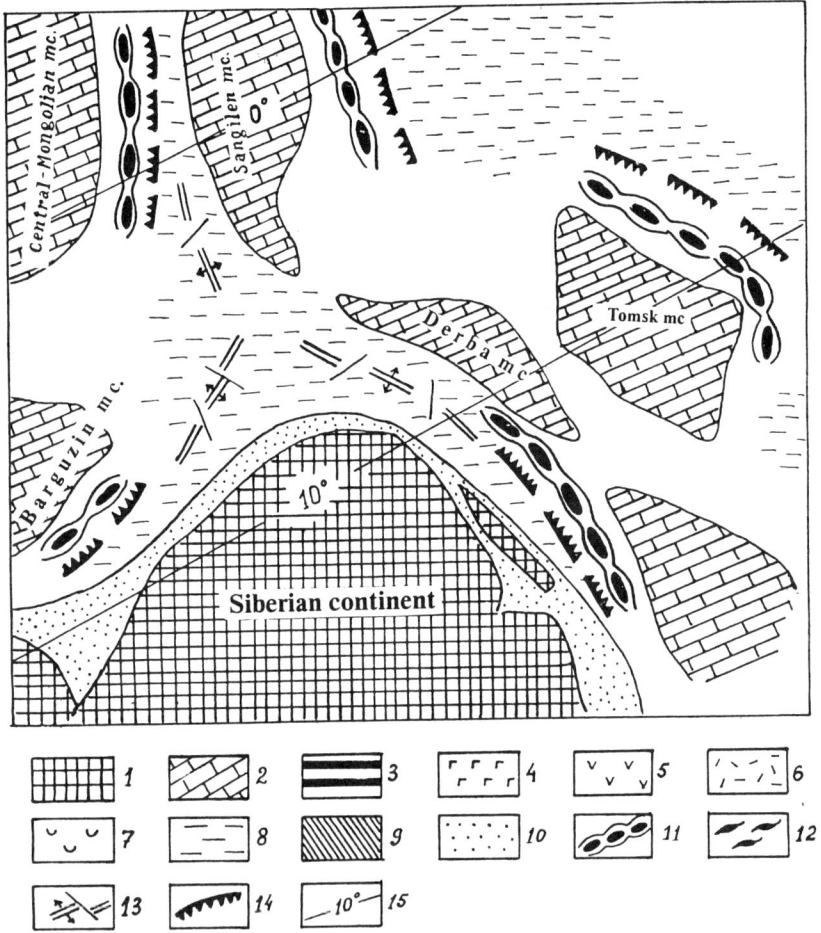

Figure 3. Reconstruction of South Siberia and West Mongolia in the Late Precambrian time (700-750 Ma). 1- Siberian continent; 2- Microcontinents of Laurasian group and newly formed continent in late Precambrian-Lower Paleozoic time with predominant carbonate cover; 3- Altai-Mongolian microcontinent (part of Gondwana passive margin); 4- Complexes of Late Precambrian island arcs, active margins and ophiolites within the structure of the newly formed continent; 5 to 7- Island arc complexes included in folded structures: 5- $V-\mathcal{C}_1$, 6- $\mathcal{C}_{1-2}$, 7- $\mathcal{C}-O$; 8- Oceanic crust; 9- Seamount; 10- Terrigenous and carbonate-terrigenous sediments of passive margins and intercontinental rifts; 11- Active island arcs; 12 - Cambrian-Ordovician accretional-collisional zone; 13- Spreading axes; 14- Subduction zone; 15- Paleolatitudes.

In the Enisey ridge they can be clearly distinguished and belong to the tectonic nappes thrust from the west to the east. According to Kuzmichev [30] the tectonic zonation is restored here, indicating formation of ophiolites within the marginal sea basin which was limited on the west by a volcanic island arc with a gradually differentiated series of calc-alkaline volcanics. Kuzmichev [30] considered this basin was localized immediately along the boundary of the Siberian craton. Using the data cited, we think that the Paleoasian ocean or basins with oceanic type crust existed in Late Proterozoic time and was surrounded by Trans-Baikal and Enisey parts of Siberian craton. It is difficult to estimate their actual dimensions, but they were not narrow rifts but represent large ocean basins formed as a result of sea-floor spreading. On the convergent boundaries of these developing oceanic plates and on the edge of the craton, island arcs and active continental margin complexes formed (Fig.3).

The occurrences of ancient paleo-oceanic and island arc complexes not only close to the Siberian craton but within the micro-continents indicates that a series of isolated continental blocks (Bargusin, Derba, Minusa, Tuva and Mongolian) existed in Late Proterozoic Paleoasian ocean near the Siberian craton. Essentially carbonate sedimentation is present on all of these blocks in Late Riphean - Vendian. Joining up of the Tuva and Mongol blocks and the accretion of Bargusin and Derba blocks to the Siberian craton occurred in the Late Precambrian or Vendian. At this time, subduction zones were inclined away from the craton. The absence of Late Proterozoic active continental margins on the craton periphery combined with formation of passive margins, or intra-continental rifts within the craton supports this subduction polarity.

Collisional processes following the Late Riphean accretion were fixed both within the micro-continent and paleooceanic zone complexes and in the passive margin rocks of Siberian continent. We cannot distinguish which passive margin complexes were connected with collision of a specific paleotectonic unit; because during the long geological history, significant horizontal displacements between the ancient craton and the surrounding accretion-collision systems occurred. As shown above there are geological and geochronological data supporting two periods of collision. One collisional event proceeded before or at the beginning of Late Riphean (1.1-0.8 Ma), the other before or at the beginning of Vendian (650-630 Ma). The first collision corresponds to the time when several micro-continents were joined up and collided with island arc, and partly (in Enisey ridge) with the craton. Simultaneously, within continental blocks and their dividing accretion-collision zones, a zonal metamorphism occurred and autochthonous granitoids were formed. On the passive margin of the Siberian craton these events are clearly exhibited in the Enisey ridge, where tectonic collision processes and granite formation are recorded 1.1-0.8 Ga. The Late Precambrian collisional granites are possibly localized within the inner (Bodaibo) zone of Baikal Patom passive margin.

Riphean collisional complexes of the south margin of Siberian craton are located within sub-meridional and north-east passive margins. In the northwest, near the Sayan plot where the ancient metamorphic sequences are widespread, Riphean age granite complexes were not found. Therefore, this plot could represent a zone along which the transform displacements predominated.

The passive margin Pre-Vendian collisional stage is characterized by unconformities and abrupt change in sedimentary environments. In West Trans Baikalia there is a change from sand-mudstone-clayish deposits of Kochargat suite to coarse-grained rocks of Ushakov suite. In Near-Sayan, there is a clear Late Riphean-Vendian boundary between Karagass and Oselok series. The Karagass suite is composed of carbonaceous-terrigenous rocks intruded by small bodies of alkaline gabbro and alkaline effusives [5]. Petrochemical data on the magmatic rocks correspond to intra-plate magmatism. Clastic debris was transported from the craton margin and according to Yu.K. Sovetov's data (oral information) the Oselok series which lies unconformably on the Karagass series consists of shallow-water sediments or alluvial sequences derived from clastic debris derived from mountains surrounding the craton.

Thus, we believe that the major collision of Tuva-Mongolian, Bargusin and other microcontinents with the Siberian craton occurred at the end of Precambrian, about 600-650 Ma. According to palinspastic reconstructions of L.P.Zonenshain and others [42], the complete joining of South Siberian and Mongolian micro continents to Siberian continent occurred at the end of the Ordovician. For a period of 200 Ma during the final collision, oceanic crust developed between the Siberian carton and these micro-continents. The problems, related to the formation and distribution of such oceanic crust is discussed below.

Vendian - Cambrian stage of active oceanic margins
The maximal opening of Paleoasian ocean took place simultaneously with accretion-collision events resulting in the joining of the above mentioned micro-continents with each other and with Siberian craton during 600-700 Ma period (Fig.4). The support for this idea comes from the occurrence of an extensive system of volcanic island arcs created on periphery of newly formed Siberian continent. Kuznetsk-Tuva and Dzhida systems were formed south-west and south-southeast (using present coordinates) from the continent, respectively.

Kuznetsk-Tuva island arc system. It was established above that the subduction zones inclined towards the continents. In the reconstructions of L.P. Zonenshain [42] these arcs together with micro-continents were located far from Siberia, and subduction zones (unlike our interpretation) had the opposite inclination. Some regularities confirm our point of view. The first one is the spatial position of fore-arc accretionary zones in which the fragments of oceanic ophiolites, island arc rocks, deposits of fore-arc troughs, serpentinite melange, ophiolite-clastic olistostromes, and other features were preserved. Almost all the outcrops of ophiolites and ultramafics are found to the west or to the south of the reconstructed island arcs.

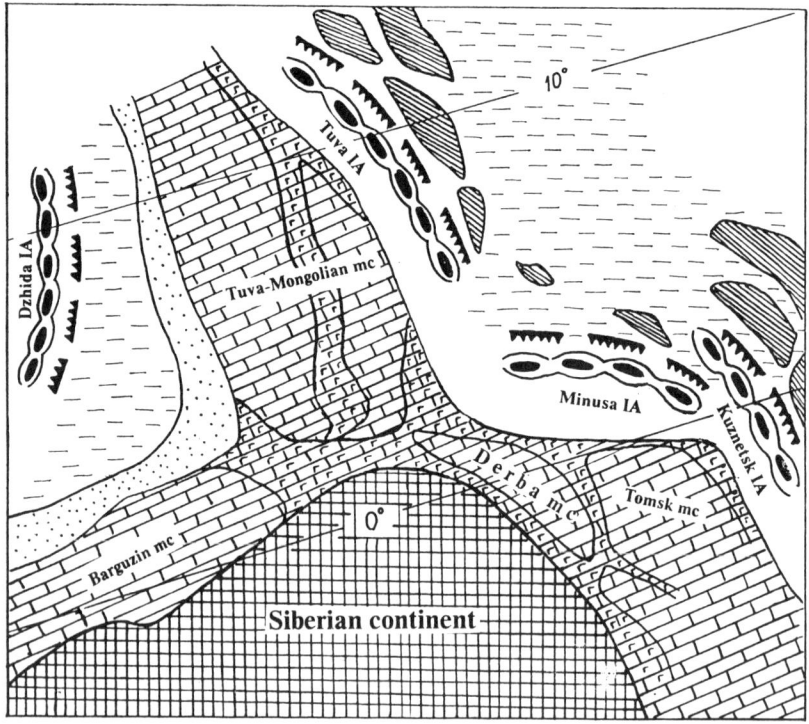

Figure 4. Reconstruction of South Siberia and West Mongolia in the Vendian - Early Cambrian time (600-550 Ma). Legend see in fig.3.

Somewhere, the inclination of subduction zones is controlled by lateral changes of effusive rock compositions (Kuznetsk Alatau). At last, the geological data show that between volcanic arcs and the Siberian continent there was no giant oceanic basin, but instead marginal seas developed. Incorporated in the basement of these seas and some arcs complexes are earlier formed accretion-collision zone rocks. Let us consider several examples, where this fact can be established.

The Agardag accretionary complex preserved between Sangilen block and Lower Cambrian island arc rocks of East Tannu-Ol is considered to be the base of the Tuva paleo-island arc (East Tannu-Ol and Khankhukhey fragments). The complexes from the backarc troughs of the preceding stage (Kharal and Okhem suites) occur beneath the Khamsarin plot of Tuva arc. They are characterized by a wide abundance of fragmental rocks high in clastic quartz, the occurrence of granite pebbles indicating their closeness to the continental blocks. The Dzhebash accretionary complex with Boruss ophiolites and serpentinite melange containing high pressure metamorphic rocks [20,21], occurs beneath North Sayan plot of Minusa arc. This complex resembles Oka-Hugein complex of Tuva-Mongolian microcontinent. Kuznetsk-Tuva arc system was established on oceanic crust (western part of West Sayan and south-eastern part of Gorny Altai). Such arcs did not reach the developed stage in Vendian-Cambrian time. They are characterized by primitive volcanism consisting of andesite-basalts, boninites and rare differentiated series.

The sequences of various composition have being accumulated in Vendian - Early Cambrian back-arc basins depending on the basement structure or remoteness of volcanic uplifts. In the place of Late Precambrian island arcs or accretionary zones mainly tuffogenic-terrigenous series with rather small lenses of lavas and carbonate rocks were formed. Carbonate sediments accumulated on the thickest sialic blocks.

In Early Cambrian within the island arcs of Kuznetsk-Tuva system the intrusions of tonalite-gabbro-plagiogranite series were formed. It is difficult to distinguish them among more widespread younger granite rocks of Altai-Sayan area corresponding to active margins or collisional environments. Plutons of the northern slope of West Sayan (Muya complex) can be considered as the most obvious subduction granites. Pebbles and boulders of such granitoids are present in conglomerates of Early-Middle Cambrian throughout all areas of extension in the Kuznetsk-Tuva system. In the back-arc situation, during subduction activity the gabbro of increased alkalinity or differentiated gabbro-wehrlite plutons were intruding. In some places basalts and contrasted differentiated series were found. Their composition proves the rift basin or marginal seas environment in the back-arc situation.

We believe that situations cited above show that the formation of Kuznetsk-Tuva island arc system was close to the Late Precambrian continent and was inclined towards the continent.

Dzhida island arc system. We have some data on this system which formed on the opposite side of Siberian continent compared the with Kuznetsk-Tuva system. Dzhida system includes Bayan-Khongor zone of Mongolia [2] and Eravnin zone of Buryati [1].

There are no reliable data on the time of Dzhida system establishment, because its appearance does not directly relate to the Late Precambrian accretion of South Siberia. We agree with L.P.Zonenshain [42] that this system formed within an oceanic basin over the subduction zone inclined away from the continent. The absence of clear signs of Cambrian magmatism and volcanic activity within the southeastern margin of the continent confirms this polarity. Likely, a passive margin existed here.

The most active volcanism in both Dzhida and Kuznetsk-Tuva systems has a Early Cambrian age. After this active period, the volcanism within the Dzhida system ceased in places where oblique collision with the continent began. Collisional environment together with volcanism existed for a long time up to Ordovician. Basalts and andesites of Barungol sequence within the Gargan block are Ordovician [18]. Oblique collision relates not only to the arcs divergence but to the horizontal displacement of sialic blocks of Bargusin microcontinent. Finally, a significantly complicated nappe-imbricated structure, is clearly manifested by zonal metamorphism [18], forming within southeastern East Sayan, Khamar-Daban and Dzhida area. When Sliudyanka-Khamardaban block was pressed into Dzhida zone, the Khubsugul-Tunka block was detached and overthrust onto Tuva-Mongolian micro-continent with continued formation of imbricated structures. The youngest signs of these giant horizontal displacements in these regions is found in the Silurian deposits [16,18].

Salair - West Ozyornaya island arc system. It was formed after Kuznetsk-Tuva system within the southwestern margin of the Siberian continent (Fig.4). Fragments of this arc system are preserved in Salair, east Gorny Altai, north West Sayan, in the Sistigkhem zone of Tuva and along the western margin of Ozernaya zone of Mongolia.

The establishment of a new island arc system can be related to the migration of subduction zones due to a roll over of subduction in the previous zone by small continental blocks and oceanic seamounts composed of intra-plate and oceanic basalts often with siliceous-carbonate cover. This situation is described in detail in the paper of Buslov and others (this volume). Seamounts from eastern Gorny Altai, Gornaya Shoriya and western Kuznetsk Alatau are overlain by siliceous-carbonate cover (Baratal suite and its analogs). In the Tuva and Ozernaya zone volcanic blocks do not contain thick carbonate covers, but only thin siliceous deposits are present.

The distance shown between the Salair-West Ozernaya subduction zone from the previous one is not equal because the different plots used show different dimensions for the collided blocks. The widest area of oceanic seamounts and accumulated blocks is located between the Salair and Kuznets arcs on the north, and between the West Ozyornaya and Tuva arcs in the south. The abrupt narrowing of inter-arc zone takes place within the West Sayan plot and its margin. It is caused by strong horizontal compression of these areas between Tuva-Mongolian and Minusa micro-continents followed by large strike-slip and overthrust dislocations. However, the subduction zone maintained its position and developed during both of the stages. In the first stage volcanism gradually or with little attenuation was followed by the second stage volcanism. In this regime the volcanic arcs of the northwest Sayan area formed.

Obviously, the collision of oceanic seamounts with the Vendian - Early Cambrian island arc systems often developed at the end of the first half of Early Cambrian resulting in a structural and sedimentary reconstruction. The collision is characterized by the volcanic activity in Kuznetsk-Tuva system and a general uplift which produced the appearance of coarse-grained rocks and unconformities. During the second half of Early Cambrian the terrigenous-carbonate deposits accumulated in the vast areas. During this time a favorable environment for development of carbonate rift rocks was created and these sediments were deposited on the newly formed island arc system and on the uplifted blocks of inter-arc zone. These sediments are widespread from Salair to southern areas of Mongolia.

The island arc volcanism of Salair- West Ozyornaya system had begun in the first half of Early Cambrian when the Kuznetsk-Tuva subduction zone was still active. Subduction was active during Middle Cambrian and ceased (Salair) in Late Cambrian. Volcanism was locally exhibited within the inter-arc accretion zone, the Kuznetsk-Tuva arc system and on the Minusa microcontinent.

Closing of South-Siberian part of Paleoasian ocean (Middle Cambrian - Early Ordovician)
From Middle Cambrian up to the Ordovician thick sequences of terrigenous sediments accumulated within the inner side of Salair-West Ozyornaya island arc system (Fig. 5). Before, Russian geoscientists believed that the sedimentation was within a relic sea basin. After the appearance of the Plate Tectonic Theory, this basin is now considered to represent a relic of a paleoocean in its final development stage. The eastern margin of Lower Paleozoic basin was thought to be part of the newly forming Siberian continent, however the question of its western margin was not considered at all nor were the causes and mechanisms of the basin termination.

At the present time, the geodynamic nature of sedimentary basin and its accumulated terrigenous sequences become clearer. Two types of sections are established in the Pre-

Ordovician deposits.

The first type relates to the central and north parts of Gorny Altai, in Salair, Sistigkhem area in Tuva close to Vendian-Cambrian island arc systems. This section is composed of polymictic and volcano-mictic sequences sometimes with tuffs and tuffites.

The second type relates to the south part of Gorny Altai, in Mongolian Altai, south-west West Sayan and south-west Tuva. Deposits consist of plagioclase quartz turbidites (sandstones and mudstones) with rather high content of clay material. Upwards in the section more polymictic and coarse-grained clastics appear and containing fragments of volcanic and intrusive rocks derived from island arc heights.

For a long time the sources of these thick sequences of plagioclase-quartz rocks remained unclear. A.B. Dergunov [14] interpreted their appearance and wide abundance as the result of weathering of ancient heights along the south margin of Siberian platform, with transport of the terrigenous clastic material along the West-Sayan trench and its deposition resulting from avalanche sedimentation within the marginal part of Late Caledonian basin. This version does not answer all the questions. For example, why are these sequences so homogenous without volcanic clasts in spite of the active development of volcanic island arcs within the eastern margin of basin? How could such terrigenous material overcome an active margin with its marginal seas, island arcs and deep-water trenches without containing any enrichment in local volcanic clasts? As a rule, we see sedimentary sequences of this age formed near volcanic arcs contain or consist of the products of their destruction (volcanic clasts). Because of this observation, Berzin suggested another source for these plagioclase-quartz terrigenous sequences. The source probably comes from a non-magmatic continental

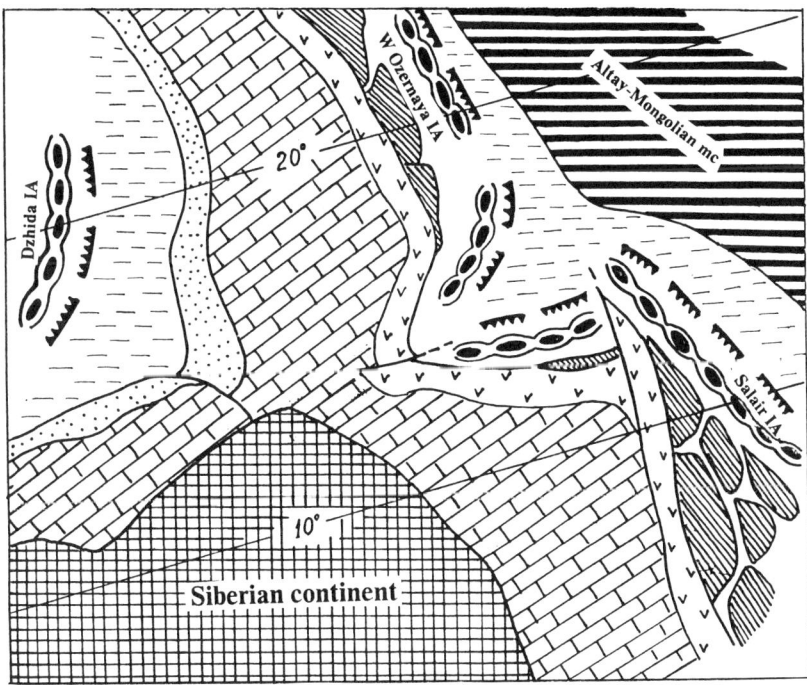

Figure 5. Reconstruction of South Siberia and West Mongolia in Middle Cambrian time (530-520 Ma). Legend see in figure.3.

blocks present on the other side of oceanic basin as it came closer to the active margin of South Siberia.

Other geoscientists came to similar conclusion independently. During the organizational meeting of the IGCP Project 283 in Urumchi (China), Prof. He Gouqi advanced and then published his ideas. He suggests that during a long period (from Sinian or Vendian up to Early Ordovician) the sequences of Gorny, Mongolian and Chinese Altai formed along a passive margin along the northern edge of the Paleo-Dzhungarian (Junggar) continental massif [26]. He Gouqi has suggested that a fragment of this massif was beneath the Dzhungarian (Junggar) trough. At this same time, the monograph of L.P. Zonenshain and others [42] was published and these sequences are considered to be part of the Altai microcontinent which forms the passive margin of the ancient Gondwana continent. Thus, the idea about these sedimentary sequences being related to passive margin of a southern continent rather than the Siberian craton is now preferable (Fig. 6).

Finally, the collision between the passive continental margin of Altai-Mongolian micro continent and Caledonian active margin commenced. This collisional process was long and its beginning coincides with the termination of the youngest subduction zones and cessation of active margin volcanism in the Late Cambrian. At that time, terrigenous material carried from these paleo-island arcs began to overlay the plagioclase-quartz turbidites. As a result, there was an upward change from turbidites to polymictic sequences sometimes containing coarse-grained rocks (conglomerates and olistostromes).

Post-Cambrian history
The latest collisional impulses are fixed by a periodical supply of coarse-grained clastic

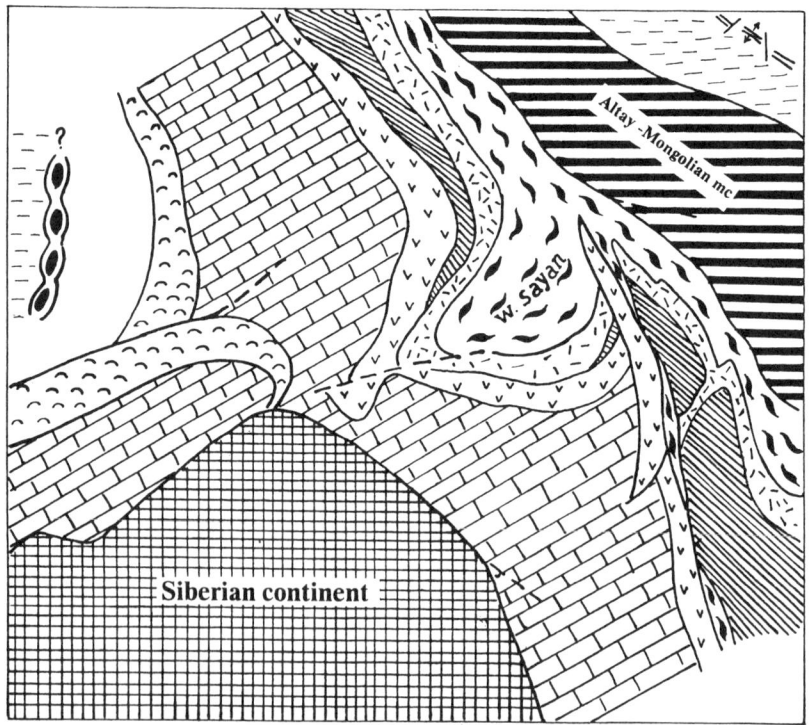

Figure 6. Reconstruction of South Siberia and West Mongolia in Late Cambrian - Ordovician time (520-440 Ma). Legend see in figure 3.

material resulting from the Earth's crust deformation. The unconformities between Cambrian and Ordovician, Early and Middle Ordovician, at the end of Ordovician and between Early and Late Silurian are the most clearly manifested and widespread. In the Ordovician and

Silurian rather shallow-water and carbonate-terrigenous deposits (sometimes resembling the molasses type) were formed on most of the Altai-Mongolian micro-continent, in the place of the earlier Cambrian island arcs and back-arc basins. In the south and southwest (in present coordinates) the deposits change to more deep-water sequences related to a passive margin of the Ordovician ocean (?). Deep-water troughs formed in the place of the West Sayan, Anui-Chuya zone of Gorny Altai and Kobdinsk zone of Mongolia. These troughs are considered as relics of the Vendian-Cambrian ocean [14]. Berzin believes them to be a series of troughs of different depth and dimension existing during the stage of divergence and collision between the Altai-Mongolian micro continent with the South Siberia island arc systems (Fig.6). The trough in the East Sayan formed in place of an oceanic bay and should be the deepest in this range. In Ordovician and Silurian times this trough was connected with the Oka zone of southeast East Sayan and also with Dzhida zone which was an active margin for another oceanic basin. Transgressive bedding of Ordovician-

Silurian sequences along southeast West Sayan confirm the presence of these troughs. Moreover, these sequences overlay more ancient complexes towards the inner part of Early Caledonian continent. Such a situation is typical for the vast territory of North Asia. These deposits result from an active spreading within the inner areas of Paleoasian ocean (Kazakhstan and South Mongolia) and generally coincide with the opening of Uralian paleoocean [8].

However, there is another interpretation for this structure as discussed in papers [16,18,42]. These authors suggests the presence of an oceanic strait or bay here with island arcs during Ordovician. This strait was "closed" by the movement of Tuva-Mongolian micro-continent with adjacent Caledonian structures of Ozyornaya zone only in the Silurian-Early Devonian. The occurrence of Ordovician-Early Silurian (?) island arc volcanic series in these structures makes such interpretation possible. These volcanics include an andesite series in the Kobdinsk zone [14] and an andesite-liparite series in the West Sayan [14,42]. In the Oka zone, the Cambrian-Vendian greenstone tuffogenous effusive series containing fragments of ophiolites and blueschists of 640 Ma age (Rb-Sr isochron) is overthrusted by Ordovician tuffogenous-terrigenous series with black shales and sills of diabases and gabbro-diabases. This series is thought to be a marginal sea sequence adjacent to an island arc. Stratigraphically upward it is overlain by Late Silurian-Early Devonian molasse-like sequence of sandstone, carbonaceous mudstones, gravellites, olistostromes with interlayered acid effusives which can be related to a neo-autochton. This sequence includes olistoliths of meta-ophiolites and blueschists [18]. There are no clear boundaries between Ordovician and Cambrian and Ordovician and Silurian in the sections mentioned. In the northern part of Dzhida zone (the Tunka'Alps) the Barungol suite (volcanic-sedimentary with andesites, andesite-basalts and siliceous interlayers) and Toltin series (carbonate-terrigenous sediments with rare andesites) form large tectonic nappes and are zonally metamorphosed in collision zone [3].

Thus, in the inner zone of this structure, the nappes and their related olistostromes are fixed in Silurian. In Middle Silurian and Early Devonian along the margins of these structures the main nappes were formed and their collision began. Zonal metamorphism and granite bodies of 420-400 Ma age mark the upper age limit of these processes.

In the Devonian, the major part of this area was an active continental margin. On the south, within South Altai and. Gobi Altai zones island arcs were located. The inter mountain depression (Tuva, Minusa, Rybinsk) with trachybasalt and trachyandesite-basalt volcanism were situated north from the zone of active volcanic activity of Cordilleran type shown in Fig.2.

While reconstructing the palinspastic environment of the South Siberia for Late Precambrian and Early Paleozoic (Fig. 3-6) we came to the conclusion that the movement of plates and terranes along the Siberian continent-Paleoasian ocean boundary had a southwest-northeast orientation for a long period (using present coordinates). This orientation resulted in various geodynamic environments along the margin of the Siberian craton and the newly-formed continent in the Late Proterozoic. Such a situation continued into the Cambrian when island arc systems and back-arc basins with local rift environments were formed under oblique subduction along a south-west margin. Oblique subduction caused a delay in volcanism from the southeast to northwest along southwest margin of the Early Caledonian continent. In the second half of Early Paleozoic the convergence of Altai-Mongolian micro continent with Caledonian active margin continued. After their collision in a wide zone of interaction of converging plates the horizontal movements were concentrated along a large northwest strike-slip and overthrust fault systems of sub-lateral and northeast strike (Fig.7). These movements intensified due to the compression of the accreted southeast sialic blocks during Devonian and Late Paleozoic. The large-scale horizontal displacements within southwest margin of Siberian continent were also related to the rotation of Siberian craton and displacement of Tarim block relative to the North Chinese craton. They caused intensive deformation of pre-collisional structural situation with a general displacement of terranes to the northwest. Those terranes which were earlier bordering the continent could be transported far to the northwest and were finally buried under the Mesozoic cover of the West Siberian platform An extended period of active margin and collisional environments southeast of the Siberian continent produced intensive granitization and metamorphism within complexes developed here.

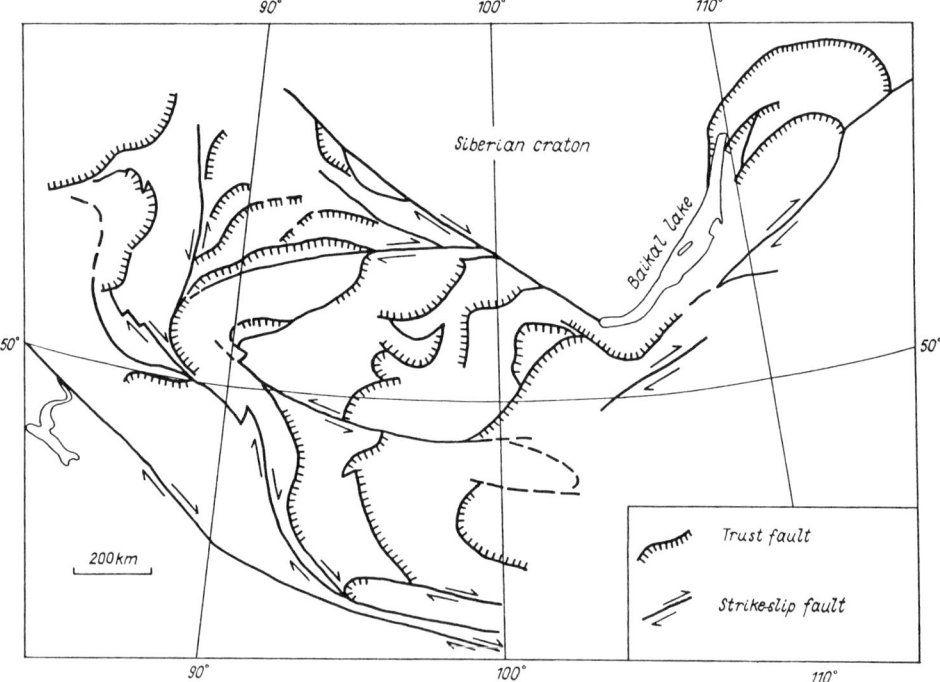

Figure 7. Scheme of the main thrusts and strike-slip faults in the modern structure.

The main period of granitoids formation with exception of local Cambrian granitoids, was in the Ordovician, Devonian and Carboniferous-Early Permian. Carboniferous-Early Permian (sometimes Pre-Jurassic) granites was post-collisional A-granites whose formation relates to the melting in the lower part of the intensively thickened sialic blocks or formed by remelting deeper in the mantle [36]. The homogenization temperature of melt inclusions in such rocks show high-temperature (1000-1200°C) and a mantle origin for the A-granites melt [35].

In Late Permian, all the territory of South Siberia and Kazakhstan was involved in the Eurasian continent and became a young platform. To the south and to the southeast the oceanic structures of Paleotethys and Mongol-Okhotsk ocean were developing. Late Permian paleoreconstructions are shown in Fig.8. As in the previous cases of Cambrian and Ordovician-Silurian ocean, Paleo-Tethys and Mongol-Okhotsk oceans can be interpreted as newly-formed oceans or a results of the continued evolution of the previous Hercynian (Devonian-Carboniferous) Paleoasian ocean. From this we conclude that the upper age limit of the Paleoasian ocean is probably Permian.

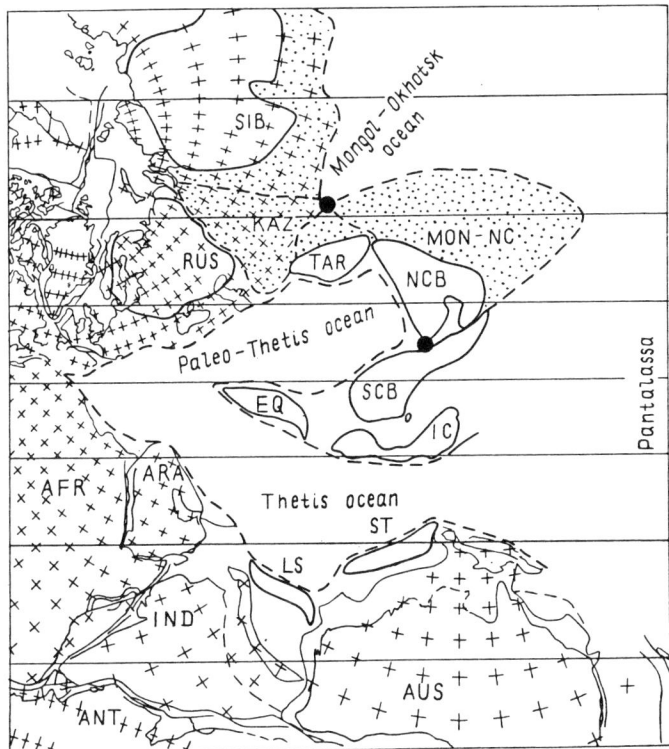

Figure 8. Schematic reconstruction of Asian and Gondwana plates in the Late Permian time, mercator projection (modified from [43]). Dotted areas correspond to folded structures originating from Paleoasian and Uralian oceans. Names of plates and blocks: **RUS**= Russian, **SIB**= Siberian, **KAZ**= Kazakhstan, **TAR**= Tarim, **NCB**= North China block, **SCB**= South China block, **EQ**= East Qiangtang, **IC**= Indoosinian, **AFR**= Africa, **ARA**= Arabian, **IND**= Indian, **LS**= Lhasa, **ST**= South Tienshan, **ANT**= Antarctica, **AUS**- Australia.

Triassic and Jurassic evolution of Paleo-Tethys and Mongol-Okhotsk oceans [13,42] was completed after formation of the Eurasian continent's modern boundaries (except for the collision of the Indian block 40 Ma ago). Which produced movement to the west of the Tarim block, the successive collision of blocks of South-East Asia (north-Chinese, South Chinese, Indo-Chinese, East Chintan, East Qiangtang), as well as the collision of the Pacific type blocks (Omolon, Okhotsk, Bureya, Hida, and others), whose history relates to the Mesozoic evolution of Pacific ocean

CONCLUSION

Most of the terranes in South Siberia and associated folded areas are related to micro-continents, island arcs and accretionary complexes (Fig.2). Ophiolites fix the spreading stage and blueschist complexes fix the late subduction-early collision stage located along boundaries of huge terranes and belong to accretionary complexes (Fig.1,2). Using the relative age dates and interrelations between these terranes, ophiolites and blueschists, we have attempted to make a paleo-reconstruction of the Paleoasian ocean within South Siberia and West Mongolia during the Late Proterozoic-Ordovician time interval (Fig. 3-6). During this evolution active spreading zones were formed and then during the Late Riphean (800-900 Ma), Vendian-Cambrian (620-500 Ma) and Ordovician-Silurian (500-400 Ma) oceanic basins were successively closed. Evolution of the Hercynian (D-P_1) Paleoasian ocean was changed by Mesozoic history of Paleotethys and Mongol-Okhotsk oceans. For example, the Cambrian ocean shows roll over of the subduction zones towards the ocean which resulted from cessation of subduction zones by submarine uplifts (choking).

All the above considered paleo-reconstructions are concordant with paleomagnetic reconstructions [27,42,43] and make these reconstructions more reliable concerning the location and displacement of subduction zone positions, back-arc basin and passive continental margins. However there is still the need for more, detailed paleomagnetic determinations in South Siberia.

Acknowledgments:
We Thank Prof. R.G. Coleman for the invitation to contribute to this volume and our colleges Prof. R. Coleman, V. Khain, L. Zonenshain, A. Dergunov for critical remarks during presentation of our material in Kyoto and in the preparation of manuscript.

REFERENCES

1. V.G. Belichenko. Early or complete Caledonides of Sayan-Baikal mountain area. *Izvestija Acadentii Nauk SSSR Geol. Ser.* **NI,** 68-75 (1983) (*in Russian*) .

2. V.G. Belichenko. Paleotectonic reconstruction of Paleozoides of southeast East Sayan, West Khamar-Daban and Near-Khubsugul, *Geology & Geophysics.* **NS,** 11-19 (1985).

3. V.G. Belichenko and R.G. Boos. Metamorphic series of Tunka hills (East Sayan). In: *The problem for stratigraphy for Middle Siberia Early-Precambrian . pp.5-21,* Nauka, Moscow (1986) (*in Russian*) .

4. V.G. Belichenko and R.G. Boos. Characteristic features of paleotectonics in the central part of the Central Asia belt in Paleozoic time. In: *IGCP Projec~ 283 Report #1* pp.11-17. Nauka, Novosibirsk (1990).

5. N A. Berzin. *Major Fault zone of the East Sayalt.* Nauka, Moscow (1967).(*in Russian*).

6. N A. Berzin. Horizontal movements during formation of the Central Asia structure. In: *IGCP Project 283 Report #1* pp.17-20. UIGGM, Novosibirsk (1990)

7. N A. Berzin. Horizontal movements during formation of Paleozoids of Altai-Sayan area and West Mongolia. In: *Geodynamics structure and metallogeny of folded structures of South Siberia.* Abstr., pp.153-155. UIGGM, Novosibirsk (1991) (*in Russia*)

8. K.B. Bogolepov, A.K. Basharin and N A. Berzin. *Tectonics and evolution of the earths crust of Siberia.* Nauka, Novosibirsk (1988) (*in Russian*).

9. Ch.B. Borukaev. *Precambrian structure and Plate Tectonics.* Nauka, Novosibirsk (1985) (*in Russian*).

10. Ch.B. Borukaev, A.K. Basharin and N.A..Berzin. *Precambrian time for continents: the features of tectonics.* Nauka, Novosibirsk (1977) (*in Russian*).

11. A A. Bukharov, V.O. Glasunov and N.M. Rybakov. Baikal-Vitim Early Proterozoic greenstone belt, *Geology & Geophysics.* **N7,** 33-39 (1985).

12. M.I. Buyakaite, A.B. Kuzmichev and D.D. Sokolov. 718 Ma Rb-Sr isochron of Sarkhoi series of the East Sayan, *Rep. Acad. Nauk USSR*. **309**, 150-154 (1989).

13. J. Dercourt, L.P. Zonenshain and others. Geological evolution of the Tethys from the Atlantic to the Pamirs since the Lias, *Tectonophysics*. **123**, 241-315 (1984).

14. A.B. Dergunov. Caledonides of the Central Asia. Nauka, Moscow (1989) (*in Russian*).

15. N.L. Dobretsov. Ophiolites and the problem of Baikal-Muya ophiolite belt. In: Magmatisn *and metamorphism of BAM zone their role in formation of useful minerals*. Nauka, Novosibirsk (1983) (*in Russian*).

16. N.L. Dobretsov. About nappe tectonics of East Sayan, Geotectonica. **Nl**, 39-50 (1985a) (*in Russian*).

17. N.L. Dobretsov, R.G. Coleman, J.G. Liou and C. Maruyama. Blueschist belts of Asia and periodicity of blueschist metamorphism, *Opltioliti*. **12**, 445-456 (1987).

18. N.L. Dobretsov and V.I. Ignatovich (Eds). *Geology and ores of the East Sayan*. Nauka, Novosibirsk (1989) (*in Russian*).

19. N.L. Dobretsov and *A.N.*Bulgato, *Geodynamic map of Trans-Baikal*. Nauka, Novosibirsk (1991) (*in Russian*).

20. N.L. Dobretsov, E.G. Konnikov and N.N. Dobretsov. Precambrian ophiolites of southern Siberia and their metallogeny, *Precambrian research* **58**, 427-446 (1992)

21. N.L. Dobretsov, E.V. Sklyarov and S.Q. Dong. Blueschist and ophiolite belts of North Asia, *J. South Asian Earth Sci*.(1991) (in Press).

22. N.L. Dobretsov, N.V. Sobolev and V.S. Shatsky. *Eclogites old blueschists in folded belts*. Nauka, Novosibirsk (1989) (*in Russian*).

23. N.F. Gabov and V.G Kartavchenko. Metamorphosed olistostrome and its relation to Precarnbrian ophiolites in North Baikal region. In: *Petrology and Mineralogy of Siberian basites*. Moscow, Nauka (1984) (*in Russian*).

24. A.S. Gibsher, A.E. Izokh and E.V. Khain. Pre-Vendian structure of Tuva-Mongol part of the Central Asia foldbelt. In: Geodynamics, structure and metallogeny of folded structures of South Siberia, Abstr. Publ. UIGGM, Novosibirsk (1991). (*in Russia*).

25. G.S. Gusev, A.I. Peskov and S.K. Sokolov. Paleo-geodynamics of Proterozoic Baikal-Vitim belt, *Geotectonics*. N2, 72-86 (1992) (*in Russian*).

26. He Guoqi. A preliminary scheme of tectonic division of the Paleo-Asian oceanic domain. In: Geodynamic evolution of Paleoasian ocean. IGCP Project 283, Rep.#1, pp.42-45, Novosibirsk (1990).

27. A.N. Khramov and L.E. Sholpo. Paleomagnetism. Nedra, Leningrad (1987) (*in Russian*).

28. M.P. Klimetz. Speculations in the Mesozoic plate tectonic evolution of Eastern China, *Tectonics*. 2, 139-166 (1983).

29. Yu A. Kosygin, N. Berzin, B.N. Krasilnikov and L.M. Parphenov. *To the interrelations of Siberian platform and geosyncline areas in Precambrian*. Nauka, Leningrad, Moscow (1964) (*in Russian*).

30. A.B. Kuzmichev. Riphean marginal-oceanic paleostructures in the Sayan-Enysei foldbelt. In: Geodynamic evolution of Paleoasian ocean, IGCP Project 283, Report #1, pp.54-59, Nauka, Novosibirsk (1990).

31. A.B. Kuzmichev. Sarkhoi series as stratotype (stratigraphy, structural position, age). In: *Late Precambrian and Early Paleozoic of Siberia*, pp.104-123, Publ. UIGGM, Novosibirsk (1990).

32. V.I. Lebedev, A.G. Vladimirov, V A. Khalilov and A.S. Gibsher. U-Pb dating and the question on stratification of Precambrian - Lower Paleozoic metamorphic and magmatic rocks of the West

Sar~gilen (South-east Tuva). *Geology & Geophysics* (1993). (in press).

33. Y. Li. An apparent polar wander path from the Tarim block, China. Tectonophysics. **181**, 3141 (1990).

34. J. Lin, M. Fuller and W. Zhang. Preliminary Phanerozoic Polar wander paths for the north and South China blocks, Nature. **313**, 444-449 (1985).

35. B A. Litvinovsky. The analyses of conditions of acid magma generation in the activized mobile belts. In: *IGCP Project* 224, Report #5, pp.45-51, Osaka (1990).

36. BA. Litvinovsky and N.L. Dobretsov. The Permian-Triassic granitoid magmatism and volcanism of the USSR Central Asian part and the Mongolian territory. In: *IGCP Project 224* Report #2, pp.44-45, Osaka (1987)

37. G.L. Mitrophanov, A.V. Sintsov and V.P. Korchagin. Geodynamic environments of south East Siberia in Riphean (based on the Plate Tectonics Theory). In: *Precanlbrian geology tectonics petrology and ores of Siberian Platform and its margins.* Geochronology: Abstr., pp.32-33, Publ. Inst. Earth's Crust, Irkutsk (1987).

38. G.V. Pinus. On the age of ultrabasites of East Sayan and some associated problems of geology, *Geology & Geophysics.* **N4**, 58-66 (1965).

39. G.V. Pinus, VA. Kuznetsov and I.M. Volokhov. *Ultrabasites of Altai-Sayan folded area.* Publ. Academii Nauk , Moscow USSR (1958) (*in Russian*).

40. A.P. Sekerin, V.A. Lashenov, B.M. Vladimirov, Yu.V. Menshagin and V.G. Domyshev. The pecularities of aulacogen magmatism of the East Near-Sayan. In: *Magmatism of rifts (petrology, evolution, geodynantics)* pp.83-89, Nauka, Moscow (1989) (*in Russian*).

41. O. Tomurtogoo. *Ophiolites and formation of folded areas of Mongolia.* Thesis doct. dissert. Publ. GIN USSR, Moscow (1989) (*in Russian).*

42. L.P. Zonenshain, M.I. Kuzmin and L.M. Natapov. *Plate tectonics and geology of the USSR.* Moscow, Nedra. (1990) (*in Russian*).

43. X. Zhao, R.S. Coe, Y. Zhou, H. Wu and J. Wang. New paleomagnetic results from northern China: collision and suturing with Siberia and Kazakhstan, Tectonophysics. **181**, 43-51 (1990).

The Geodynamic Evolution Of The Mobile Fold Belts Of The Territory Of Mongolia

D. DORJNAMJAA, G. BADARCH & D. OROLMAA

Geological Institute of the Mongolian Academy of Sciences, Ulaanbaatar 210351, Mongolia

Abstract - The paper deals with the geodynamic environments of various stages, and changes in the tectonic regime and style of tectonic deformations in Mongolia. New data on kinematics of heterogeneous tectonic structures are presented and illustrated on the geodynamic map of Mongolia (1:2,500,000). The map portrays rock complexes and structures in their present position (Fig.1). Analysis of the Earth's crustal structure and its dynamics of formation in the territory of Mongolia suggests that the geodynamic settings existing at the present stage has emerged largely in the Riphean-Phanerozoic. The map shows that Mongolia consists of a composite region of a mosaic of folded mountains and linear nappe-fold belts. These structures were formed as a result of long term divergence and convergence of lithospheric plates. Within the territory of Mongolia two important and distinctive types of evolutionary process can be found within the oceanic segments. In the Indo-Atlantic segment, an evolution of the geodynamic setting reveals replacement of previously existing mosaic systems by linear systems accompanied by replacement of an autochthonous mechanism in the formation of granite-metamorphic layer by the allochthonous mechanism. In the Pacific segment, the granite-metamorphic layer forms mainly as a result of granite intrusion and imbrication of the mantle substance, namely oceanic crust and island-arc complexes. The early tectonic pattern of the Precambrian-Paleozoic folded structures controls the general arcuate shape of the major tectonic structures in Mongolia. The mosaic of folded mountains and linear-fold belts within the territory of Mongolia developed as a result of successive closure of three oceans: 1). Early Paleozoic Paleo-Asian, 2). early-Middle Paleozoic south Mongolian and 3). late Paleozoic Inner Mongolian. Later, during the continental stage of development which extends through the Middle Carboniferous to Permian and Triassic periods, volcano plutonic complexes and molasse were deposited in large depressions and intermountane troughs. In Mesozoic time, a structural reconstruction took place and is marked by block-uplift movements accompanied by intrusive activity with continental molasse zeolite- and coal-bearing deposits interbedded with significant calc-alkaline and alkaline volcanism.

Key Words - Geodynamic, tectonic, terrane, structure, ophiolite, granite, sediments

INTRODUCTION

The territory of Mongolia belongs to the eastern part of the huge Urals-Mongolian orogenic belt and is composed of a mosaic of fold mountains and linear nappe-fold belts. A great number of monographs and papers are devoted to the description of tectonics and magmatism of Mongolia [1,2,9,10,11,19,20,21,25,27,28,29,31]. Most of these authors discuss the geological history of Mongolia in terms of geosynclinal, platform, and post-platform (so-called tectono-magmatic activization) stages. In this paper we will show, that oceans often become the most reliable source of information about the Earth's geological history within the Mongolian mobile fold belts. The formation of a new Mongolian ocean began with the

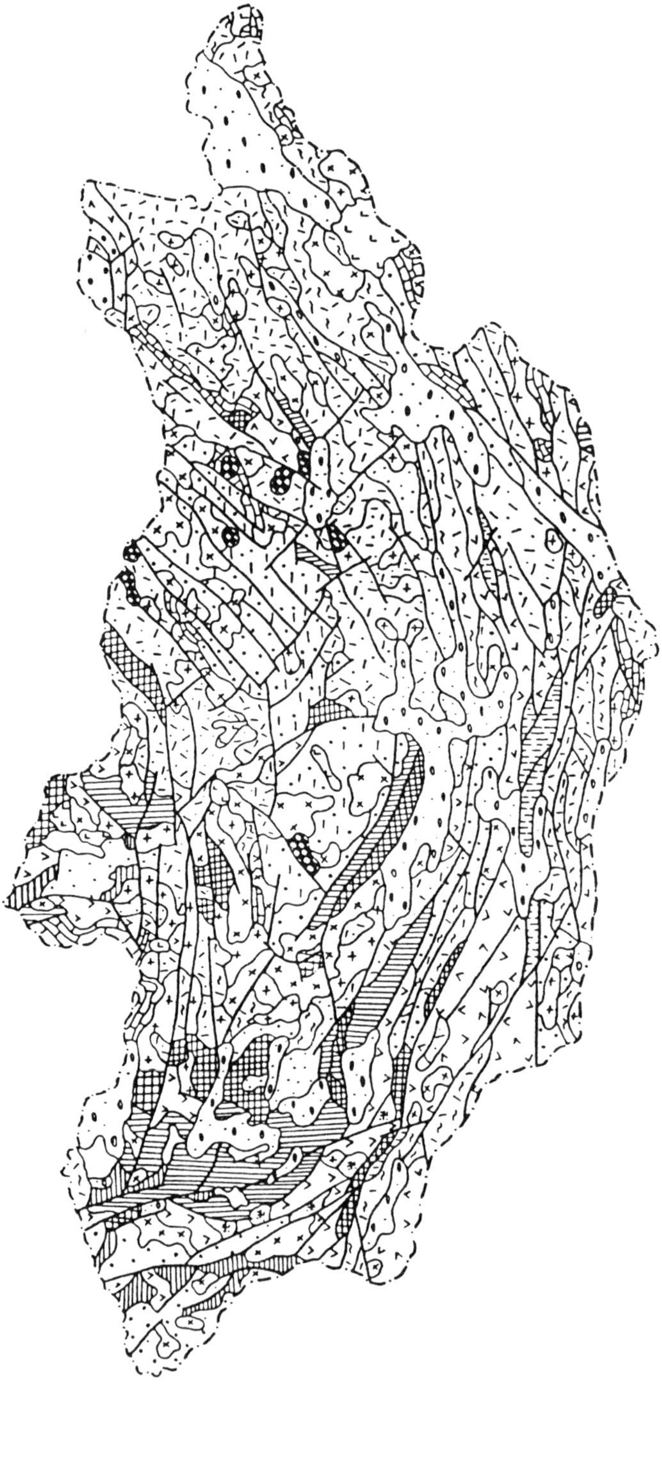

Figure 1 Geodynamic scheme of Mongolia

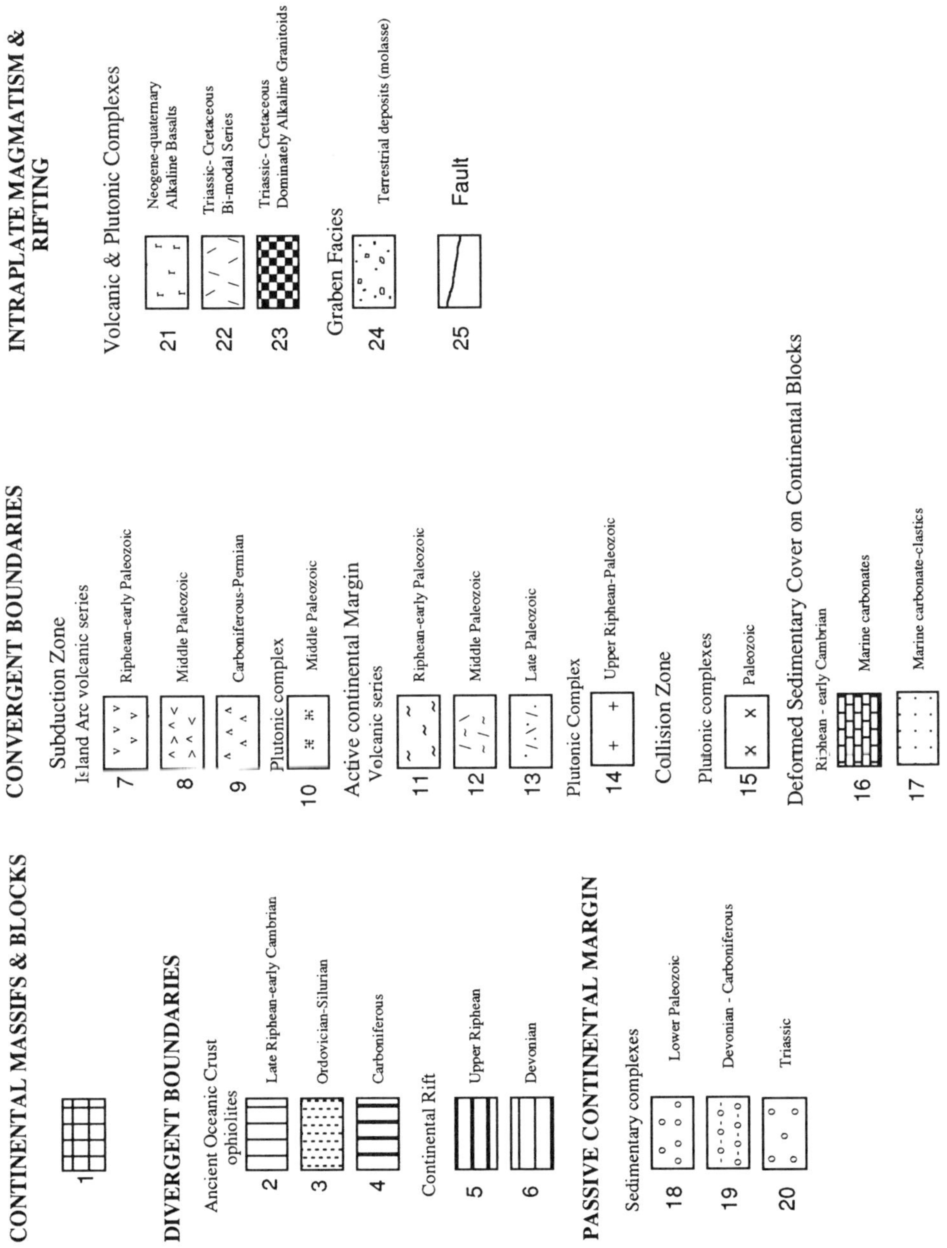

Geologic Evolution of Mongolia

CONTINENTAL MASSIFS & BLOCKS

1

DIVERGENT BOUNDARIES

Ancient Oceanic Crust ophiolites

2 — Late Riphean-early Cambrian
3 — Ordovician-Silurian
4 — Carboniferous

Continental Rift

5 — Upper Riphean
6 — Devonian

PASSIVE CONTINENTAL MARGIN

Sedimentary complexes

18 — Lower Paleozoic
19 — Devonian - Carboniferous
20 — Triassic

CONVERGENT BOUNDARIES

Subduction Zone
Island Arc volcanic series

7 — Riphean-early Paleozoic
8 — Middle Paleozoic
9 — Carboniferous-Permian

Plutonic complex

10 — Middle Paleozoic

Active continental Margin
Volcanic series

11 — Riphean-early Paleozoic
12 — Middle Paleozoic
13 — Late Paleozoic

Plutonic Complex

14 — Upper Riphean-Paleozoic

Collision Zone
Plutonic complexes

15 — Paleozoic

Deformed Sedimentary Cover on Continental Blocks
Riphean - early Cambrian

16 — Marine carbonates
17 — Marine carbonate-clastics

INTRAPLATE MAGMATISM & RIFTING

Volcanic & Plutonic Complexes

21 — Neogene-quaternary Alkaline Basalts
22 — Triassic- Cretaceous Bi-modal Series
23 — Triassic- Cretaceous Dominately Alkaline Granitoids

Graben Facies

24 — Terrestrial deposits (molasse)

25 — Fault

Figure 1 Geodynamic scheme of Mongolia

formation of rifts and rifto-genetic complexes of late Riphean and Vendian age. In northern and western Mongolia, late Riphean bimodal volcano-plutonic complexes belong to continental rifting. These older rifts were distinctly more mature. Everything indicates that the rift system was compound and new ocean crust has developed by expansion of the older oceanic crust. The final intricate construction of the Asian Paleoocean was completed in the Vendian-lower Cambrian and produced an extremely heterogeneous assemblage of rocks. The undifferentiated basalts, cherts and pelagic limestones were formed within the oceanic basin, and andesite-greywacke and tuffaceous formations were deposited within the island arcs and marginal seas. The closing of the Asian Paleoocean was followed by the obduction and subduction which produced imbricated blocks and terranes of ophiolite now exposed at the present day surface. The closing also produced ancient ensimatic and ensialic island arcs. Finally, the collisional and the post collisional processes produced orogenic zones. The destructional processes resulted in granite intrusion and metamorphism of the island arc complexes. Prevailing structural elements forming in late Precambrian-early Paleozoic have been manifested during the Phanerozoic stages of development in the heterogeneous geological structures of Mongolia. These structures are sharply expressed in the recent deformation, and especially in the geomorphological shape of the region.

I. FOLD MOUNTAINS-MOSAIC AREAS

The northern part of Mongolia contains the fold mountain-mosaic areas where they usually correspond to the Caledonides. Within the mosaic areas, the Tuva-Mongolian and central Mongolian massifs, smaller blocks have been recognized. These massifs and blocks consist Archean and early Proterozoic basement of high-grademetamorphic rocks covered by the Riphean to Vendian protoplatform rocks and the upper Riphean rift-related volcanics [22, 16, 9, 10] The Vendian-lower Cambrian calcareous and terrigenous sediments with phosphorites unconformably cover all of the older formations. The Archean-early Proterozoic stage(2.8-1.6 Ga) was characterized by the development of protocratons later united by various small granulite-gneissic blocks (Hanhuhiy, Tarvagatay, Uriygol, Buteeliynnuruu, Songino, Baydrag, Tsogt-Tseel, and others) (Fig. 2). These events occurred by remobilization of a tonalite substratum and formation of normal granitoids. Early Precambrian complexes (Baydrag and Buteeliynnuruu complexes, U-Pb ages 2800-2650 Ma) consist largely of recrystallized supracrustal rocks such as quartzite, high alumina gneiss, amphibole-biotite grey gneiss, amphibolite, garnet-biotite gneiss, marble and enderbite, which form the sialic basement. Most of the early Precambrian granulite-gneiss areas contain some lenses and layers of metamorphosed iron-bearing sediments. They are usually a few metres thick and typically are interbedded with magnetite quartzite in supracrustal sequences with mica schists, marbles, amphibolites and paragneisses. Along with these sequences an anorthositic pluton (so-called Mustuliyn massif-U-Pb ages 3050-2500 and 2450 -1880 Ma) is present within the north Hangay highland (Tarvagatay block). The anorthosites commonly intrude the gneiss-amphibolite complex following their high-grade regional metamorphism. These anorthosites contain more than 90% plagioclase and are associated with a suite of granitic rocks and probably represent products of a bimodal magmatic processes. The late Precambrian stage (1.6-0.57 Ga) involved quartzite-carbonate, effusive-carbonate and schist-quartzite units. These metamorphic complexes are derived from a uniform sedimentary cover of the pre-Riphean crystalline basement. Vendian-lower Cambrian carbonate-terrigenous complexes are covered by middle-upper Cambrian molasse-flysch formation which consists of a terrigenous melange containing separate boulders of the Vendian limestone and granite pebbles. Probably, these characteristic formations can be considered as accretionary complexes formed during the collision of the small ancient sialic blocks.

The extension of sialic crust and generation of narrow oceanic rifts as well as formation of broad oceanic areas developed within the recent ophiolite (Nuuriyn, Bayanhongor, Dzida and Herlen) zones. As shown in Figs.1 and 2, the divergent boundaries are divided into oceanic

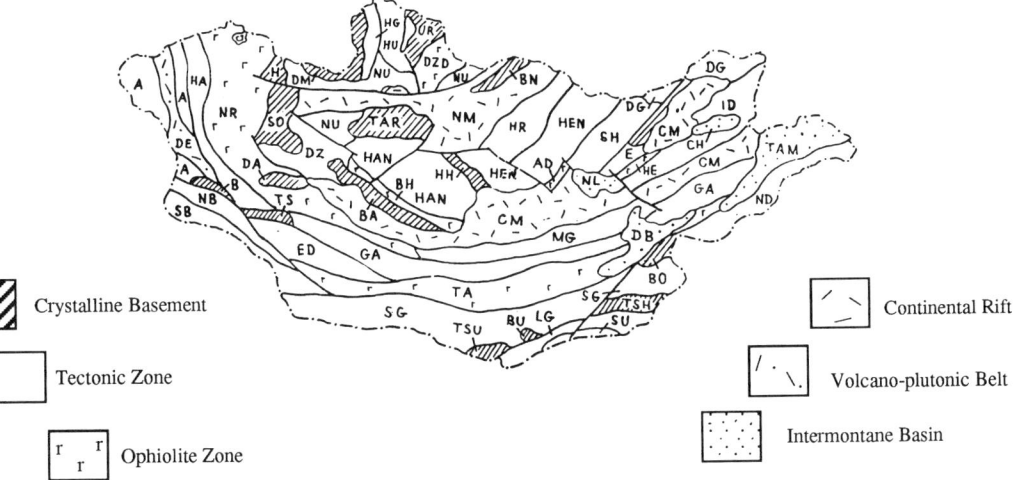

Continental Massifs & Blocks

- B- Bodonoch
- BA- Baydrag
- BN- Buteeliynnruu
- BO- Bargiynovoo
- BU- Bulgannuul
- DA- Dariv
- DM - Delgermoron
- E- Ereendavaa
- H- Hanhuhiy
- HH- Harhorin
- SO- Songino
- TAR- Tarvagatay
- TS- Tseel
- TSH- Totoshan
- TSU- Tsagaanuul
- UR- Uriygol

Continental Rift

- DE- Deluun

Volcano-plutonic Belt

- CM- Central Mongolian
- NM- North Mongolian

Intramountain Basins

- CH- Choybalsan
- DB- Dzununbayan
- NL- Nyalga
- TAM- Tamtsag

Tectonic Zones

Fold-Mountains & Mosaic Areas

- A- Altay
- AD- Adaatsag
- BH- Bayanhongor
- DG- Dochgol
- DZ- Dzavhan
- DZD- Dzida
- MG- Middle Gobi
- HAN- Hangay
- HA- Harhiraa
- HR- Haraa
- HEN- Hentey
- SH- South Hentey
- HE- Herlen
- HU- Hubsugul
- HG- Hug
- ID- Idermeg
- MG- Middle Gobi
- NU- Numreg
- NR- Nuur

Linear Nappe-Fold Belts

- NB- North Baruunhuuray
- SB- South Baruunhuuray
- ED- Edren
- GA- Gobi Altay
- SG- South Gobi
- LG- Lugiyngol
- ND- Nuhutdavaa
- SU- Sulinheer
- TA- Transaltay

Figure 2 Scheme of tectonic subdivisions of Mongolia

and continental rift zones. The geodynamic setting of oceanic rift zone is characterized by imbricated ophiolite complexes (or tectonic klippe) and the continental rift zone is characterized by bimodal volcanic and plutonic complexes. The ophiolite complexes are stratigraphic units ideally comprising the following upward sequence: serpentinized dunite-wehrlite-clinopyroxenite-gabbroic banded complex, gabbro, "sheeted" basic dike zone, basic pillowed volcanic rocks, chert, pelagic limestone and argillite [10, 26,32]. The petrochemical study of the metavolcanics of ophiolite complexes which are located in the Dzida zone have revealed varying types of volcanic rocks, tholeiitic, calc-alkaline, alkaline and marianite-boninitic series. Their composition and differentiation trends correspond best of all to geodynamic conditions of a frontal arc-marginal basin-continental margin. Other authors [17, 16, 32] are of an opposite opinion. After their study of the ophiolite rocks of the Hantayshir area, they came to the conclusion that the formation of the ophiolites of the Lake (Nuuriyn) district formed in a back-island arc as a result of the back-arc spreading. Neodymium-samarium whole rock ages on the gabbros from the adjacent Dariv-Nuruu ophiolite give a Vendian age [6] . From this point of view, this system of island arc-marginal sea produced oceanic type crust not far from the Siberian margin. Ophiolites of late Riphean-lower Cambrian age are shown in Fig.1 as formed in rift zones. The rifting of marginal parts of the Siberian continent started in the upper Riphean [16] . It was accompanied with siliceous magmatism dated 820 m.y. The further extension of these old continental rifts is associated with a younger magmatism, dated 752 m.y. and expressed by bimodal volcanics, and comagmatic alkaline granite. Clearly, the Darhat-Hubsugul rift in the Hubsugul zone was once part of the old rift system, which was filled with clastics of arkose composition with some rhyolites and minor basalts. Another later continental rift zone is found at Deluun including the Achitnuur and Tsagaanshiveet grabens in western Mongolia.

Dergunov et al [9] suggest that the Deluun zone is a destructional trough-monogeosynclinal. This zone is located in western Mongolia between Altay and Harhiraa zones (Fig.2), extending about 450 km in a NW direction. The Deluun zone was filled with orogenic molasse, clastic sediments (shales, sandstones and conglomerates) and bimodal terrestrial volcanic rocks. Probably, the alkaline granites (for instance, the Halzanburged rare-metal bearing granite massif not far from Hovd city) are associated closely with the Devonian rift zone, which is characterized by bimodal volcanics.

Rock complexes related to convergent plate boundaries are variable and are subdivided into subduction and collision complexes. Zones of calc-alkaline magmatism (Nuur, Dzida, Herlen and others.) are very good indicators of former subduction zones related to island-arc and active continental margin environments. The oceanic stage of the late Riphean-early Cambrian in the early Paleozoic is replaced by a transitional stage. Indicators of which are tonalite-plagiogranite, granodiorite, and granodiorite-granite magmatic intrusions. The first is the M-type granite concentrated in Nuur, Dzida, Bayanhongor, and Herlen zones invading earlier formed oceanic crust. Another type is mainly I-type granites within the marginal area of continental crust blocks bordering the imbricated ophiolites. A magmatic transitional stage within the Nuur zone and others during the early Paleozoic changes the magmatic setting of the developed island arc and continental block magmatism to a setting of mature island arc and active continental margin producing belts of batholith-like granitoids. Indicative, of this magmatic associations in the early Paleozoic, is the complete cessation of extension within the former basins where ophiolites formed and were replaced by conditions of compression. The island arc calc-alkaline volcanism of these newly compressional areas are characterized by basalt-andesite-rhyolite, andesite-basalt, dacite-andesite, and andesite-dacite associations. The most characteristic complexes are established in the Dzida zone [16] ,Tariat-Selenge trough [15] , and Nuur zone [8, 9] . The continental stage came to an end in early Paleozoic (Ordovician) throughout the whole territory of the northern mountain-mosaic area. The passive continental margin in north and central Mongolia was changed into an active one in which a volcanic arc was developed and characterized by calc-alkaline and basic-intermediate volcanics. The North Mongolian (Selenge-Vitim) rift-type volcano-plutonic belt is characterized by the calc-alkaline volcanics (especially andesite-rhyolite type), subalkalic and alkalic comagmatic complexes with extensive development of the intraplate A-type granites as shown below.

The Numreg active continental margin in space corresponds to the Ider volcanic (V- $\mathcal{C}_1$) zone in north Mongolia. As shown in Figs.1 and 2, this zone extends as a stripe 800 km long and 120 km wide. In this zone, the Devonian volcano-plutonic rocks are very extensive. Recent work [15, 13] in this area shows that the Devonian volcano-plutonic complex is a typical bimodal association. This complex is considered to have formed by active continental margin or continental extensional tectonism. Sediments (thickness of more than 10 km), form along an active continental margin of the Andean-type and contain submarine volcanic rocks of acid and intermediate composition (for instance, in the Orhon, Selenge, and Zelter depressions). Analogous complexes are found in the central Mongolian belt (or zone) (Fig. 2) which forms on an active continental margin that can be traced nearly 1500 km with widths of 300-350 km. This complex consists of several subzones, Hantayshir-Buutsagaan (west part), north Gobi and the country around R. Herlen (eastern part). These parts are characterized by the wide development of Permian continental sedimentary and volcanic units accompanied by contemporaneous granitoid intrusions. The Hantayshir-Buutsagaan subzone extends as a strip 10-15 km wide for a distance of 500 km and is represented by numerous grabens and troughs. These structures were filled with bimodal alkaline volcanics associated closely with subalkaline intrusions. Magmatic rocks of this area are high K-Na-type containing rare-metals. The north Gobi and Herlen subzones are characterized mainly by rhyolitic and andesite-rhyolitic series products of aerial volcanism. These volcanics are up to 1000 km long and tens to hundreds of km wide. In this area the intrusive rocks are represented mainly by batholiths and stocks consisting of granodiorite and granosyenite. Thus, the western part of the central Mongolian volcano-plutonic belt is characterized by the alkaline, mostly bimodal volcanic association and intrusion of granosyenite-syenitic and alkali-granitic rock. They are mainly located in tilted-block rift-valley or trough-like structures. The east part of this belt includes the normal and subalkaline, often differentiated volcanic series and intrusive rocks of granodiorite-granitic and granite-granosyenitic composition. Volcanic rock is interbedded with various terrigenous rocks. Within this area terrigenous rocks are deposited in the intracontinental depressions. During the Permian and lower Triassic periods both north and central Mongolian active continental margins underwent intrusion of granite into the terrigenous and volcanogenic-terrigenous rocks. In these intrusive zones, A-type granites are distributed as stocks and batholiths and are characterized by an K-feldspar granite, alkali granite and hornblende syenite. Each period is characterized by a unique magmatism and metallogeny. Numerous deposits and occurrences of nepheline syenite are established within the west Hubsugul Devonian alkaline intrusion. There are the Dochiyngol, Uvurmaraatgol, Beltesgol and other deposits with rich metal reserves. The main metallogeny of the active continental margin in the late Paleozoic time is characterized by the fact that some deposits of molybdenum and copper (for instance, Erdenet and Tsagaansuvraga) are genetically related to the porphyry copper ore formation. It is possible that late Paleozoic porphyry copper-molybdenum associated with andesite volcanoes and granite plutons formed along the continental margin. Volcanic-plutonic arcs [20] suggest that a large belt of calc-alkaline granite stretches from SW to NE along the Hangay-Hentey arched uplift and further to the Hinggan-Bureyan massif, then it turns to the south and reaches the south Primorye. We are persuaded that in the Mongolian territory a boundary zone exists between calc-alkaline latite magmatic series [14] and alkaline granite belts along the passive continental margin of Hangay-Hentey zones (uplift) forms a ring shaped area consisting of late Permian subalkaline granites, quartz syenites, monzonites, and monzodiorites. According to Gerel et al [14] the complex polymetallic mineralization (Gobi-Ugtaal ore-magmatic system) is the result of coincidence between the granitoid and latite magmatism.

Magmatic complexes are sensitive indicators of their geodynamic regime, and can be used to reconstruct a paleogeodynamic setting. By the end of the Cambrian age rather intensive processes of tectonic imbrication occurred and this was accompanied by metamorphism and granitoid magmatism. These processes led to the formation of granitoids in early (middle and upper Cambrian), middle (lower and middle Ordovician) and late(upper Silurian-Devonian) Caledonides, reflecting the successive growth of the granite-metamorphic crustal layer [7] . A number of continents and microcontinents (Tuva-Mongolian, central Mongolian), small continental blocks, island volcanic arches, marginal sea structures were converted into a shelf that represented the continental slope and rise of the Siberian continent which formed the Paleoasian ocean basin margin [4] . Probably at the end of lower Paleozoic age this whole oceanic area with ophiolites and volcanic arcs was consumed along zones of subduction and tectonic collision which resulted in amalgamation of the above microcontinents and other ancient tectonic blocks (or terrenes) on to the Siberian continental margin. These processes were accompanied by formation of Caledonian intrusions mainly of S-type and I-type granites (for instance,Telmen, Tohtohinshil and Herlen complexes of C_{2-3} age) which are large isolated batholiths or stocks. They are mainly found in the Altay, Nuur, Numreg, Hubsugul, Dzida, Hangay, Herlen, and Haraa zones and are characterized by tonalite, granodiorite, and quartz-diorite, usually associated with calc-alkaline volcanics. In the territory of the Hangay and Hentey zones (or arched uplift) during the Carboniferous-early Permian time calc-alkaline (I-type) diorite-granodiorite-granite series was formed. Granitoids of this series are also found in the west, in the Altay; the central part of Dzavhan; and to the east, in the Xerlen areas. Syncollisional S-type and Caledonian I-type granites were formed in the Hangay, Hentey, Hubsugul, Altay, middle Gobi, Nuhutdavaa, and other zones. At the same time, a marine and terrestrial molasse was deposited in these zones. S-type granites are represented by large batholiths or stocks containing abundant K-feldspar, two-mica granite or granodiorite (for instance,Tarvagatay,Hangay, Sharusgol intrusive complexes in the Hangay zone; Bagahentey, Bayanulaan, Delgerhaan, Burentsogt intrusive complexes in the Hentey and south Herlen areas). Pegmatitic veins or blocks are common. Enclaves are predominated by sub-micaeous rocks. Such features are genetically related to Sn, W (Be), and fluorite mineralization.

The accumulation of the coarse clastic molasse sediments in the collision stage of development is particularly rapid. Molasse is known in the Vendian, early, middle, and late Paleozoic, Mesozoic, and Cenozoic. In northern Mongolia, a thickness of marine, shallow-marine, and continental molasse deposits are estimated to be 3.5-4 km thick. The largest superimposed depressions are the Orhon-Selenge and Noen Sumiyn areas filled with molasse sediments which are up to 5-7 km in thickness. The molasse sediments are also widely developed in the Mesozoic and Cenozoic. The maximum thickness of molasse in the Mesozoic depressions may exceed 2 km, sometimes up to 4-5 km (for instance, Sayhanovoo trough in central Mongolia).

The convergence and collision between the various landmasses is not simple but has developed a compound-collision or compound-suture zone. The Caledonian suture zones as Nuur, Dzida, Bayanhongor and Herlen are marked by remnants of eroded island arcs, highly tectonized ophiolitic melange and high P/T metamorphics. Nappes, thrusts, granite batholiths, and chaotic complexes can be regarded to result from different styles of collision. High-pressure exotic components such as blueschist and eclogites (Dariv Nuruu) probably occurred during the late Proterozoic. The blueschists (crossite and winchite bearing) were first described within the western Xubsugul region in the Ilugiyn zone [25] (Fig.2) . These higher-pressure associations were established in the greenschist volcano-clastics of the early Riphean Hugeins formation. The mineral and chemical characteristics indicate that the blueschists were developed from mafic volcanic rocks. The Rb-Sr age of the glaucophane-schist metamorphism and the neighboring East Sayan area is 640 Ma. In our opinion, blueschists were formed at an early stage of accretion and tectonic imbrication. Rocks of the oceanic stage (ophiolites) are associated with zones of serpentinitic melange and form separate blocks. For instance, the so-called Ichetiyngol, an olistostrome tectonic melange in the Hanhuhiyn zone is several kilometres long and tens of metres wide and has a matrix of chlorite and talc slate. The melange blocks consist of highly deformed, partly metamorphosed, disrupted rocks of serpentinite, pyroxenite, greenschist, amphibolite, jasper, gabbro, marble, chert and volcanics. The following main stages can be recognized in the tectonic history of the northern fold mountains-mosaic areas: (1) the stage of the initiation of a proto-metamorphic layer, which was granitized by the end of the early Precambrian; (2) stage of partial destruction of the developed continental crust in the late Precambrian and formation of the continental margin with some elements of collision; (3) stage of the generation and formation of a broad oceanic basin in the late Riphean-early Cambrian; (4) stage of the closing of the Asian paleoocean and formation of the collisional and post-collisional orogens in middle, late Paleozoic and Mesozoic.

II. LINEAR NAPPE-FOLD BELTS

The linear nappe-fold belts lie to the south within the Mongolian territory and include the Junggar Variscan fold belt, the south Mongolian Variscan fold belt and the Inner Mongolian Indosinian fold belt (Fig-2).

The Junggar belt is situated on the western part of south Mongolia. The northern side of the belt (north Baruun Huuray zone) is characterized by Devonian and lower Carboniferous alkaline volcanics which consist of trachyandesite, quartz latite, trachydacite, rhyodacite and shoshonite. These volcanic rocks have been formed on the sialic block along an active continental margin. In addition, to the north of the block along the Bulgan fault, an extended narrow tectonic wedge exists and is characterized by the lower to middle Devonian tholeiitic basalt, tuff, chert, sandstone and siltstone. The Bizh zone is analogous to the Ertysh zone of northern Junggar in China [28] . The southern side of the belt (south Baruun Huuray zone) consists of Devonian tholeiitic pillow lava, calc-alkaline basalt, basaltic andesite, dacite, tuff, silicified rock and volcano-clastic turbidite. This sequence of rocks is covered by the lower Carboniferous flysch-like deposits. The paleotectonic setting for the southern Baruun Huuray zone may be a combination of frontal arc, fore-arc and back-arc basin.

The south Mongolian belt lies in the vast territory of the linear nappe-fold areas. The belt is divided into the Gobi Altay zone, Edren zone, Trans-Altay zone, south Gobi zone, and Nuhutdavaa zone (Fig.2). The Gobi Altay zone is characterized by the presence of linear structures, numerous longitudinal faults and by a variable facies of Ordovician to Carboniferous volcano-sedimentary deposits. In this zone, a system of troughs and uplifts is present. These troughs consist of a thick Ordovician to Devonian sequence of sandstone, shale, chert and turbidite. These rocks have suffered a greenschist and amphibolite-grade metamorphism and are intruded by concordant sheet-like bodies of leucocratic granite. Ultramafic bodies found in these troughs are considered fragments of melanocratic basement. The uplifts are covered by Ordovician and Silurian carbonate-terrigenous deposits. Crystalline basement is exposed only on the western end of the Gobi Altay zone, Tseel and Tsogt districts. These small blocks are represented by high alumina gneiss, hypersthene granulite, amphibolite, marble with mica schist and magnetite quartzite. They are overlain by lower-middle Riphean Dzargalantiyn formation, which consists of sandstone, siltstone, minor quartz-carbonaceous schist and porphyrite. This formation contain microfossils[22]. The pre-Riphean metamorphic rocks are correlated with the early Precambrian complex of the Uench and Bodonch areas on the southern end of the Mongolian Altay. The amphibolized hypersthene gneisses of this complex has a Pb-Pb zircon age of 2200 Ma [19].

We suggest that the system of troughs and uplifts was formed on the northern continental shelf of the south Mongolian ocean. Along the southern margin of the Gobi Altay zone, the upper Ordovician and Silurian fossiliferous sediments are overlain unconformably by the lower and middle Devonian consisting of predominantly bioaccumulated limestones and carbonaceous sandstones with occasional interlayered rhyolite. This sequence is conformably overlain by the middle-upper Devonian and lower Carboniferous volcanic and sedimentary sequences. The middle-upper Devonian volcanics are andesite-basalt, andesite, breccia, aquagene tuff, minor pillow lava, volcanoclastic and chert with radiolaria.and conodonts. The lower Carboniferous sedimentary formation consists of sandstone, siltstone, tuff with coquinoid limestone. These Devonian and lower Carboniferous rocks were formed in a volcanic arc environment.

In contrast to the southern margin, the northern side of the Gobi Altay zone is characterized by relatively deep marine sediments, consisting of dark colored middle-upper Devonian slate, sandstone and pyroclastic material deposited in a back-arc basin. One outstanding feature of the Gobi Altay zone is the presence of Carboniferous and Permian granite and molasse sediment. The Edren zone contains thick Devonian calc-alkaline basalt, andesite and volcanoclastic sediment. Lower Carboniferous intermediate volcanic rocks and terrigenous deposits lie unconformably on the Devonian volcanic rocks These deposits were accumulated on the volcanic uplifts, but the basement is unknown. The Trans-Altay zone lies on the axial portion of south Mongolian Variscan fold belt and contains widespread Paleozoic ophiolites. This zone consists of several structural-formational complexes, as follows: Dzoolon, Hadatuul, Berhuul and Gurvansayhan [23]. The lower part of these complexes is composed of siliceous rocks and pillow basalt of upper Silurian and lower Devonian age. The radiolarian cherts with volcano-clastic rocks in the lowermost parts are regarded as deep-water deposits. Gravel consisting of ultramafic rock has been found in the volcano clastic sandstone intercalated with radiolarian chert. The upper part consists of middle-upper Devonian tuff-terrigenous rock with minor basalt and chert. All of these complexes were formed on the melanocratic basement under special geodynamic conditions.

The basement is largely composed of serpentinized peridotite and serpentinite melange. The blocks in the melange include gabbro, diabase, pyroxenite, plagiogranite, locally amphibolite with basic dikes intruding the melange. The south Gobi and Nuhutdavaa zones are distributed south of the Trans Altay zone. Their metamorphic basement is composed of Proterozoic crystalline schists, garnet-biotite gneiss, amphibolite and marble. The sedimentary cover is upper Riphean-lower Cambrian fossiliferous carbonate, quartzite, intermediate-acid volcanic and clastic rock up to 2 km in thickness, intruded mainly by Devonian batholiths. The oldest rocks exposed in numerous small blocks such as Tsagaanuul, Bulganuul, Totoshan continue eastward into Arligemiao-Xilinhote landmass in Chinese Inner Mongolia [3]. The middle Ordovican to early Devonian carbonate-terrigenous and minor volcanic -terrigenous sequences were deposited on the northern passive continental margin of the south Gobi microcontinent. The middle Devonian molasse exposed only in the western end of the zone consists of conglomerate, sandstone, siltstone, and tuff resting directly on the granite and Silurian folded strata. However, the Carboniferous molasse more extensively distributed in the south Gobi zone is a result of the main collision stage.

The Inner Mongolian Indosinian fold belt occurs along the southeastern border of Mongolia and includes two tectonic zones Lugiyngol in the north and Sulinheer to the south. The Lugiyngol zone is characterized by the occurrence of Permian volcano-sedimentary rocks. The lower-upper Permian strata rest on the upper Riphean rocks of the south Gobi zone. They are exposed only in a few localities and consist of andesite, dacite, rhyolite, greywacke, conglomerate, sandstone and chert with fossiliferous limestone lenses. These rocks are unconformably overlain by thick (5 km) tightly folded upper Permian sediments which are widely developed and extend in a northeastward direction. The Permian sediments consist of conglomerate, sandstone, minor siltstone and limestone, intruded by the Triassic-lower Jurassic leucocratic granite. The Permian strata of the Lugiyngol zone, probably, accumulated in a marginal basin. The Sulinheer zone consists of several nappes of various Carboniferous and Permian sequences [24]. The lower nappe is represented by Permian pillow basalt, volcanoclastic rock, chert and terrigenous flysch with olistostrome layers. Carboniferous age fossiliferous carbonate sequences are present in the lower slices of the middle nappe. The upper slices are composed of middle Carboniferous to lower Permian andesite, tuff, volcanoclastic sandstone with limestone lenses. The upper nappe consists of serpentinized peridotite, dunite, wehrlite, websterite, listvenites and minor gabbro. Environments of each individual member of the nappe complex can be reconstructed. The basalts of the lower nappe were formed in an intra-arc basin. The carbonate and volcanic rocks of the middle nappe suggest an ensimatic volcanic arc and its slopes. The ultramafic rocks of the upper nappe are considered to be relicts of oceanic crust from the Inner Mongolian ocean.

The above mentioned history of the linear-nappe belts can be summarized as follows. The south Mongolian basin oceanic crust formed as a result of extension within an older continental crust that existed in the early Paleozoic. A system of troughs and uplifts of the Gobi Altay zone is regarded as structural indications of the continental destruction. During the upper Ordovican and Silurian the south Mongolian ocean expanded and probably connected with the Junggar basin [5]. However, at the beginning of the Devonian, when the oceanic lithosphere extension gave way to compression, volcanic arcs were formed on the axial (Trans Altay) zone of the ocean. In the middle-upper Devonian, the south Gobi block moved northward and collided with these volcanic arcs. Obviously, the middle Devonian molasse in the south Gobi zone was formed in the foreland basin following collision. At the same time an extensive volcanic belt was being formed along the southern margin of Gobi Altay continental shelf. These continental margin volcanic belts indicate that the oceanic lithosphere plate was sliding beneath the continent from the south. During the middle Carboniferous, convergent movement of the south Gobi microcontinent and north Asian continent caused closing of the south Mongolian oceanic basin. In the late Paleozoic the Inner Mongolian ocean opened south of the south Gobi micro-continent. The opening of the new ocean led to concurrent closing of the south Mongolian ocean. In late Carboniferous and early Permian a volcanic arc was formed by subduction of oceanic crust within south Inner Mongolian basin. During the middle-upper Triassic, convergent movement of Siberian and Sino-Korean plates caused rapid closing of the Inner Mongolian oceanic basin.

III. INTRACONTINENTAL MAGMATISM AND RIFTING

In Mesozoic and Cenozoic time an intraplate rifting took place mainly in eastern Mongolia with block-faulting movements, accompanied by significant intrusive activity accompanied by continental molasse, containing zeolite, coal and oil-bearing depressions (for instance, Choybalsan, Tamsag, Dzuunbayan and others.). After the formation of the Hangay, Nentey, Altay and Hamardavaa orogenic belts intracontinental rifting occurred at the end of Paleozoic and the beginning of Mesozoic-Cenozoic producing mainly A-type rare-metal and agpaite granites. These granites followed the late orogenic granitic magmatism and are considered post orogenic A-type, often rich in rare-metals and rare earth elements indicating a transition from compressive to tensional regimes. Predominantly Mesozoic-Cenozoic subalkaline and alkaline volcanics are bordered by the south Hentey zone and the main Mongolian lineament on the north and the south respectively [11]. Within the areal volcanic field there are local areas of complicated volcanic successions [27]. The alkaline granite magmatism of the east Mongolian belt can be subdivided into two stages. The first spans the Triassic to middle Jurassic with K-Ar ages 175-220 Ma. The typical intrusive complexes of these granites are Janchivlan, Sharhad, Egudzur, Avdarharaat, Halzanuul and Dulaanhan. They are largely distributed in northeastern Mongolia (or south Hentey, Uldz, south Herlen, and Nuhutdavaa zones). The early intrusive stage includes subalkaline, alkaline granite, syenite and W, Be, Mo, Sn-bearing granite. The second intrusive stage or latest formed in upper Jurassic to lower Cretaceous with K-Ar ages 120-170 Ma. The representative complexes are Tsav, Bayanuul, Bayandun and Haydelgerhaan. These complexes are placed within the Dashbalbar, Choybalsan, and south Herlen troughs. The latest stage includes lithium-fluorine mineralization mainly in the latite or standard granite and also in plumasite granite. Differential blocky deformational elements consisting of both compression and extension are present during the Jurassic and in the early Cretaceous in the Mongolian territory.

The character of the magmatic activity is considered intracontinental and is related to development of the eastern Mongolian rift zone which stretches from SW to NE along the Hentey zone. The rift zone is up to 250-300 km wide and is represented by a system of parallel and divergent grabens, filled with basalt and trachy-rhyolite, sometimes with latite, trachyte, carbonatite or rare-metal-bearing ongonite. This late Mesozoic volcanic area based on its composition and structural position is similar to rift zones found in the intra-continental

Alpine-Himalaya collision belt. The majority of the known economic deposits of gold, tin, tungsten, copper, molybdenum, fluorspar and rare metals occur in central and eastern Mongolia, and is related to the early and late Mesozoic period of tectono-magmatic activity. The metallogenic zones form a pattern related to the sub latitudinal and NE trending structural zones that are complicated by domal structures. The cores of the domal and arcuate uplifts found in Altay, Hiangay, Hubsugul, Hentey, south Herlen and Gobi-Nuhutdavaa apparently control the mineralization of tin, tungsten and molybdenum. Gold occurs around the Hangay and Hentey dome uplift and within its NW-, S- and SE-margins. Khutorsky [18], established important stages of mineralization that group in the late Paleozoic, early Mesozoic, late Mesozoic and Cenozoic cycles of the post-Devonian tectono-magmatic history of Mongolia. The first cycle is related to development of the central Asian segment of the Paleotethys, the second and third cycles are related to the development of the Mongolo-Okhotsk belt, and the fourth cycle to the formation of the Baikal rift zone.

Possibly at the end of the Permian, a complete closing of the Paleotethys occurred and the regime of an active continental margin was replaced by a setting of a more mature continental collision with continuing intracontinental magmatism and rifting. A new cycle of tectono-magmatic events began in the Triassic in eastern Mongolia where the geodynamic setting was related to movement of the Paleotethyan spreading zone under a basement of the Mongolo-Okhotsk belt. On the whole, considering the intracontinental character of these events, we suggest that the geodynamic setting at this time was similar to that of continental collision characteristic of the Mediterranean segment of the Alpine mobile belt. The magmatic activity of Mongolia in the Cenozoic manifests itself as basic magma activity along a meridional stripe from Hubsugul Lake to the Gobi Altay through the lost volcanoes of the central Hangay. The last eruptions of the "Horgo" volcano in the central Hangay occurred 5-6 thousand years ago. This group of volcanic manifestations is related to the Cenozoic Baikal rift zone. The age of these distinctive events has been confirmed by helium isotopic analysis. Another area of recent volcanic activity is located on the Dariganga plateau where basalt outpourings are related to the rift grabens of the Shansu and Layakhe of Chinese Inner Mongolia. A number of Cenozoic basaltic fields are distributed within the Lake District and south Mongolian zone. These basalts are controlled by a sub-latitudinal rift system. The basites of the Hubsugul-Hangay area are characterized by leucite basalts, olivine-trachybasalt, subalkali andesite-basalt and on the Dariganga plateau hawaiite, limburgite, nephelinic mugearites and olivine-basalts predominate. In all probability hot spot activity within the Hubsugul, Hangay, Dariganga and southern Gobi regions is marked by these widespread alkaline and flood basalts and alkaline granite pluton intrusions.

IV. CONCLUSIONS

In light of new mobilistic views we have analyzed the evolution and structural-stratigraphic interpretations of Mongolia. It is suggested that different accretional and collisional structural complexes are marked by orogenic and rift zones which were born in the environment of the Paleoasian and Paleotethys Oceans and can be related to plate tectonic schemes.

Acknowledgments
The authors express their special thanks to Academician Barsbold, Director of Geological Institute of the Mongolian Academy of Sciences for giving permission to present this paper in the book "Reconstruction of the Paleoasian Ocean", 1993.

REFERENCES

1. *Tectonics of the Mongolian People's Republic.* Vol. 9., Nauka, Moscow. (1974) (*in Russian*).

2. G. Badarch and D. Dorjnamjaa. *Main features of tectonics of the southern Gobi zone of Mongolia.* International Geologic Correlation Program, Project 283: Geodynamic evolution and main sutures of the Paleoasian Ocean (1991) Conference in Shenyang, China

3. H. Baoguan and Z. Shigian. *Characteristics of the composite-type suture zone and its tectonic development in the northern part of Inner Mongolia, China.* International Geologic Correlation Program, Project 283: Geodynamic evolution and main sutures of the Paleoasian Ocean: (1991) Conference in Shenyang, China, Report 2

4. A.N. Bulgatov and I.V. Gordienko. *The Pre-Jurassic Geodynamical evolution of the Transbaikalia.* in *International Geologic Correlation Program. Project 283: Geodynamic evolution and main sutures of the Paleoasian Ocean.* (1990). Conference in Ulan Ude, Russia, Report 1

5. A.R. Carroll, S.A. Graham, S.A. Hendrix, J. Chu, C.L. McKnight, X. Xiao, and Y. Liang. Junggar basin, northwest China: Trapped late Paleozoic ocean, *Tectonophysics.* **181**, 1-14.(1990).

6. R.G. Coleman. Can international cooperation help reconstruct the paleo-Asian ocean?, *Episodes.* **13**(3), 184-185.(1990).

7. A.B. Dergunaov. *The tectonics of central Asian Caledonides: The Altai-Sayan region and west Mongolia.* International Geologic Correlation Program, Project 283: Geodynamic evolution and main sutures of the Paleoasian Ocean. (1990) Conference in Ulan Ude, Russia, Report 1

8. A.B. Dergunov, B. Luvsandanzan, and V.S. Pavlenko, *The geology of west Mongolia.* Vol. 31., Nauka, Moscow. (1990) (*in Russian*).

8 a A.B. Dergunov and B. Luvsandanzan, Caledonides of central Asia; Some general results and problems of studies, In: *Evolution of geological processes and metallogenesis of Mongolia.*(Ed.), Vol. 49, pp. 55-72, Nauka, Moscow. (1990). (*in Russian*).

9. D. Dorjnamjaa. *Precambrian and Cambrian deposits in Mongolia and Latest Precambrian phosphorites.* in International Correlation Program, Project 303: Precambrian and Cambrian event stratigraphy (1991) Conference in Calgary, Alberta, Canada

10. D. Dorjnamjaa, G. Badrach, J. Badamgarav, D. Orolmaa, Z. Dashdavaa, and G. Eenjin. *Geodynamic conditions of tectonic structures on Mongolia, based on Remote Sensing image interpretation.* Proceedings of the 13th Asian conference on remote sensing. I-9-1-6 (1992).

11. D. Dorjnamjaa and Y. Bat-Ireedui, *The Precambrian of Mongolia.* Ulan Bator. (1991) (in Russian)

12. S.P. Garvrilova, A.I. Luchitskaya, D.I. Frich-Char, D. Orolmass, and J. Badamgarav, *Volcano-plutonic associations of the central Mongolia.* Nauka, Moscow. (1991) (*in Russian*).

13. O. Gerel, S. Batzhargal, and V. Balzhinnyam. *Magmatism and polymetal mineralization in Mongolia.* International Geologic Correlation Program, Project 283: Geodynamic evolution and main sutures of the Paleoasian Ocean (1990) Conference in Ulan Ude, Russia, Report 1

14. I.V. Gordienko, *The Paleozoic magmatism and geodynamic of the central-Asiatic orogenc belt.* Nauka, Moscow. (1987) (*in Russian*).

15. A.B. Ilyin, *Geological development of south Siberia and Mongolia in the Late Precambrian and Cambrian.* Nauka, Moscow. (1982) (*in Russian*).

16. K.B. Kepezhinskas, V.V. Kepezhinskas, Tomurkhuu, and D. Dorjnamjaa, Riphean-lower Paleozoic ophiolites of northern Mongolia., In: *Riphean-Paleozoic ophiolites of north Eurasia*(Ed.), pp. 19-34, Nauka, Novosibrsk. (1985) (*in Russian*).

17. M.D. Khutorsky and V.V. Yarmolyuk, Heat flow, structure and lithospheric evolution of Mongolia, In: *Evolution of geological processes and metallogenesis of Mongolia*(Ed.), Vol. 49, pp. Nauka, Moscow. (1990) (*in Russian*).

18. I.K. Kozakov, *Precambrian infrastructural complexes of the Paleozoides of Mongolia.* Nauka, Lennigrad. (1986. (*in Russian*).

19. B.A. Litvinovsky. *Late Paleozoic magmatism of the North-Eastern Asia.* International Geologic Correlation Program, Project 283: Geodynamic evolution and main sutures of the Paleoasian Ocean (1990) Conference in Ulan Ude, Russia Report 1

20. M.E. Markov, A.S. Perfiliev, Y.M. Pushcharovsky, and V.S. Fedorovsky, Evolution of tectonogenesis and geodynamics of the territory of the USSR., In: *The evolution of geological processes*(Ed.), pp. 77-86, Nauka, Moscow. (1989) (*in Russian*).

21. F.P. Mitrofanov, I.K. Kozakov, and I.P. Palei, *The Precambrian of western Mongolia and southern Tuva.* Nauka, Leningrad. (1981) (*in Russian*).

22. S.V. Ruzhentsev, G. Badarch, T.A. Voznesenskya, and N.G. Markova, Tectonics of the southern Mongolia, In: *Evolution of Geologic processes and metallogenesis of Mongolia*(Ed.), Vol. 49, pp. 111-117, Nauka, Moscow. (1990 (*in Russian*).

23. S.V. Ruzhentsev, I.L. Pospelov, and G. Badarch. Tectonics of Mongolian Indosinides, *Geotectonics.* **6**, 13-27.(1989) (*in Russian*).

24. E.V. Skylarov. *Ophiolites and blueschists of the South-East Sayan, Guide book for IGCP 283 meeting in Ulan Ude.* in *Reconstruction of Paleoasian Oceans.* (1990).

25. O. Tomurtogoo. *Ophiolites of Mongolia and problem of the tectonic evolution of Central Asia.* International Geologic Correlation Program, Project 283: Geodynamic evolution and main sutures of the Paleoasian Ocean (1991). Conference in Shenyang, China, Report 2 (65-66)

26. I.K. Volchanskya, E.N. Sapozhnikova, and Y.A. Baskin, Morphostructural regularities in the distribution of endogenous mineraliztion in Mongolia, In: *Global tectonics and metallogeny*, Vol. 2, pp. 95-110, (1983) (*in Russian*).

27. C. Wenjun, Preliminary study on Plate Tectonics on NE Junggar in Xinjiang, Uygar Autonomous Region, In: *Contributions to Plate Tectonics, North China, No. 1*, Vol. pp. 1-21, Publishing House, Beijing. (1983) (*in Chinese*).

28. A.L. Yanshin, *et al.*, The evolution of sediments accumulation in the Precambrian and Phanerozic, In: *The evolution of geological processes*, pp. 114-126, Nauka, Moscow. (1989) (*in Russian*).

29. N.S. Zaitsev, Tectonics of Mongolia, In: *Evolution of geological processes and metallogenesis of Mongolia*, Vol. 49, pp. 15-22, Nauka, Moscow. (1990. (*in Russian*).

30. L.P. Zonenshain, *Geosynclines and their implication to Central-Asian fold belt.* Nedra, Moscow. (1972) (*in Russian*).

31. L.P. Zonenshain, M.T. Kuzmin, O. Tomurtogo, and V.V. Kopteva, Ophiolites of the west Mongolia, In: *Riphean-Paleozoic ophiolites of north Eurasia*, N.L. Dobretsov (Ed.), Vol. pp. 7-19, Nauka, Novosibrsk. (1985) (*in Russian*).

32. L.P. Zonenshain and L.A. Savostin, *An introduction to geodynamics.* Nedra, Moscow. (1979) (*in Russian*).

# Ophiolites of the southern Siberia and northern Mongolia

E.V. SKLYAROV[1], V.A. SIMONOV[2] and M.M. BUSLOV[2]
[1]*Institute of the Earth's crust, Russian Academy of Sciences, Lermontov St., 128, Irkutsk, RUSSIA.*
[2]*United Institute of Geology, Geophysics and Mineralogy, Russian Academy of Sciences, Universitetsky Av., 3, Novosibirsk, RUSSIA.*

Abstract. Folded systems of the southern Siberia and northern Mongolia include Precambrian (800-1300 Ma) and Late Precambrian - Early Paleozoic (500-700 Ma) ophiolites. According to age and peculiarities of the ophiolites five areas of their distribution or terranes are distinguished. The first area adjacent to the Siberian craton includes the most ancient (1100-1300 Ma) Baikal-Muya and Enisey ophiolite belts. These belts consist of slices of peridotite, gabbro and volcano-sedimentary series, incorporated in assemblages of Late Precambrian island arc, back-arc and passive margin formations. The presence of komatiites in effusive section is the essential feature of the ophiolites. Late Precambrian (700-1100 Ma) ophiolites of the second terrane, known in literature as Dzhabhan-Hubsugul microcontinent, constitute sometimes complete sequences. These ophiolites correspond to very low-Ti geochemical type and are rich in dikes and lava flows of boninites. The presence of metamorphic sole and some other features support emplacement of the ophiolites onto passive margin. Three terrains - Dzhida-Bayan Hongor, West Sayan, and Gorny Altai - incorporate Late Precambrian - Early Paleozoic ophiolites. The ophiolites associate with volcanic, volcano-sedimentary, and sedimentary series of the same age, formed in developed or primitive (with boninites) island arc or back-arc settings. For the western terrains HP-complexes (eclogites and blueschists) are typical, being absent in the eastern terrane. Cordillerian type of ophiolite accretion is supposed these terranes.

Keywords: ophiolite, southern Siberia, northern Mongolia, tectonics.

INTRODUCTION

Folded systems of southern Siberia and northern Mongolia are the adjacent to the Siberian platform part of the Central Asian foldbelt. This part has a complex linear-mosaic structure, comprising Early Precambrian metamorphic massifs, Late Precambrian and Early Paleozoic sedimentary, volcanic and igneous complexes of various origin. Ophiolite belts are relatively widespread in the area under consideration. Sometimes they compose linear belts, marking suture zones between terranes, in other cases ophiolites are incorporated in complex thrust-folded structures, occurring sporadically in wide areas.

The great interest to ophiolites as to relics of paleooceanic lithosphere triggered a "rolling stones" of publications about geology and petrology of discrete ophiolite massifs and belts, but till now there isn't a review, summarizing geological and geochemical features of the ophiolites of the all region under consideration, as well as of the Central Asian foldbelt. More over, any attempt of such a work will collide with difficulties caused by several reasons. First of all, the usual practice was to study the separate components of ophiolite sequences - ultrabasites [23], gabbro [19], or volcano-sedimentary sequences. The second reason is the usual absence of combination of structural and petrological investigations of ophiolites. So, good structural schemes are not supported by petrological investigations, and detailed petrological studies have no correct geological base. And the third reason is connected with

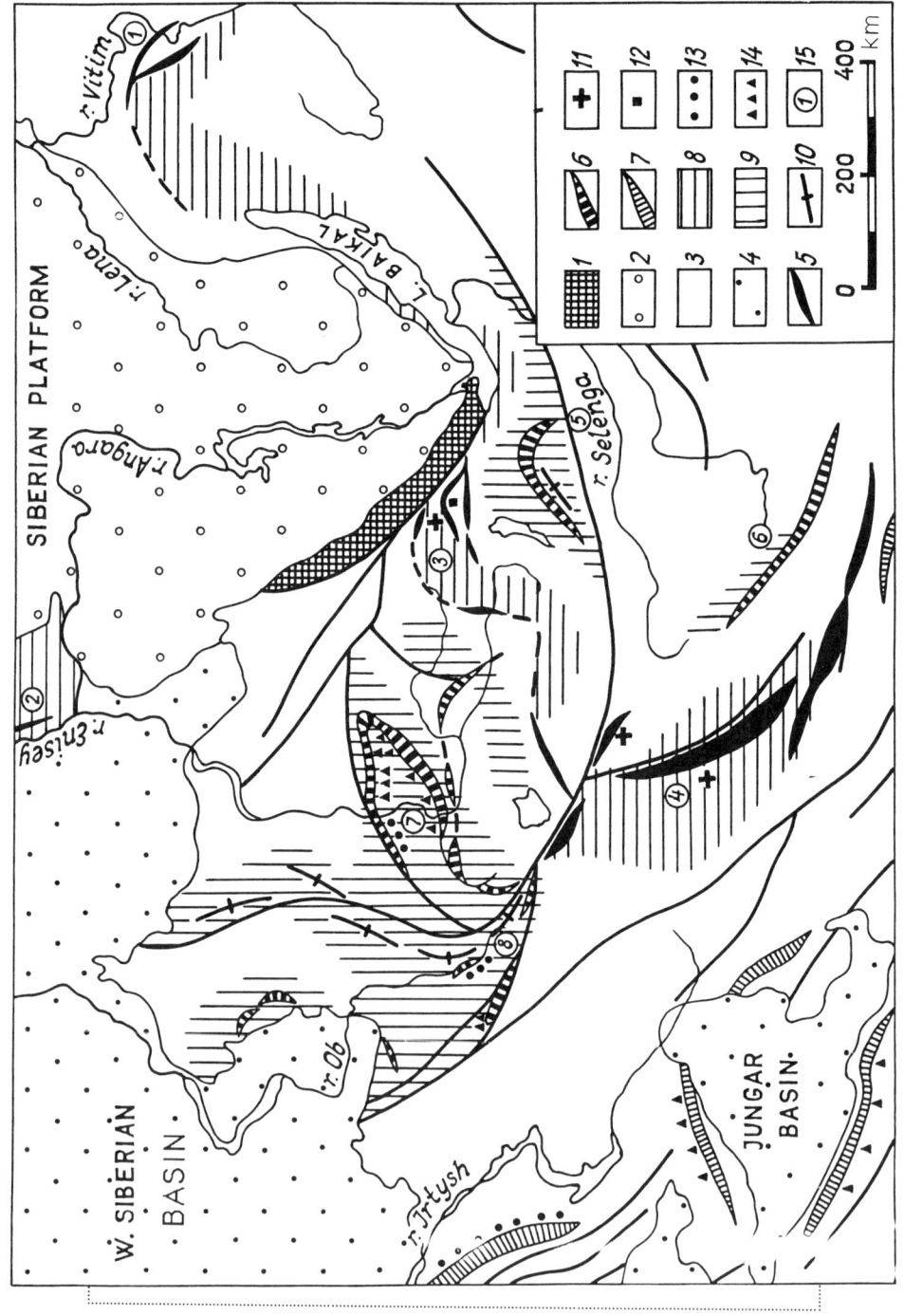

Figure 1. Ophiolite belts of southern Siberia and northern Mongolia. 1- Early Precambrian basement of the Siberian platform; 2- cover of the platform; 3- Late Precambrian-Paleozoic foldbelts surrounding the south Siberian platform; 4- Cenozoic-Mesozoic cover; 5 to 7- ophiolite belts of Late Precambrian (5), Vendian-Cambrian (6), and Early Paleozoic (7) age; areas of wide occurence of Late Precambrian(8) and Early Paleozoic(9) island arc and back-arc complexes; primitive oceanic crust and island arc complexes (10), occurrence of boninites in ophiolites (11), presence of metamorphic sole (12), eclogites and jadeitic rocks (13), blueschists (14); 15- ophiolite belts and systems of belts described in the text: ① - Baikal-Muya belt; ② - Enisey belt; ③ - East Sayan system; ④ - Lake system; ⑤ - Dzhida belt; ⑥ - Bayan Hongor belt; ⑦ - West Sayan system; ⑧ - Gorny Altay

absence or invalidity of isotopic and precise geochemical investigations. So, the ages of ophiolites are determined usually in wide range, using indirect data, and ophiolites are transported in time in accordance with points of view of researchers. Only some belts, such as Kurtushibin belt of West Sayan [27], North Ilchir belt of East Sayan [7, 26], Han Taishir [28], and some others are relatively well investigated geologically and petrologically.

The goal of this paper is to provide a brief review of the south Siberia and northern Mongolia ophiolites with emphasis on their most essential features.

DESCRIPTION OF THE OPHIOLITES

Ophiolite belts and systems of belts of Late Precambrian and Late Precambrian-Early Paleozoic age are recognized in the southern Siberia and northern Mongolia (Fig.1). Baikal-Muya and Enisey belts, East Sayan and Lake systems are referred to the first group and Dzhida and Bayan-Hongor belts, West Sayan and Gorny Altai systems - to the second group.

Baikal-Muya belt
This belt (1 in Fig.1), adjacent to the margin of the Siberian platform, has Riphean age and is believed to be the most ancient in the region under consideration [4, 17]. Indirect data (ages of intruding granite and associated olistostrome) sustain 1100-1300 Ma age of the ophiolites [4]. Sedimentary and volcano-sedimentary series of similar age, considered as fragments of island arcs and back-arc basins, associate with ophiolites. The complete and tectonically undismembered sequences are unknown here, but all constituents of ophiolite sequence in various combinations are distinguishable in descrete tectonic blocks or slices. There are three types of such blocks consisting of: a) serpentinites, serpentinised dunites and harzburgites; b) massive or layered gabbro, sometimes with cumulate pyroxenites and peridotites at the base or with discrete dikes at the top of sequence; c) metavolcanic series with associated sedimentary rocks. Tectonic dismembering and later high-grade metamorphism in amphibolite facies offer not high assurance about real ophiolite nature of the many gabbroic massifs and, especially, of amphibolite series. In petrologically well studied cases [14, 17] metabasalts with MORB, IAT and WPB affinities are distinguished in the same localities, but their interrelationships are unknown. Gabbro and metavolcanics with typical island arc affinities predominate in this region. The presence of komatiites in ophiolitic volcanics [17] (see analyses 10 and 11 in Table 1) is the essential feature of these, as well as of the other, described in the literature on ancient ophiolites [24]

The Enisey belt (2 in Fig.1), adjacent to the western margin of the Siberian platform, consists of slices of serpentinised peridotites, metavolcanics, and rarely metagabbro incorporated in complex structure, including Late Precambrian volcanic, terrigenous, and volcano-sedimentary series of island arc origin [18]. The belt is not well investigated petrologically.

East Sayan system

Relying on recent investigations [6, 7, 25, 26] this area is characterized by complex juxtaposition of Precambrian and Early Paleozoic formations. The main units are: 1) Precambrian metamorphic massifs, mantled sometimes by package of slices, including Upper Precambrian carbonate-terrigenous series and Early Paleozoic carbonate, essentially terrigenous and volcano-terrigenous series; 2) ophiolites, composed of ultrabasite and gabbro-ultrabasite massifs or complete ophiolite sequences; 3) Late Precambrian (R_2) terrigene-volcanic or essentially terrigenous series of turbidite type, supposed to originate in back-arc or for-arc tectonic settings. HP-metamorphism, corresponding to transitional between greenschist and blueschist facies, is described at the base of volcano-terrigenous sequence [8, 25]; 4) calc-alkaline volcanics with associated sedimentary series (R_3), supposed to form in environment

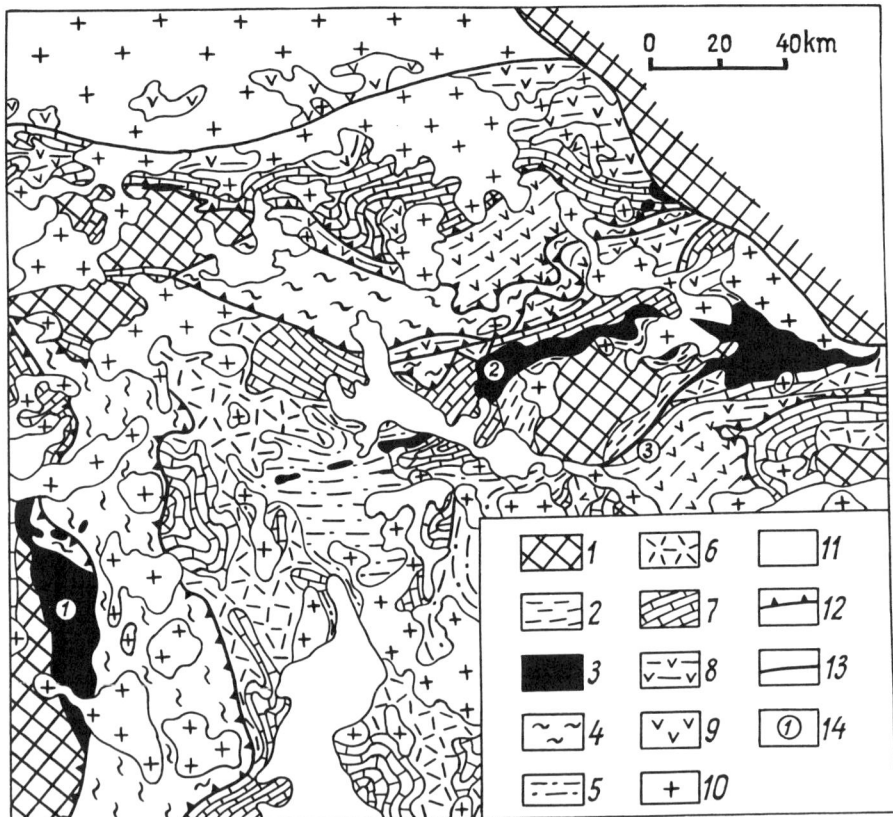

Figure 2. Tectonic sketch map of South-East Sayan ophiolite system (3 in Fig.1).
1 - Precambrian metamorphic series; 2 - package of Late Precambrian and Early Paleozoic carbonate-terrigenous and volcano-sedimentary series; 3 - ophiolite; 4 - Late Precambrian (R_2) basalt- sedimentary series with blueschists at the base; 5 - Late Precambrian (R_2) essentially sedimentary series of flysch type; 6 - Late Precambrian (V) rhyolite-andesite-basalt-sedimentary series; 7 - Vendian - Cambrian essentially carbonate series; 8 - Early Paleozoic (O-S) carbonate-volcano-terrigenous series; 9 - Devonian volcanic series; 10 - granitiods undivided; 11 - Quaternary sediments and basalt; 12 - thrusts; 13 - other faults; 14 - Shishgid (1), North Ilchir (2) and South Ilchir (3) ophiolite belts, described in the text.

of a developed island arc; 5) Essentially carbonate series (V-Cm), regarded as sedimentary cover of the microcontinent or as fragments of passive margin; 6) complex package of Early Paleozoic (O-S) series, including carbonate-siliceous, carbonate-volcanic, essentially terrigenous, and sedimentary-volcanic sequences. Some of them, exposing in the south-east section of the region are believed to belong to Early Paleozoic Dzhida zone. The origin of others is supposed to be connected with intracontinental extension and the Gargan metamorphic core complex formation.

There are three ophiolite belts with sublatitude strike in the eastern part of the region and submeredional - in the western part (Fig.2). North-western or Shishgid belt (1 in Fig.2) comprise the large ultrabasite massif of the same name and several small slices of serpentinites and pyroxenites. The Shishgid massif consists mostly of metaperidotites, predominantly of dunites and harzburgites (60-70%) with less lherzolites and Cpx-bearing harzburgites (30-40%) [20, 21]. The thickness of peridotites is estimated as exceeding 3.5-4 km. Rocks of cumulate zone occur sporadically and include wehrlites, clinopyroxenites and layered gabbro. Upward they grade to layered and massive gabbro and then to diabases with uncertain structure. The thickness of diabases exceeds 800 m. The uppermost part of ophiolite

Table 1.
Selected analyses of boninites and komatiites.

	1	2	3	4	5	6	7	8	9	10	11
SiO_2	53.91	53.28	59.78	51.28	50.89	50.98	54.62	54.70	54.66	45.42	48.02
TiO_2	0.12	0.18	0.24	0.14	0.14	0.17	0.13	0.23	0.34	0.47	0.35
Al_2O_3	8.29	10.10	11.34	9.52	12.07	11.20	9.85	11.28	9.41	9.02	7.44
Fe_2O_3	1.64	2.04	1.51	4.01	9.54*	1.44	1.93	10.57*	10.85*	10.50*	9.70*
FeO	6.57	6.19	4.98	4.61	-	6.43	5.77	-	-	-	-
MnO	0.15	0.15	0.11	0.12	0.14	0.14	0.15	0.21	0.25	-	-
MgO	17.10	14.65	10.14	13.60	14.32	17.02	14.08	10.22	10.23	17.80	17.36
CaO	5.68	4.50	4.05	6.72	6.91	5.19	8.36	7.89	9.78	10.70	11.96
Na_2O	0.42	1.64	3.35	3.01	1.28	1.41	1.00	1.13	2.15	0.88	0.92
K_2O	0.01	0.28	0.15	0.65	0.51	0.46	0.14	0.54	0.32	0.30	0.20
P_2O_5	0.03	0.04	0.04	0.02	0.03	0.04	0.04	0.03	0.03	0.18	0.04
LOI	5.46	6.55	3.95	5.97	4.67	5.51	3.91	3.04	1.92	-	-
Total	99.37	99.60	99.65	99.65	100.50	99.99	99.98	99.84	99.94	-	-
Sr	6	49	68	57	n.a.	69	100	104	110	220	170
Cr	800	1100	430	650	n.a.	969	846	800	1050	560	830
Co	16	44	31	49	n.a.	53	41	40	46	51	62
Ni	370	300	110	590	n.a.	425	200	82	91	510	930
V	120	25	100	130	n.a.	400	100	210	290	220	55
La	2.4	4.9	4.3	2.0	n.a.	0.19	3.1	n.a.	n.a.	5.7	2.6
Ce	5.2	8.2	8.8	4.4	n.a.	<1	6.0	n.a.	n.a.	15.0	7.9
Nd	2.3	4.5	4.8	2.3	n.a.	<1	3.0	n.a.	n.a.	7.0	5.6
Sm	tr	1.20	0.95	0.60	n.a.	0.18	0.70	n.a.	n.a.	2.40	1.60
Eu	0.14	0.14	0.19	0.15	n.a.	0.07	0.20	n.a.	n.a.	0.49	0.42
Gd	0.48	0.74	0.84	1.05	n.a.	<1	<2	n.a.	n.a.	n.a.	n.a.
Yb	0.39	0.54	0.36	0.70	n.a.	0.98	0.63	n.a.	n.a.	1.00	0.90
Y	3.60	4.50	3.20	4.20	n.a.	n.a.	n.a.	n.a.	n.a.	10.0	9.8

Comment: 1-4 - North Ilchir belt (after [25]): 1 - norite, 2, 3 - dikes, 4 - upper-most lava; 5-6 - dikes in volcanics of the Dzhida belt (after [16]); 7 - dike of the Han Taishir belt (after [16]); 8-6 dikes in volcanics of the Gorny Altai system (after [9]); 10-11 - komattites of the Baikal-Muya belt (after [17]). * - all iron in Fe_2O_3 form; n.a. - not analysed.

sequence consists of plagiogranites.

North Ilchir belt (2 in Fig.2) contain complete ophiolite sequence, including restite dunite-harzburgite complex, transitional zone with complex or layered structure, composed of wehrlites, websterites, and gabbro, massive gabbro, sheeted dike complex, and ophiolitic lavas with overlying turbidites [6, 26]. The dikes and lava have unusual chemical composition (Table 2). They correspond to andesite with relatively high MgO content and are referred to very low-Ti type according to [1]. The second feature is wide occurrence of boninite in ophiolite sequence. They are met as screens in sheeted dike complex, late dikes cutting parallel ones, and lava flows in the low as well as in uppermost parts of the effusive unit. More over, some gabbro also have boninite affinities, but it is difficult to distinguish the contours of their bodies in gabbroic section. The calculations using computer program "Crystallization" [11] show possibility of derivation of all components of the ophiolite sequence except for restites from primary boninite magma. The composition of boninite magma (4 in Table 2) after 30-40% of crystallization is identical to average composition of dikes and lavas (1 and 3 in Table 2). Estimated relative thickness of the critical zone as well as the composition of this zone also are in a good agreement with calculations. One more essential feature of this belt is presence of metamorphic sole at the base of ophiolite package. It consists of amphibolite, garnet amphibolite, and minor two-mica gneiss. This sole with thickness, ranging from tens to hundreds meters, was traced at the distance more than 60 km. Mineral parageneses correspond to T=650-700 °C and P=5-7 kbar [25]. Preliminary isotopic investigations provide the age of the ophiolites about 1000 Ma.

The South Ilchir belt (3 in Fig.2), consisting of slices of serpentinites, pyroxenites and basalts intercalated with cherts, is highly dismembered. Basalts usually have island arc affinities and are similar to the main type volcanics of Dzhida zone.

Lake system

Lake zone of north-western Mongolia (4 in Fig.1) is characterized by wide distribution of Late Precambrian island arc and ophiolite complexes. It is very similar to East Sayan system and may be regarded as its south-western continuation. In most cases ophiolites occur as small slices of serpentinites or gabbro among volcanic and sedimentary series. But large ophiolite slabs with relatively complete sequence also occur. Isotopic investigations (Sm-Nd mineral isochrone) show 830 Ma age of layered gabbro [13]. Well studied Han Taishir belt includes all ophiolite sequence [28] and is completely identical structurally and petrologically to North Ilchir belt. In both belts there are numerous boninite dikes, and basites correspond to very low-Ti type.

Dzhida belt

This sliced belt (5 in Fig.1) is located to the east from described above area. Numerous slices of serpentinised dunites and harzburgites, layered metaperidotite-pyroxenite-gabbro massifs, diabases of uncertain origin are incorporated in complex structure with predominance of volcano-sedimentary and sedimentary series, formed in island arc setting [12, 16]. Age of the ophiolites is supposed to be Late Precambrian - Early Cambrian, relying upon fossils in sedimentary rocks, associated with ophiolite basalts. Petrological investigations [12, 16] show juxtaposition of basites with MORB, IAT and WPB-affinities. In the western part of the area boninite dikes and flow among island arc volcanics are described [16]. Voluminous gabbro-tonalite massifs of Middle-Late Cambrian, supposed to form in island arc or active margin settings are widespread in this area.

Bayan-Hongor belt

This belt (6 in Fig.1) compose narrow suture zone of north-west strike, separating Riphean metamorphic series from Early-Middle Paleozoic terrigenous and volcano-sedimentary sequences [28]. Two units are distinguished in ophiolites. The lower unit comprise the package of slices of serpentinites, metagabbro and amphibolites. According to description [28] some amphibolite slices may be the relics of metamorphic sole. The thickness of this package is 300-500 m. The upper unit consists of stratified ophiolite sequence with cumulate wehrlite and pyroxenite at the base, grading upward to coarse-grain gabbro. All this section is cut by dikes of diabases. Sheeted dike complex above gabbro have a thickness about 1000 m. The uppermost part of ophiolite sequences consists of massive and pillow lavas, corresponding by chemical composition to IAT-tholeiites or calc-alkaline basalt [28]. The age of gabbro is about 650 Ma [16].

West Sayan system

Three pre-Silurian series are distinguished here [5, 27]. Metamorphic series of the Dzhebash anticlinorium, separating the ophiolite belts are predominantly represented by greenschist, more rare low amphibolite facies blastomylonites. The metamorphic series are usually regarded as the Early-Middle Proterozoic basement for ophiolite nappes, but metamorphic schists are very similar to "schistes lustres" of Tauern window [27]. The latter now are regarded as metamorphic core complex [10]; 2) Ophiolite sequence, including gabbro-ultrabasite and basalt-chert slabs or package of slices. The associated with basalts sedimentary rocks have Vendian - Early Cambrian age. 3) Cambrian-Ordovician sedimentary series of flysch type with subordinate volcanics of andesite-basalt composition.

All three ophiolite belts have north-east strike (Fig.3). The Kurtushibin belt (1 in Fig.3) stretches on the distance more than 200 km and includes the most complete sequences [27].

Table 2.
Average volatile-free analyses of ophiolite dikes and lavas.

	1	2	3	4	5	6	7	8	9	10	11	12	13	14
SiO_2	58.30	58.37	58.23	58.87	56.64	52.16	53.65	49.68	56.76	56.83	51.00	50.26	52.77	50.51
TiO_2	0.42	0.21	0.44	0.33	0.64	2.29	1.31	1.12	0.18	0.41	1.92	0.89	1.40	1.98
Al_2O_3	15.65	10.36	16.62	15.34	14.37	14.90	15.15	13.98	12.98	14.98	15.16	21.64	14.39	13.54
FeO^{tot}	8.39	7.94	8.18	7.86	9.92	12.34	8.79	10.89	8.55	10.80	11.43	5.91	11.20	13.85
MnO	0.15	0.16	0.15	-	0.10	0.18	0.23	0.21	0.17	0.19	0.19	0.10	0.19	0.23
MgO	6.20	15.43	6.57	6.42	8.17	7.12	6.32	10.27	10.56	6.51	6.84	4.94	6.53	7.39
CaO	6.19	5.59	4.66	8.27	7.92	7.61	10.01	9.98	7.36	5.59	10.07	12.88	10.01	9.41
Na_2O	4.07	1.26	4.53	2.17	2.10	2.37	3.61	2.98	3.16	4.56	3.13	2.62	3.03	2.94
K_2O	0.63	0.68	0.62	0.74	0.14	1.03	0.93	0.89	0.28	0.13	0.26	0.76	0.48	0.15
Sr	110	48	120	-	-	-	-	-	64	64	165	240	-	-
Cr	220	860	170	-	-	-	-	-	750	249	125	187	160	-
Co	36	35	32	-	-	-	-	-	68	31	40	24	32	-
Ni	50	240	65	-	-	-	-	-	150	75	81	62	64	-
V	80	50	100	-	-	-	-	-	220	208	294	163	240	-

Comment: 1-4 - North Ilchir belt: 1,2 - dikes parallel (1) and cutting (2), 3 - lavas, 4 - calculated after [11] composition of melt after 35% crystallization of boninite magma; 5-6 - Kurtushibin belt: dikes (5) and lavas (6) [27]; 7 - lavas of Boruss belt [27]; 8 - lavas of North Sayan belt [27]; 9-10 - Han Taishir belt: dikes (9) and lavas (10) [28]; 11-12 - Bayan Hongor belt: dikes (11) and lava (12) [28]; 13 - Dzhida belt effusives [16]; 14 - Gorny Altai lava.

In the north-western part the package of thick slabs is exposed. The low slab is composed of volcano-sedimentary series, regarded as the uppermost unit of ophiolite sequence. The upper slab comprise complete ophiolite sequence: serpentinised dunite-harzburgite complex grading upward to cumulate harzburgite-pyroxenite-gabbro complex, massive gabbro, sometimes changed by sheeted dike complex. Both slabs are underlain by serpentinite melange. The general dip of this structure to the south-east. The blueschist unit of volcanic-sedimentary composition is exposed at the base of ophiolite package. In blueschists mineral parageneses of glaucophane with lawsonite are typical [8, 27]. The protholite of blueschists is considered to be the uppermost part of ophiolite sequence, but metabasalts usually have higher TiO_2 content and WPB-affinities. The western part of the belt is tectonically dismembered and consists of separate small slices of serpentinites and metavolcanics.

Boruss belt (2 in Fig.3) consists of package of slices of ultrabasite, basalt with metacherts, and island arc volcano-sedimentary sequences, intruded by gabbro and granites [5, 27]. Utrabasites, composing large slabs, are serpentinised dunites and harzburgites with minor lherzolites (less than 20% of the volume). At the base of ultrabasite slabs thick zones of serpentinite melange are exposed. According to composition of "knockers" two types of

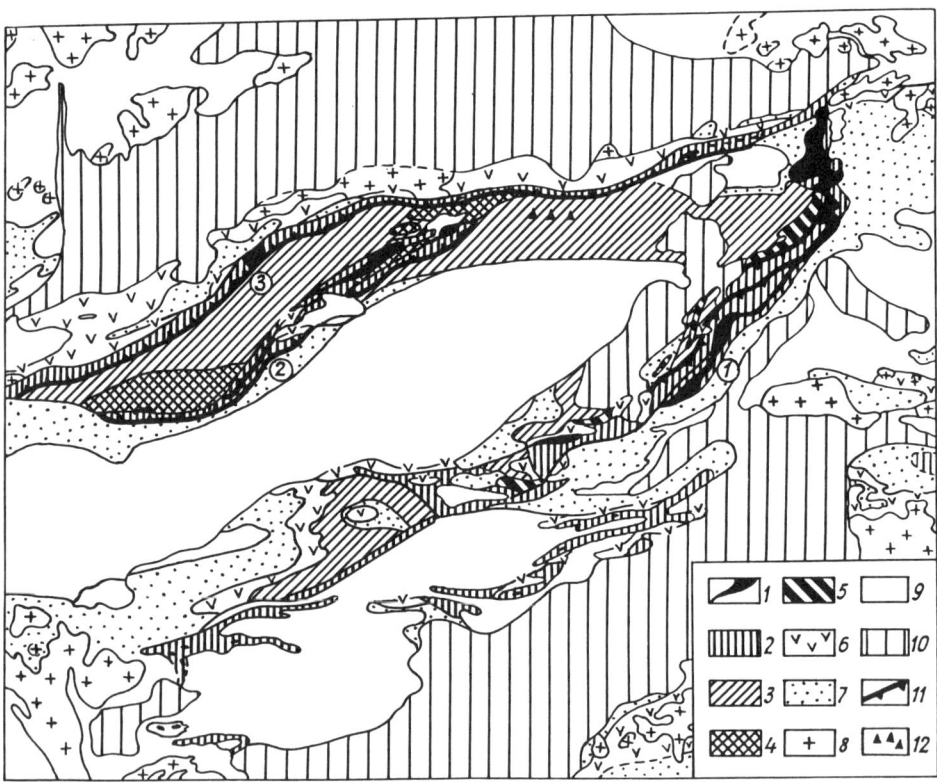

Figure 3. Tectonic sketch map of West Sayan ophiolite system (7 in Fig.1) (after [27]).
1 - ophiolite ultrabasites and gabbro; 2 - basalt-sedimentary upper unit of ophiolite sequence (V-Cm); 3-4 - Proterozoic blastomylonites of greenschist (3) and amphibolite (4) facies; 5 - blueschist unit; 6 - Early Cambrian andesite-basalt series; 7 - essentially sedimentary series of flysch type (Cm_{2-3}); 8 - granites (Cm_3-O); 9 - Late Ordovician - Early Silurian series; 10 - superimposed volcanic and coal-bearing depressions (D-C); 11 - most important thrusts; 12 - occurrences of blueschists. Number in the figure correspond to Kurtushibin (1), Boruss (2) and North Sayan (3) ophiolite belts.

melange are identified [5]. Melange of the first type includes eclogites, garnet amphibolites and jadeitic rocks. PT-conditions of metamorphism of these rocks are estimated as T=600-700 °C and P=10-16 kbar. Melange of the second type contain rodingites of various mineral composition corresponding to low or moderate pressures of metamorphism.

North Sayan slice belt (3 in Fig.3) is poorly studied. Small slices of serpentinites are incorporated in complex package of volcano-sedimentary series of different origin. Most of the volcanics have IAT or WPB-affinities, but some are believed to belong to ophiolite sequence (8 in Table 2) and are compared with ophiolitic basalt-sedimentary unit of the Kurtushibin belt [27].

Gorny Altai system

This system is characterised by complex structure, including Late Precambrian, Early and Middle Paleozoic volcanic, volcano-sedimentary and sedimentary series of perioceanic origin. Ophiolites participate in complex thrust-folded structure with west dipping (Fig.4). The compiled cross-section of this structure comprise (downward) [9]:

1. Upper nappe, composed of thick siliceous-carbonate series, sliced at the base (R_2);

2. Package of slices including chert-bearing olistostrome (Cm_1), volcanic-carbonate series (V) and volcanic-siliceous series (V-Cm_1). The thickness of this unit sometimes reach 3 km.

3. Package of slices of essentially volcanic (V-Cm_1) and volcano-sedimentary (Cm_1) composition. The essentially volcanic sequence consists of pillow and variolitic basalts with more rare andesite flows and minor terrigenous, tuffaceous, and calcareous rocks, cut by dikes and sills of gabbro-diabases and diabases of tholeiitic and subalcaline basalt composition. This sequences are regarded as fragments of back-arc basin with seamounts. The volcanic-sedimentary series are predominantly composed of limestones, slates, sandstones, tuffaceous rocks with rare flows of basalt.

4-5. The Chagan-Uzun ophiolite massif, consisting of two large slices. The upper slice is composed of serpentinised harzburgites, changing downward to massive serpentinites. The zone of serpentinite melange with massive serpentinite knockers and small slices of diaphthorized eclogites, garnet amphibolites, amphibolites, minor cherts, and black schists is exposed at the base of the upper slice. Low slice consists of massive or foliated serpentinites with boudinated and deformed dikes of gabbro, gabbro-diabases and diabases, sometimes rodingitised. Basalt, regarded as upper unit of the ophiolite sequence, underly low slice. Metamorphic sole represented by garnet-free amphibolites with the thickness to hundreds meters is distinguished. PT-conditions of metamorphism of the upper slice correspond to high pressures (7-9 kbar), and low unit - to low pressures (2-4 kbar) [9].

To the north from described area the package of slices, including essentially volcanic series of island arc origin is exposed. They consist mostly of basalt with minor andesites, tuffs and pyroclastics, cut by dikes and sills of gabbro-diabases and diabases. The essential feature of these series is a presence of boninite, occurring as dikes and lava flows. Usually they are situated near the southern contact of volcanics and possibly compose separate slice.

DISCUSSION AND CONCLUSION

Features of ophiolites

Two families of the features should be taken into account [22]. The first one is connected with environment of ophiolite generation and include the composition of restitic ultrabasites (lherzolite and harzburgite types) (a), presence of boninite in ophiolites or associated island arc series (b), and geochemical affinities of ophiolite basites (c). The second family comprise features connected with emplacement or tectonic evolution of ophiolites. It includes presence

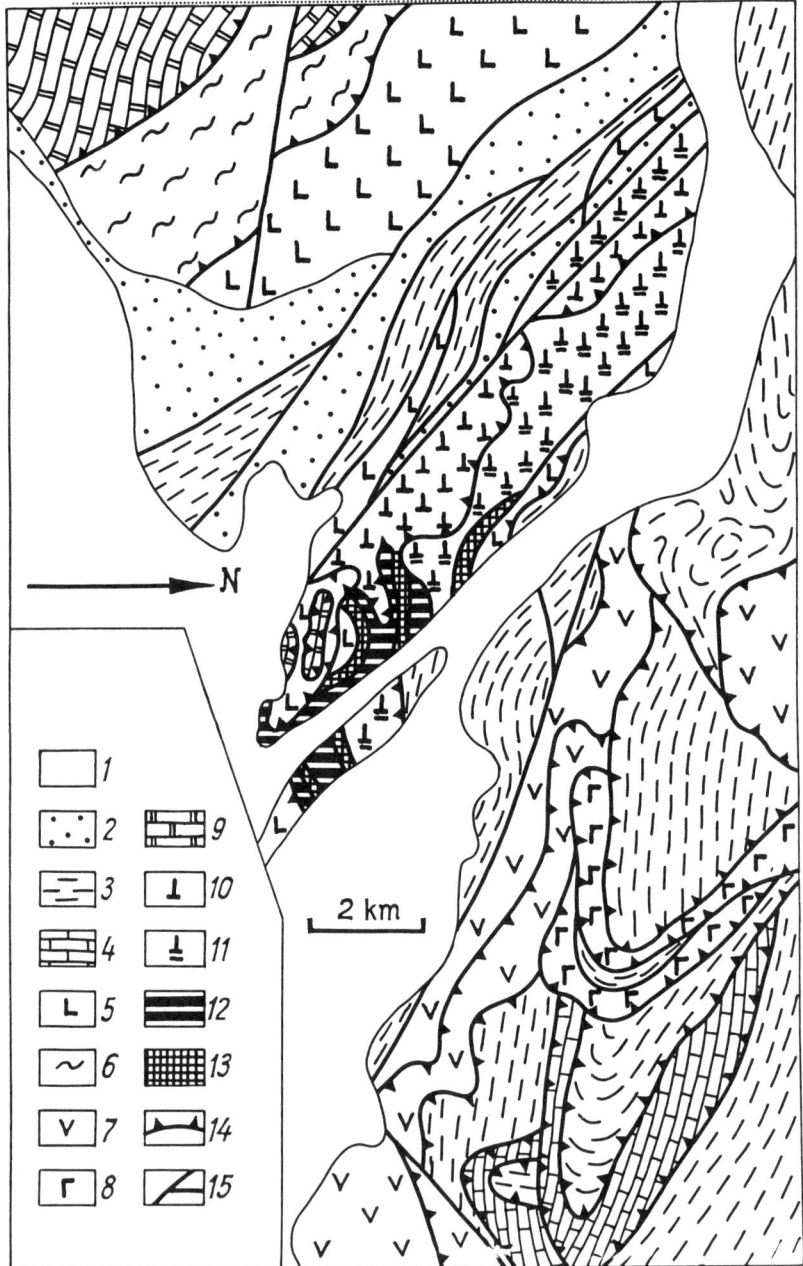

Figure 4. Tectonic sketch map of Gorny Altai system (8 in Fig.1).
1 - Quaternary deposits; 2 - Devonian volcano-sedimentary series; 3 - Cambrian sandstone-slate series; 4 - Early-Middle Cambrian limestones; 5 - Vendian-Cambrian basalt-sedimentary series of seamounts; 6 - package of Vendian-Cambrian siliceous, volcanic, and sedimentary series with olistostrome; 7-8 - island arc series (V-Cm): basalt-andesite volcanics and tuffs with boninite (7) and sills of gabbro (8); 9 - Upper Riphean carbonate-siliceous series; 10-13 - ophiolite: harzburgite and massive serpentinite (10), massive and foliated serpentinite (11), serpentinite melange (12), slices of eclogite, garnet amphibolie and amphibolie (13); 14 - thrusts; 15 - post-Devonian normal faults and strike-slips.

of HT metamorphic sole at the base of ophiolite slabs (d) presence of HP-metamorphic rocks in melange or in slices of ophiolite packages (e) and degree of preservation of ophiolite sequences (f).

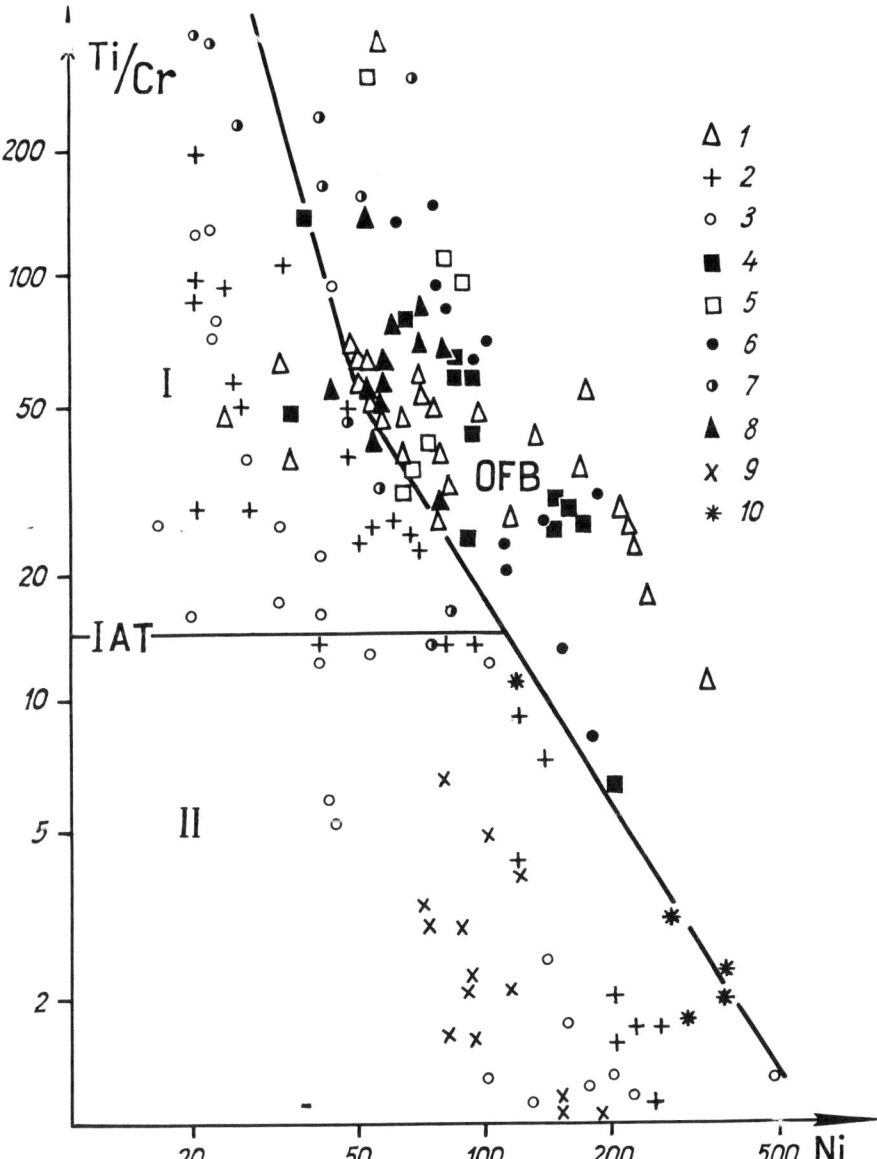

Figure 5. Ti/Cr-Ni diagram for ophiolite volcanics.
1 - Baikal-Muya belt [14]; 2 - North Ilchir belt of East Sayan system; 3 - Hanaishir belt of Lake system [16]; 4 - Dzhida belt [16]; 5 - Bayan-Hongor belt [28]; 6-7 - eastern (6) and western (7) parts of the Kurtushibin belt of West Sayan system [27]; 8-9 - amphibolites (8) and boninites (9) of Gorny Altay system; 10 - boninites of Dzhida belt [16].

(a) All ophiolites described have dunite-harzburgite composition of restites, only sometimes with subordinate lherzolites (Shishgid belt of the East Sayan system, two small massifs of the Lake zone [23] and Boruss belt of the West Sayan system) and refer to harzburgite type [22].

(b) Boninites are very common in ophiolites of the North Ilchir and Han Taishir belts, occurring as dikes and lavas. In Dzhida belt and Gorny Altai system boninites are described as dikes and lava flows among island arc volcanics, associated with dismembered ophiolites. The latter case suggest existence of primitive island arc in Vendian - Early Paleozoic, similar to modern ones. Representative analyses of boninites are shown in Table 1.

(c) Ophiolites of the East Sayan and Lake systems have peculiar composition of basites, corresponding to andesite or andesite-basalts with relatively high MgO content (4-9%). They are reffered to very low-Ti type according [1]. For other ophiolites juxtaposition of basites with MORB-, IAT-, and WPB-affinities in usual. Interrelationships of these types are usually poorly investigated. MORB-type basites are described almost in the all belts (Table 2), and some discriminate diagrams may illustrate this (Fig.5). But usage of precise geochemical data (REE-elements and some others) show that ophiolite basalts differ from typical oceanic ones as in most ophiolites described. After R.G. Coleman [2] we believe, that described ophiolites derived in settings of small oceanic basins.

(d) High-pressure metamorphic complexes are relatively usual in western part of the area under consideration. There are two types of HP mineral associations. The first one comprise blocks of eclogites, garnet amphibolites and jadeitic rocks in serpentinite melange (Boruss belt of the West Sayan and Gorny Altai). HP-metamorphism is supposed to occur in accretionary wedges [9]. The second type of HP-metamorphic rocks, including low temperature mineral associations with glaucophane [8] is usual in volcano-sedimentary series underlying ophiolite packages, regarded as upper units of ophiolite succession (Kurtushibin and North Sayan belts of the West Sayan, Gorny Altai).

(e) Metamorphic sole is described only in two cases - in the North Ilchir belt and in the Gorny Altai system. In the first case metamorphic sole underlies relatively complete ophiolite sequence and correspond to classical examples, described in Mediterranian ophiolites [22]. So it is correct suggestion about emplacement of the ophiolites on passive margin. In Gorny Altai all other data support Cordilleran type accretion of ophiolites. We assume that in this case amphibolites are only indicator of intraoceanic detachment [15].

(f) Most of the ophiolites are strongly dismembered. Only three belts - North Ilchir, Han Taishir, and Kurtushibin - preserve relatively complete and thick ophiolite successions. So, the emplacement of these ophiolites on passive margin is possible. As for the other belts, their mordern structure may be the result of accretion in active margin setting as well as of later collision. In most cases there are no correct evidence to distinguish accretional or collisional settings.

Tectonic implications

According to age, mode of occurrence and petrological features of ophiolites five areas of their distribution or terranes may be identified in the region. Two terranes include Precambrian ophiolites and three - Late Precambrian-Early Paleozoic ones.

1. Folded systems adjacent to Siberian platform contain the most ancient (1100-1300 Ma) ophiolites (Baikal-Muya and Enisey belts), usually strongly dismembered. The presence of komatiites is their essential feature as well as in some other ancient ophiolites [24]. These ophiolites fix early stages of evolution of the Paleoasian ocean.

2. The terrane, described before as a Dzabhan-Hubsugul microcontinent [7], includes Late Precambrian ophiolites with supposed age 800-1100 Ma. Other constituents of the terrane are Early Precambrian metamorphic massifs, Late Precambrian igneous, volcanic, volcano-

sedimentary and sedimentary series, supposingly formed in island arc, back-arc, and passive margin settings. The ophiolites (North Ilchir and Han Taishir belts) are characterized by distinctive composition (boninite type) and by wide occurrence of boninites. Relatively preserved successions and occurrence of metamorphic sole support point of view about thrusting of ophiolites onto passive margin. Ophiolites as well as other constituents of this microcontinent were amalgamated before Late Vendian. The time of accretion of the microcontinent to Siberian paleocontinent is supposed to be Early Paleozoic.

3. The terrane relatively poor in ophiolites occupy the territory to the east from the previous one. Ophiolites with Late Precambrian - Cambrian age (650-500 Ma) usually are strongly dismembered and are incorporated in complex structures, only sometimes revealing relatively preserved successions (part of Bayan-Hongor belt). Association of island arc fragments with boninite is usual. Ophiolites are supposed to be incorporated in accretionary structure in active margin setting.

4. West Sayan terrane with wide distribution of Early Paleozoic island arc and back-arc complexes include Late Precambrian - Early Paleozoic ophiolites (500-650 Ma). The ophiolites are regarded to be the fragments of back-arc basins. They are mostly dismembered (North Sayan belt), or preserve relatively complete sequence (Kurtushibin belt). So it is difficult to say doubtlessly about mode of emplacement of the ophiolites. The main difference of ophiolites of this terrane with previous one is association with HP-metamorphic rocks.

5. Gorny Altay terrane contain ophiolites of the same age and in many general features and ophiolite peculiarities (for example presence of HP-complexes) is similar to West Sayan terrane. Here it is more clear, that ophiolites were accreted in active margin setting. It shouldn't be excluded that both terranes have to be united in single one.

Acknowledgement. The recent work was carried out as a part of IGCP Project 283 "Geodynamic evolution of Paleoasian ocean". We express our gratitude to prof. N.L. Dobretsov for help and guidance and prof. V.G. Belichenko for helpful discussions.

REFERENCES

1. L. Beccaluva, P. Di Girolamo, G. Macciota, and V. Morra. Magma affinities and fractionation trends in ophiolites, *Ofioliti*. **8**, 307-324 (1983).
2. R.G. Coleman. The diversity of ophiolites. In: *Ophiolites and ultramafic rocks - a tribute to Emile den Tex.* H.J. Zwart, P. Hartman and A.C. Tobi (Eds). Geol. Mijnbouw, 63, 141-150 (1984).
3. A.B. Dergunov. *Caledonides of the Central Asia*. Nauka, Moscow (1989) *(in Russian)*.
4. N.L. Dobretsov. Problems of tectonics and ophiolite belts of Central Asia, South Siberia and North China. In: *Problems of magmatism and metamorphism of the Eastern Asia*. N.L. Dobretsov and B.A. Litvinovsky (Eds). pp. 7-25, Nauka, Novosibirsk. (1990) *(in Russian)*.
5. N.L. Dobretsov, and A.V. Tatarinov. *Jadeite and nephrite in ophiolites*. Nauka, Novosibirsk (1983) *(in Russian)*.
6. N.L. Dobretsov, E.G. Konnikov, V.N. Medvedev, and E.V. Sklyarov. Ophiolites and olistostromes of the East Sayan. In: *Riphean-Paleozoic ophiolites of the Northern Eurasia*. N.L. Dobretsov (Ed.) p. 34-57, Nauka, Novosibirsk (1985) *(in Russian)*.
7. N.L. Dobretsov, and V.I. Ignatovich (Eds). *Geology and ore deposits of the East Sayan*. Nauka, Novosibirsk (1989) *(in Russian)*.
8. N.L. Dobretsov, and E.V. Sklyarov. Blueschist belts of the South Siberia and Northern China. In: N.L. Dobretsov, N.V. Sobolev, V.S. Shatsky et al. *Eclogites and blueschists in the folded systems*. pp. 132-157, Nauka, Novosibirsk (1989) *(in Russian)*.
9. N.L. Dobretsov, M.M. Buslov, and V.A. Simonov. Associated ophiolites, blueschists and eclogites of the Gorny Altai. *Dokl. Acad. Sci. USSR*. **318**, 413-417 (1991) *(in Russian)*.
10. M. Fournnier, L. Jolivet, B. Goffe, and E. Turco. Alpine Corsica metamorphic core complex. *Tectonics*, **10**, 1173-1188 (1991).
11. M.Ya. Frenkel and A.A.Ariskin. Algorithm of computational solution of the equilibrium problem for crystallization of basalt melt. *Geochemistry*. **5**, 679-690 (1984) *(in Russian)*.

12. I.V. Gordienko. *Paleozoic magmatism and geodynamics of the Central Asian foldbelt*. Nauka, Moscow (1987) *(in Russian)*.
13. A.S. Gibsher, A.E. Izokh, and E.V. Khain. Pre-Middle Ordovician structure of Tuva-Mongolian segment of Central Asian foldbelt. In: *Report No 2 on the IGCP Project 283 "Geodynamic evolution of Paleoasian ocean"*. Beijing, p. 25-28 (1991).
14. G.S. Gusev, A.I. Peskov, and S.K. Sokolov. Paleogeodynamics of the Muisky segment of the Proterozoic Baikal-Vitim belt. *Geotektonika*, N2, 72-86 (1992) *(in Russian)*.
15. S. Karamata. Metamorphism beneath obducted ophiolite slabs. In: *Ophiolites - Proc. Int. Ophiol. Symp. 1979*. A. Panayiotov (Ed.). p. 219-227, Geol. Surv. Dept. Ministry of Agriculture and Natural Resources, Nocosia (1980).
16. K.B. Kepezhinskas, V.V. Kepezhinskas, and N.S. Zaitsev. *Earth's crust evolution of Mongolia in the Precambrian - Cambrian*. Nauka, Moscow (1987) *(in Russian)*.
17. T.G. Konnikov, and A.A.Tsygankov. To a nature of the Baikal-Muiskii belt ophiolites. In: *Report No 1 on the IGCP Project 283 "Evolution of the Paleoasian ocean"*. Novosibirsk, p. 49-53 (1990).
18. A.B. Kuzmichev. Riphean marginal-oceanic paleostructures in the Sayany-Enisey fold belt. *Ibid*, pp. 54-59 (1990).
19. V.A. Kutolin (Ed.) *Gabbroic complexes of the Western Mongolia*. Nauka, Novosibirsk (1990) *(in Russian)*.
20. A.A. Melyakhovetsky. *Metamorphism of the East Tuva ultrabasites*. Nauka, Novosibirsk (1982) *(in Russian)*.
21. A.A. Melyakhovetsky, and E.V. Sklyarov. Ophiolites and olistostromes of the West Sayan. In: *Riphean-Paleozoic ophiolites of the Northern Eurasia*. N.L. Dobretsov (Ed.). p. 58-70, Nauka, Novosibirsk (1985) *(in Russian)*.
22. A. Nicolas. *Structures of ophiolites and dynamics of oceanic lithosphere*. Kluwer Academic Publisher, Dordrecht/Boston/London (1989).
23. G.V. Pinus, L.V. Agafonov, and F.P. Lesnov. *Alpine-type ultrabasites of Mongolia*. Nauka, Moscow (1984) *(in Russian)*.
24. D.G. Scott, M.R. St-Onge, S.B. Lucas, and H. Helmstaedt. The 1988 Ma Purtunk ophiolite: imbricated and metamorphosed oceanic crust in the Cape Smith Thrust Belt, northern Quebec. *Geoscience Canada*. 13, 144-147 (1989).
25. E.V. Sklyarov. *Ophiolites and blueschists of the South-East Sayan*. Guidebook to field excursion. Ulan-Ude (1990).
26. E.V. Sklyarov, N.L. Dobretsov, E.G.Konnikov et.al. Petrochemistry and geochemistry of the ophiolites. In: *Geology and metamorphism of the East Sayan*. N.L. Dobretsov (Ed.). Nauka, Novosibirsk (1988) *(in Russian)*.
27. V.S. Sobolev, and N.L. Dobretsov (Eds). *Petrology and metamorphism of ancient ophiolites*. Nauka, Novosibirsk (1977) *(in Russian)*.
28. L.P. Zonenshain, M.I. Kuzmin, O. Tomurtogoo, and V.V. Kopteva. Ophiolites of the Western Mongolia. In: *Riphean-Paleozoic ophiolites of the Northern Eurasia*. N.L. Dobretsov (Ed.). pp. 7-18, Nauka, Novosibirsk (1985) *(in Russian)*.

Blueschist belts of North Asia and models of subduction-accretion wedge

NIKOLAI L. DOBRETSOV and ANATOLY G. KIRKDYASHKIN

United Institute of Geology, Geophysics and Mineralogy, Siberian Branch of Ruassian Academy of Sciences, Novosibirsk, Russia, 630090

Abstract - A large number of the Paleozoic and Late Precambrian blueschist belts are located in South Siberia and North China. They show the most important sutures between micro-continents and terrains formed during the closing of the Paleozoic ocean. The majority of these belts are characterized by a transition between blueschist and greenschist pressure conditions of 6-7 kbar and have formed long before the last accretion. Therefore it is unlikely that such accretionary complexes have been formed by underplating according to the model of Platt [31]. But they are essentially different in structure and peculiarities of metamorphism. The high-pressure eclogite-blueschists complexes of the same age (520-640 Ma) are very common in the western part of Asia. The complexes with extremely high pressures carrying diamond (Kokchetav) and coesite (South Ural) occur here. The coincidence and periodical repetition of collision and accretion processes are observed in many complexes.

Our subduction wedge model [14] permits us to evaluate the conditions of return flow under normal subduction and accelerated return flow under transition from subduction to accretion when the descending flow into the mantle disappears. The rapid and intensive tectonic transportation from the mantle's depths is only possible after cessation of subduction and at the prolongation of the convergence process with additional buoyancy effect of the sialic block that terminates the subduction process. Correlation of this model with the general model of convective flow in the mantle confirms the observed periodicity of observed blueschist ages and transportation of the most deep-seated blocks.

The two-layer model of mantle convection shows a strong correlation with stable subduction zones and the descending flows in the lower mantle convection cell. The transitional or unstable relations between the lower mantle and upper mantle convective cells and mantle plume pulses are reflected in the cyclic impulses of the tectonic processes.

Key Words - subduction, blueschist, eclogite, metamorphism, convection, tectonic model

INTRODUCTION

According to Plate Tectonics Theory glaucophane-schist belts together with associated ophiolites fix the ancient subduction zones [8, 9, 11, 20, 21]. In the modern tectonic position they are localized along the greatest tectonic sutures showing the collision zone of continents (micro-continents) and/or island arcs and are often involved in subduction-accretion complexes that are moved away along structural sutures. The most important elements of accretionary complexes are the nappes of ophiolites and glaucophane schists, melange zones and olistostromes complexes. Often all these elements are regularly combined as shown below.

BLUESCHIST BELTS AND COMPLEXES IN ASIA

On the largest continent of Eurasia the young belts (Late Paleozoic-Mesozoic) are a specific feature of Asia. Their localization along the greatest suture zones [13]. Figure 1 illustrates the most important peculiarity of Eurasia as a composite continent or collage of blocks and terrains of various dimensions including continents, micro-continents and island arc of Vendian-Paleozoic age. Among those continents and microcontinents we can distinguish: 1. Laurasian type (Russian or East-European, Siberian, North-China platforms and a number of smaller blocks), 2. Gondwana type (Indian, Indo-China, South China platforms), 3. Blocks of Pacific type or doubtful origin (Omolon, Hida, Kazakhstan and others).

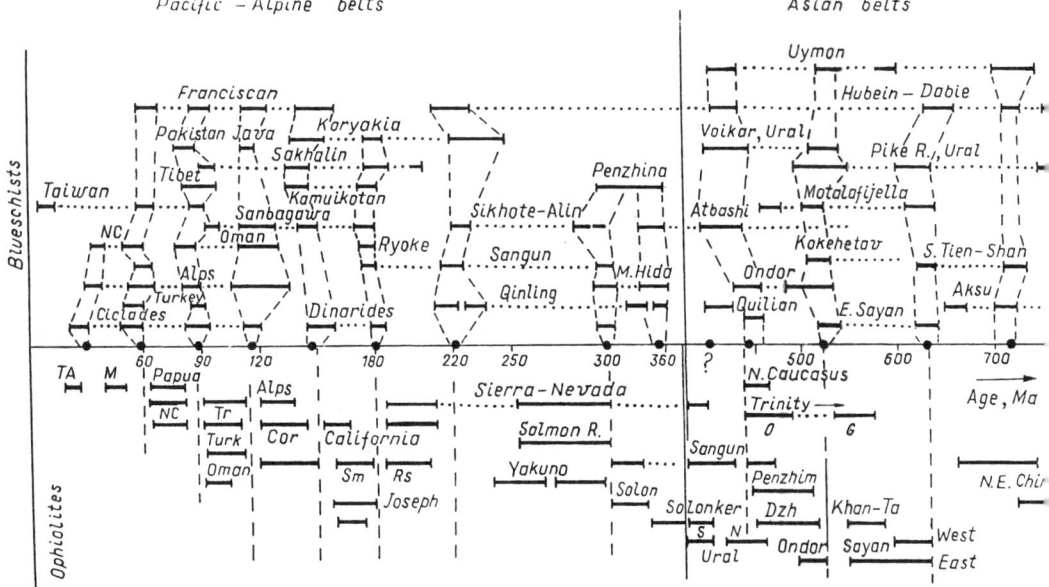

Figure 1 Blueschist belts of Asia and their selected isotopic ages [13]. For details see Fig. 2 and Tables 1 and 2. J-Jurassic, K-Cretaceous, T-Triassic, P-Permian.

Late Cambrian and Paleozoic glaucophane schist belts are widespread in the inner Asian folded belts in the Urals, South Tianshan, South Siberia and North China (Fig. 2). All of them fix the most important sutures between micro-continents, fragments of island arcs or other terrains formed during the closing of the Paleoasian ocean. The age of glaucophane metamorphism stages vary from 1.0 - 1.2 to 0.45 Ga (Table 1) In all cases, the metamorphic age is more ancient than accretion and collision of the related fold belts. The Urals represent the Hercynian fold belt where the accretion and collision processes have started in the Late Devonian (360 Ma) and finished in the Permian. The age of the Uralian eclogites and glaucophane schists is over 400 Ma. The most abundant ages are about 450 Ma, but more ancient ages of 1.0 - 1.1 Ga are found in both the Maksutov complex of the south Urals and in the Polar Urals. An analogous situation is found in the Hercynides of South Tianshan and in the Ob-Zaisan folded system.

In the Caledonides of the Central Kazakhstan and South Siberia, the collision terminated in the Ordovician and Silurian, whereas the age of the eclogites and glaucophane schists are predominately Late Precambrian (1.0- 1.1 Ga, 740-850 Ma), and the youngest are Cambrian (about 520-540 Ma, see Table 1). So, the Late Precambrian age is well

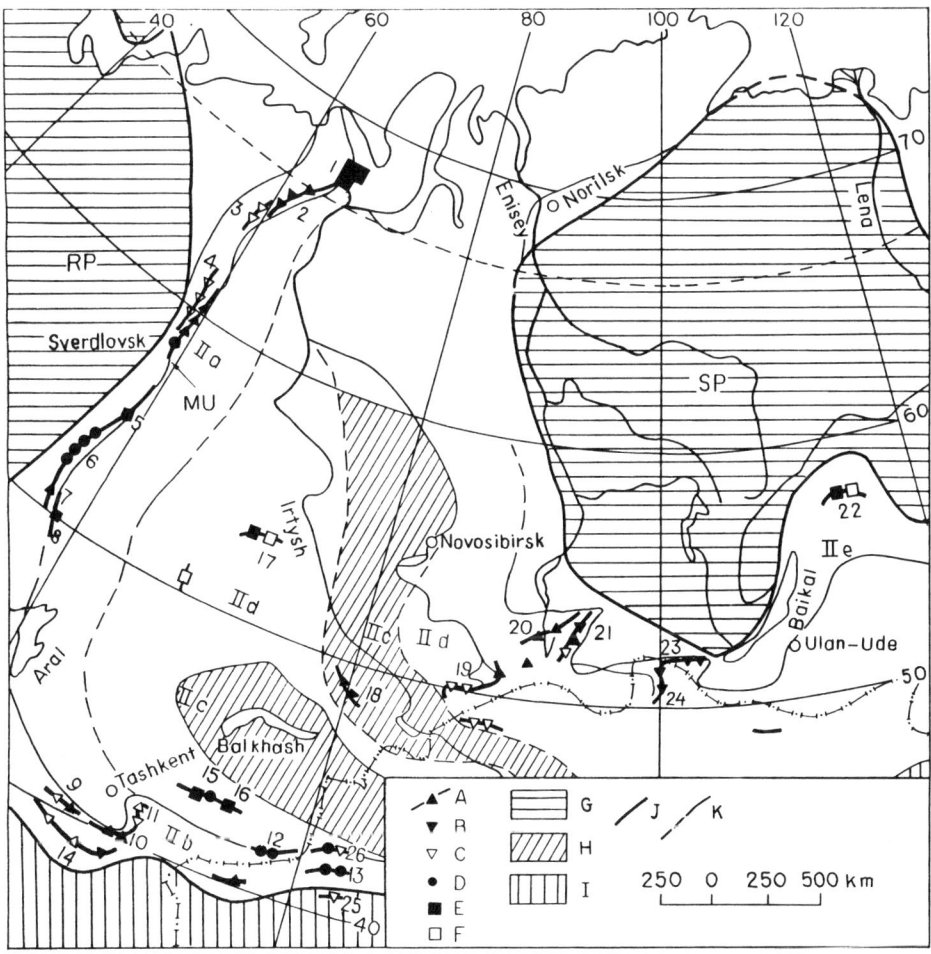

Figure 2 The glaucophane schists and eclogites of the Paleoasian ocean (Ural-Mongol system). See Table 1 and [35] for locality numbers. A-F: outcrop settings of the glaucophane schists and eclogites. (A) in a melange; (B) glaucophane schists at the base of an ophiolite nappe, or (C) without any relation to nappes; (D) eclogite-glaucophane schists; (E) eclogites in blastomylonites, and in usual gneisses (F). Platforms are shown as (G), including the Russian (RP) and Siberian (SP). Hercynian systems (H) include the Dzhungar(Junggar)-Balkhash and the Ob'-Zaisan areas. The Mesozoic and Alpine folded system (I) with blocks of the ancient continental crust. The heavy line (J) marks the boundaries of the eugeosyncline zones and the main structural lines (MU-the main Ural line). The lighter line (K) is the approximate boundary of the Hercynides of the Urals (IIa), Southern Tianshan (IIb), Dzhungar(Junggar)-Balkhash (IIc), Kazakhstan and Altai-Sayan Caledonic folded system (IId) and Late-Precambrian Baikal folded system (IIc).

documented (about 700-720 Ma) by geological and isotopic methods in the Aksu complex on the northern edge of the Tarim massif [27] (Table 1). The ages of 620-640 Ma using Rb-Sr isochron and K-Ar dates have been established in many South Siberian belts (Oka, Uymon, and Kurtushibin) (Table 1). We think, the conclusion about intracontinental belts in Eurasia being products of continent-continent collision [30] or A-type subduction [19] is erroneous. Our compilation [9, 17, 26] indicates that Paleozoic and Late Precambrian intracontinental blueschists and ophiolites in Eurasia formed before the final closing between cratons, hence, continental collision may not be the cause of blueschist metamorphism [11,26] and it is unlikely that these complexes have formed by underplating processes according to Platt's model [31].

Table 1 Ages of eclogites and blueschists of domains of Paleoasian oceam. Data from [12,17,23,34,38,].

Number (fig.2)	Belts, complexes	Rock	Age(Ma) method
Paleozoic belts			
6	Maksutov	Eclogite, BS Mu-schists	1110(?) Rb-U-Pb,400-460 K-Ar
12	Tien-Shan Atbashi	BS, eclogites	1150(?), 620 K-Ar
13	Hanyalak	BS	729(a),634(b)Rb-Sr
17	Kokchetav massif	Bi-Mu gneiss	516-517+-5
		gneiss, eclogite with diamond	530+-7U-Pb zircon
		eclogite, diamondiferous rocks	525-528 Sm-Nd
		gneiss with diamond	549+-8 internal isochron
		gneiss, Py-carb rock	2.0-2.65, 2.3-2.4 Nd model
		eclogite	0.8-1.0 Nd model
18	Chara	BS,E-Melange inclusions	540-580(a), K-Ar 460-477 (b)
19	Uymon, Altai	Blueschist (BS,BGS)	620-570(a),520-570(b),K-Ar
20	Borus, W. Sayan	E,BS melange inclusions	>460 (520?)
21	Kurtushibin	BS in the base of Ophiolite	>460 (520? 580?)
Pre-Cambrian complexes			
22	North Muya	eclogite	1700(a),1200(b) zircon
23	Oka, E.Sayan	blue-greenschists	620-640 Rb-Sr isochron
24	Hugain	blueschists	>490, Rb-Sr granites
25	Aksu (China)	blue-greenschists	a)720-740 Rb-Sr,K-Ar b)650-676 Rb-Sr

The diamond-bearing gneisses and eclogites of the Kokchetav massif are included in the Caledonides of North Kazakhstan and their ages were studied in detail by various methods[23, 34,36] (Table 1). These age data show that the major stage of ultra-high pressure metamorphism had taken place in 530-540 Ma, with the quenching and cooling proceeding very fast judging by the mica age (520-517 Ma), but more ancient whole rock dates are available (about 1.0 and 2,0 -2.6 Ga). In this case, as in the other examples, for ancient dates noted above the following mechanism is suggested [10]. Micro-continents or other terrains

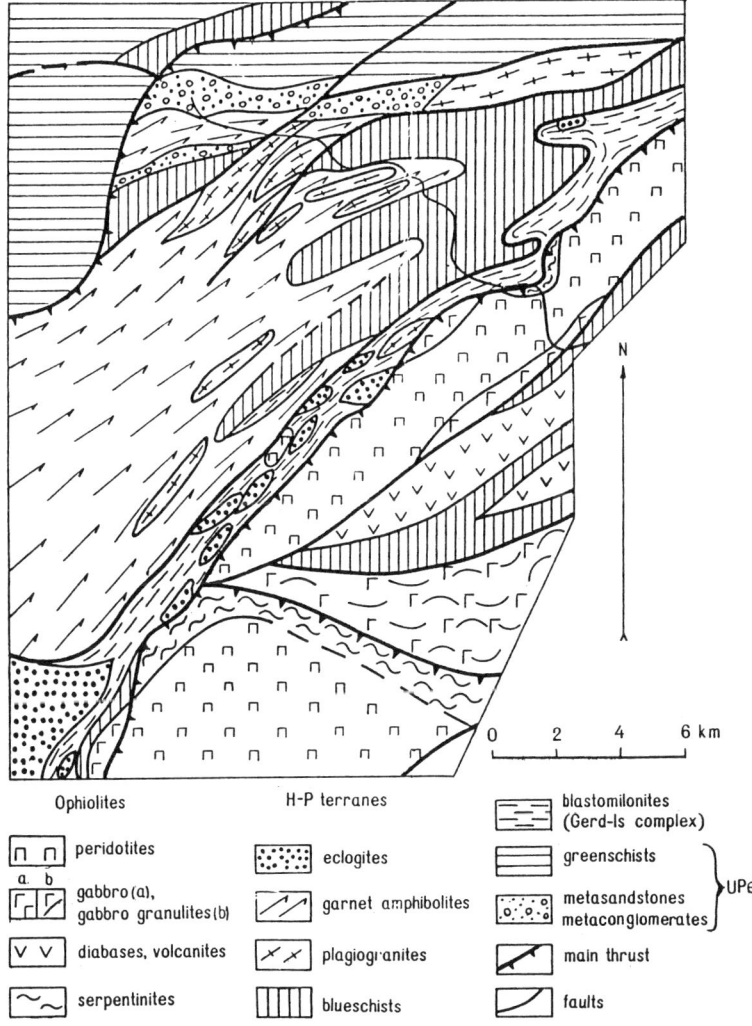

Figure 3 Structure of the Syum-Keu belt in the Polar Urals (upper part) and cross section along A-B line (below) [9, 11]. Ophiolites, ultramafics (upper units) and high pressure (H-P) terrains (lower units) are allochthonus and the Upper Precambrian, partly Ordovician schists on continental basement is autochtonous

(seamount, fragment of an ancient island arc) with ancient dates (2.0 - 2.6 Ma in Kokchetav terrain, 1.0 - 1.1 Ga in other cases) have choked the subduction zone causing its termination or displacement ocean ward. This process results in rapid tectonic transport of eclogites and/or blueschists out of the subduction zone as suggested by the isotopic and geologic data. The mechanism of this fast tectonic transportation is discussed later.Tectonic transportation of rocks from deeper parts of the subduction zone proceeds either as blocks or as a system of sheets, whose upper part is a hanging wall of the subduction zone, with the lower part consisting of weakly metamorphosed rocks produced before termination of the subduction zone. Sometimes, a combination of these forms is observed, for example the Sium-Keu belt in the Polar Urals and Chagan massif of Gorny Altai.

In the Polar Urals (Fig. 3) the upper (south-east) unit is composed of a big mass of high-temperature ultramafics which may represent the hanging-wall mantle and metamorphosed ophiolite. Below the ultramafics there is a melange of blastomylonite zone with inclusions of ultramafics and high pressure eclogites followed by a high pressure unit in the lowest position. The high pressure unit consists of three zones in overturned position (from south to north): eclogites, garent amphibolites and blueschists. The outer northern subzone of blueschist formation corresponds to metamorphosed olistostrome [9, 11]. These metamorphic zones and subzones may represent the primary metamorphic zonation within the subducted plate.

A similar structure is found in Chagan-Uzun massif (Gorny Altai) (Fig. 6). The uppermost unit (metabasalt of subalkaline affinities and siliceous limestone) may be a part of a guyot (seamount) which terminated the subduction process. The second tectonic unit consists of two subunits: a) large mass of high temperature lherzolite similar to Syum-Keu hanging wall mantle rocks and b) serpentinitic melange with inclusions of eclogite, garnet amphibolite, and pyroxenite (altered to actinolite facies). The pressure estimations for eclogites are equal to 13-15 kbar [12]. The lower tectonic unit consists of dismembered ophiolites in an overturned position (serpentinites with inclusions of gabbro-diabase and rodingite with dikes and pillow lava) metamorphosed in the epidote-amphibolite-greenschist facies. The Lower Cambrian olistostrome with fragments of ophiolitic rock forms the sole of the lower plate.

The highest pressures in rocks transported from the subduction zone are established in melange zones or sialic terrains being parts of a micro-continent squeezed into subduction zones. They are diamond-bearing gneisses and eclogites of Kokchetav massif (Table 1), the siliceous lower sheet in Maksutov complex in the South Urals containing eclogites with psuedomorphs after coesite, quartz-almandine-jadeite rock with jadeite-kyanite-quartz associations [4, 9, 11] and pyrope quartzites with coesite in Dora-Maira, the Alps [5], which may be a fragment of the European craton involved in subduction.

Nevertheless, the complexes of transitional type (glaucophane-greenschist) with the longest and most complicated geological history are present in many belts of South Siberia and in the Pacific (Japan, Sakhalin, Kamchatka, Koryakia). It was shown that the earlier blueschists of Upper Paleozoic and Mesozoic belts surrounding the Pacific were developed in subduction zones closed by seamounts. They are the Akiyosky complex in the Sangun belt, Bieu complex in the Kamuikotan belt, Devonian guyot on Penzhina belt in Koryakia. In such cases, the accretionary prism took part in all of the future events often followed by analogous environments in new parallel subduction zones. This suggestion was initially applied to the Franciscan complex in California as a model of polystage obduction [10] that can explain the multi-stage character and regular periodicity of metamorphism in these belts (Table 2) and repeated diaphthoresis of blueschists and eclogites. This type of belt was called Alpine or collisional [11, 21] according to a regular change from blueschist metamorphism to greenschist after cessation of subduction and beginning of collision. The long time and mutli-stage metamorphism in these belts (in the Alps 140-150 to 60 Ma) is typical and Hsu [22] interpreted this fact by long-term uplift of rocks from the subduction zone. However, the necessity of rapid tectonic transportation of blueschists and eclogites from subduction zone contradicts such a scheme. Regular periodicity of blueschist data (Fig. 5) in the world correlate with determined data of blueschist tectonic transportation of rocks from the deep parts of the subduction zone.

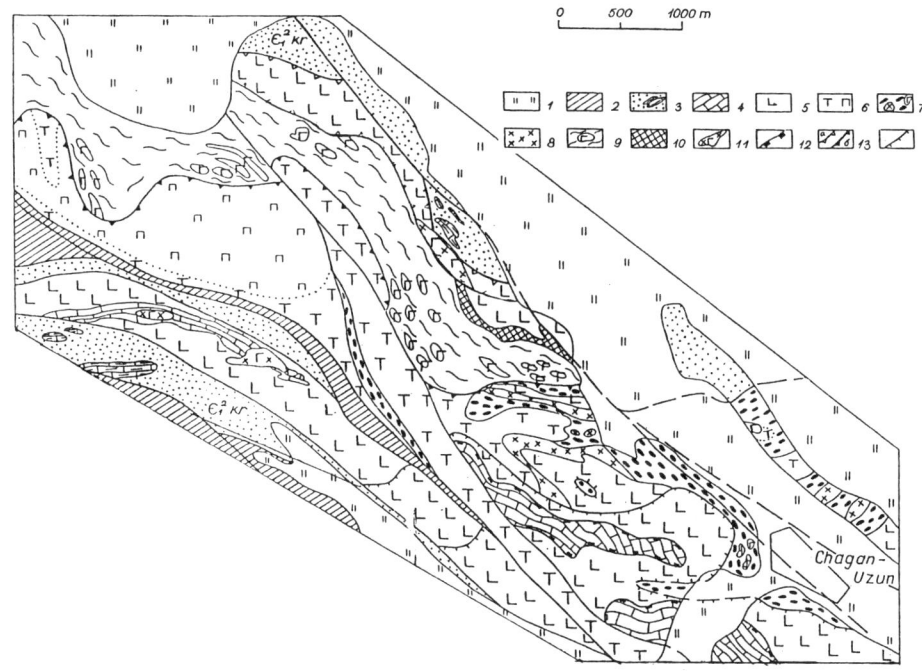

Figure 4. Structure of the Chagan-Uzrun ophiolite massif in the Altai Mountains [12]. 1- Neogene-Quaternary deposits; 2- Devonian Neo-autochton rocks; 3- Cambrian turbidites with olistostrome lenses; 4 & 5- first tectonic unit consists of Vendian-Lower Cambrian sea mount terrain, (4- siliceous limestones; 5- basalts); 6 to 8. Second tectonic unit composed of peridotites and massive serpentinite (6), serpentinitic melange with eclogite and garnet amphibolite inclusions (7), metabasalts transformed into garnet amphibolites (8); 9-11 - lower tectonic unit composed of serpentinitic melange with inclusions of gabbro, diabase, rodingite (9), basaltic pillow lava and dikes transformed into amphibolites (10), lenses of gabbro partly transformed into rodingite (11); 12 - sedimentary thrust contact; 13 - thrust boundary between upper unit (q) and lower unit (b); 14 - other thrusts.

Mineralogical and experimental data concerning the preservation of such easily converted minerals such as aragonite, lawsonite and coesite in glaucophane schist or eclogite [2, 3, 11] indicate that the velocity of the reverse movement of nappes or melange containing inclusions of glaucophane schists and eclogites with minerals noted above should be higher than subduction velocity. Cloos's model [6, 7] shows that the coincidence of the direct and reverse curves of glaucophane metamorphism PT-evolution is possible when the velocities of direct and return flow in a wedge coincide, and that these velocities may be lower than the subduction velocity. According to the kinetics such slow velocity relations are not rapid enough to preserve blueschists because during the slow movement, temperatures changes overstep the preservation temperatures of the metastable minerals [9, 11, 17]

Table 2 Ages of metamorphic belts in Japan, Sakhalin, Kamchtka and California. * new data from [18]. ** - abbreviations, oph- ophiolite (oceanic metamorphism), e- eclogite (and blueschists), bs- blueschists, d- diaphoresis, grg- granitogneisses, granulites, g- granites (g1, g2, g3 phases). *** block in Kurosegawa zone.

Belts	Stages (I-X), ages (Ma), rock types **									
	I	II	III	IV	V	VI	VII	VIII	IX	X
Susunai, Sakhalin	-	55-77 d	90-96 gs	?	133-145 e	175 oph				
Kamuikotan, Hokkaido	-	55-70 d	90-95 gs	109-120 bs	133-145 e	180 oph				
Hidaka, Hokkaido	9-15 g1 19-33 g2	55-40 g1 52-70 grg	?	110 oph						
Aniva Gomon, Sakhalin	17-32 g2	55-44 g1	?	110-120 oph						
Tokoru, Hokkaido			85-90 bs	?						
Franciscan, California	-	55-70 d	85-95 bs,gs	120 bs	145-155 e	150-180 oph				
W. Kamchatka	-	-	-	-	-	180 d	210 bs,gs			
Sanbogawa and Ryuku			85-90 d	110-120 gs	145-150 bs	175-180 e	208-240 bs ***			
Sangun, Japan						180 d	220 bs,gs	250 bs (block)	300 bs	
Omi-Range, Japan									300 bs	340-350 bs
Penzhina, Kamchatka					150 d	180 d	?	?	300 bs	330-340 bs

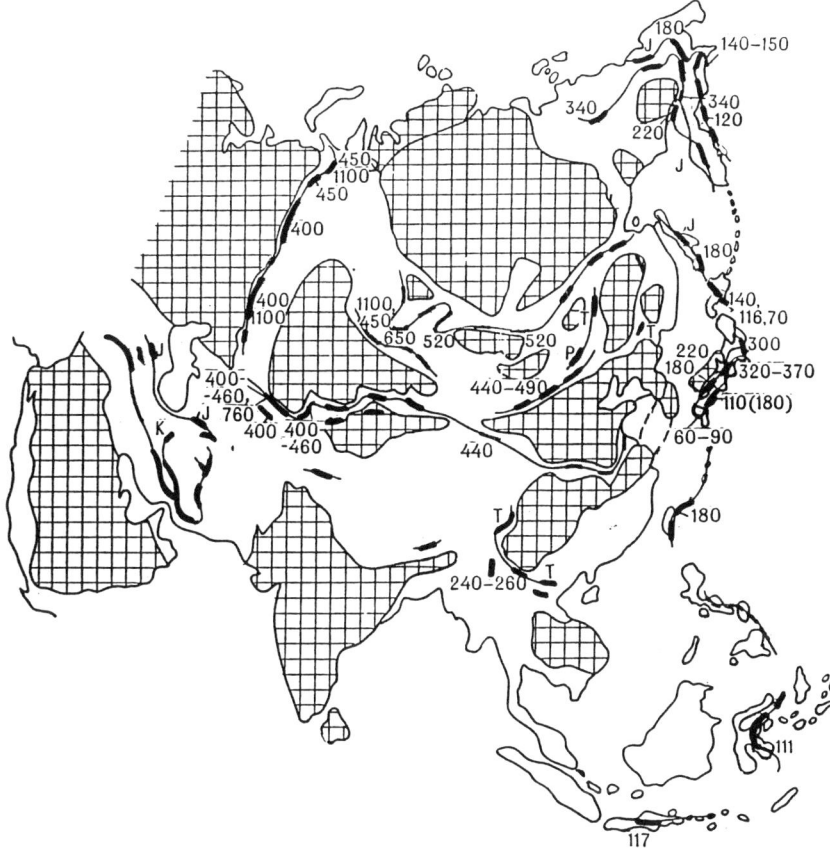

Figure 5. Isotopic ages of blueschists and ophiolites in the world [13] with additions.

MODEL OF SUBDUCTION-ACCRETION WEDGE

Our model of oceanic lithosphere/continent interaction [14, 15] shows that appearance of a viscous wedge between the subducting lithospheric plate and continent (island arc) is possible. This wedge is able to self-regulate by change of wedge shape and viscosity of the intermediate media (Fig. 6). A material comparable with a viscous liquid may consist of sediment and volcanic rocks involved in the subduction process. The viscosity od sedimentary and mixed rock ranges from 10^{15} to 10^{19} poises whereas the viscosity of lithospheric plates is 10^{21} - 10^{23} poises [14, 15]. In reality a wedge consists of a series of sheets which can be recognized in the lower part of an arc-trench slope at depths of 6.5 - 3.5 km [24, 32].

Such media of viscous interaction between continent and oceanic lithosphere significantly decrease the friction between them and stipulate the stable existence of plate subduction. Analytical solutions [14, 15] under a stationary plate's submergence and the presence of normal velocities of ocean lithosphere and continental surfaces provide the velocity of uplift movement to be less than the subduction velocity with μ = constant.

The possible stability of a viscous intermediate wedge within a subduction zone may be as follows [15]:

1. The resistance to subduction due to the pressure between the oceanic plate and the continent (island arc).

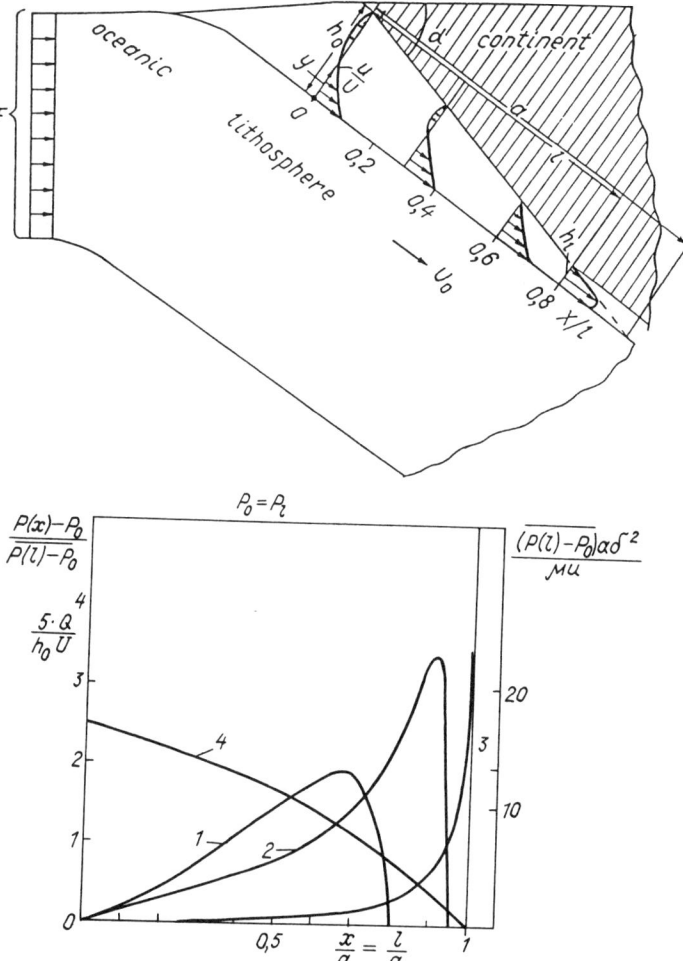

Figure 6. Accretion wedge model: (A) markings when l=0.8, $P_0 = P_l$; (B) calculations results, when $P_0 = P_l$, local prerssure changess along wedge length: 1) when l= 0.89a; 2) when l=0.95a; 3) when l=1.0a; 4) relative outlay $5Q/h_0 U$ at different values of l. See for details [14, 15]. Inlet section is h_0 when l=0, and outlet section is h_l when l=a

2. The hydrodynamic pressure increases in a viscous wedge counteracting the push force F between converging plates and preventing them from coming in contact,. In general for wedge stability and subduction steadiness, plate pressure against a viscous wedge must be equal to the internal pressure in a wedge.

3. Considerable violations of stability and stationariness lead to "corking" and cessation of subduction process. Of course, these points need to be stated in a more rigorous manner and here we show only the minimal scheme.

Let us suppose than an oceanic plate moves downward at the rate U_0 plate F, and a wedge is thin and the wedge thickness h(x) changes linearity (Fig. 6A).

$$h(x)=\delta(a-x), h(x)/l \ll 1 \tag{1}$$

Results of some calculations for the case of inlet and outlet pressure equality ($P_0=P_1$) which are cited on Figure 6, show that maximum relative pressure relates to the outlet section h_1 and increases with the decrease of outlet section h_1. Pressure of x surrounding (x=l) increases together with the relative wedge length (l/a) and at $l \geq 1$, $P_1 \rightarrow \infty$.

This relationship means that when an outlet section (in Fig. 6 at l=a) is closed, pressure can rise up to the limit of firmness and cause the tectonic erosion of a continental plate in a region of narrowing h_1. It also means that when a wedge angle increases, return flow velocity rises. This ascending flow can transport the rock blocks which have been involved in the subduction zone. But in stable conditions the return flow rate is lower than that of subduction.

But the subduction-accretion wedge conditions may be unstable as a result of the entrance of a large block in subduction zone closing h_1, or a quick change in interacting forces and geometry of the wedge, or a change in viscosity of the subduction wedge. In those cases the convergence of the lithbosphere plates will proceed only up to such wedge dimensions, when the pressure force in a wedge will be in disequilibrium with the force of the plate interaction. So, the squeezing out of wedge rocks will occur and return flow velocity could exceed the subduction velocity. In the extreme case, when the internal pressure in the viscous layer is significantly lower than that of the lithosphere plate pressure we were able to find the approximate solution for a quasi-stationary state obtained under boundary conditions at $h = H_0$, $T = 0$, $U_0 = 0$, $y = h$.

$$\bar{U} = \frac{v_0 x}{h} = \frac{F x h_0^2}{2F t h_0^2 + \mu l^3} \tag{2}$$

$$\bar{U}_{max} = F h_0^2 / 2\mu l^2 = P h_0^2 / 2\mu l \tag{3}$$

Where $\bar{U}$ indicates an average velocity of return flow through the thickness of the layer and $\bar{U}_{max}$ is the average maximum velocity when $t = 0$ and $x = 0.5 l$, h_0 = initial layer thickness, t = time, μ = viscosity. Figure 7 shows $\bar{U}_{max}$ calculated according to [3] at variable P, h_0 and μ at $l = 10^5$ m.

The initial conditions on Figure 7 can be defined in the following way. In normal subduction conditions the force of the interaction between lithosphere is comparable with the force appearing due to friction on the asthenosphere/lithosphere boundary resulting from convective flows in the asthenosphere. Estimation of Kirdyashin [24] shows in this approximation $F = 0.5 - 1.0 \times 10^{12}$ N/m and average plate pressure $P = F/l = 5 - 10 \times 10^6$ N/m^2 = 50 - 100 bar under $l = 10^5$ m. Under wider admitted pressures P = 50 - 500 bars the average viscosity expected in a wedge would be $\mu = 10^{17}-10^{118}$ poises. So, the maximal velocity of return flow $\bar{U}_{max} = 10$ cm/year under $h_0 = 2 \times 10^4$ m (Fig. 7). To obtain $U_{max} \gg U_0 = 10$ cm/yr there are two limiting variants: increase of P at constant μ and h_0 as a result of closing of the outlet section h_1 by a large block or change of interaction force (F). In this case, F and P increase can be due to the same reason, and P can increase up to 2-10 kbar. In a second variant, $\bar{U}_{max}$ increases at constant internal pressure as a result of decrease in rock viscosity of one or more orders within the subduction wedge and/or increase h_0. Decrease μ and increase h_0 may have the same effect: for example the change of plate interaction geometry and shales of low-viscosity rock (saturated with fluids) within the wedge.

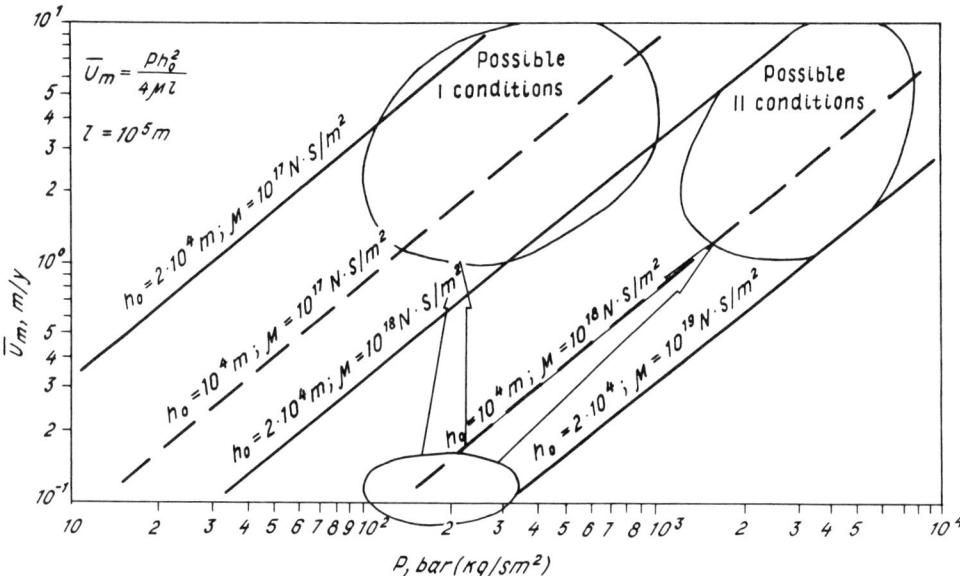

Figure 7. Calculated average maximal rate of return flow in as wedge (U_{max}) at variable external pressure p, viscosity μ and inlet section h_o.

The velocities calculated (Fig. 7) relate to t = 0 (beginning of squeezing or closing). In time, U_{max} decreases due to decrease in the viscous layer thickness and at a later time equal to:

$$t = \frac{\mu \, l^2}{2P \, h_o^2} \qquad (4)$$

$\overline{U}_{max}$ is decreased by two-fold. For instance, at $\mu = 10^{18}$ Ns/m^2, P = 500 bar, $h_o = 2 \times 10^4$ m it will occur at = 2.5 $\times 10^{11}$ sec or 10^4 years. In our approximation we supposed $P_l + P_0$, under real conditions the pressure increase may occur close to the outlet section and pressure may become equal throughout all the layers with the outer pressure of the lithospheric plates marking the transition to a stable state. However the short-term effects are great enough to obtain a situation of return flow velocity in a "viscous wedge" that is much higher than the usual subduction velocity.

DISCUSSION

Recently, the numerous data about global regularities and periodicity of geoprocesses including blueschist metamorphism, ophioilite formation (Fig. 5) and their inter

-relationships with a general system of convective flow in the Earth's mantle and core have appeared. Thus, we note blueschist maximums of 30 Ma intervals in the Mesozoic (30, 60, 90, 120, 150, 180, and 220 Ma) and 50 to 60 Ma intervals in the Paleozoic (300, 340, 400, 460, 520, Ma) (Fig. 5) coincide with the greatest tectonic phases and usually with stages of acceleration and reconstruction of lithospheric plate movement[1,13,17,39]. In turn, these maxima can be grouped into two epochs of the most preserved (or widely manifested glaucophane metamorphism) between 60-150 and 400-520 Ma [1, 20, 26, 28]. The third epoch of 700-800 Ma [26] and a periodicity of 120 Ma (520, 640, 760, 880, 1000, 1150 Ma) in the Late Precambrian [17] can be distinguished. The first two epochs correspond to a rise in sea level during Phanerozoic times of 40-125 Ma and 380-540 Ma [37] and this change in sea level has been correlated with the production rate of new ocean crust on a global scale [33]. In other words, the higher spreading rates of 60-130 (150), 400-520 and 700-850 Ma indicate that higher subduction rates favored blueschist origin along island arcs and continental margins[11, 12] including their tectonic transportation and preservation.

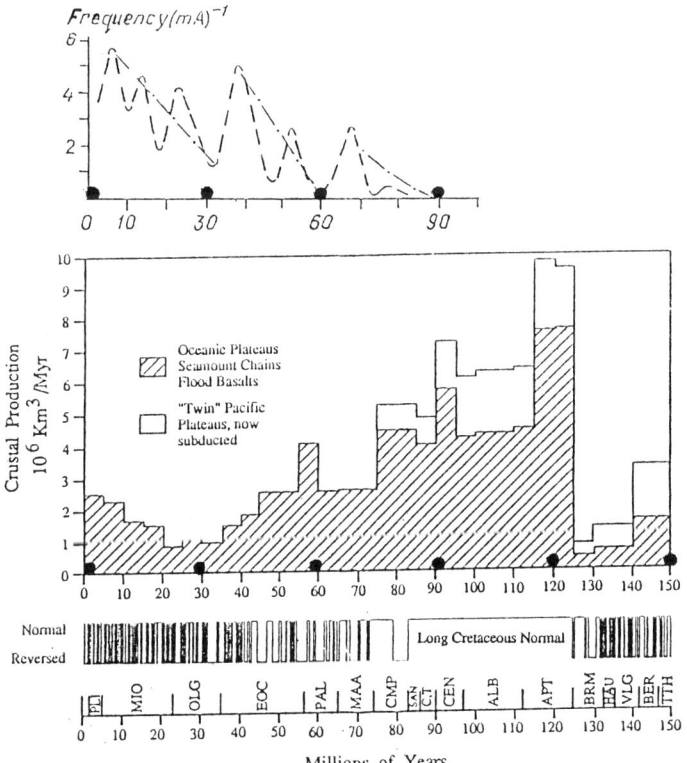

Figure 8. World-wide distribution of mantle magmatism plotted with the magnetic reversal time scale as a function of geologic time according to [25]. Each 5 Myr histogram level represents the amount of crust produced per million years during this 5 Myr time span. Upper frequency diagram shows the periodicity of magnetic reversal [29]. Black dots from both diagrams are from Table 2 and Figure 5 and correspond with the peak of blueschist metamorphism on a world scale.

According to our model of the subduction wedge [14, 15] higher subduction rates lead to ultrahigh return transportation during unstable conditions caused by the closing effect produced by sea mounts or continental margins entering subduction zones. So ultra-fast subduction is very important not only for ultra-high pressure metamorphism and production of geochemical anomalies in the upper mantle, but also for high speed tectonic transportation of blocks up to the surface and preservation of such minerals as diamond, coesite, lawsonite, aragonite and jadeite [17].

Recently, the correlation between periodicity of tectonic an petrological processes and intensity of magnetic reversals was shown, of course, only for Mesozoic and Cenozoic history. The last intense epoch of blueschist metamorphism of 60-120 Ma reflects the high rate of spreading and subduction processes that coincides with the Cretaceous "megachron" of 80-120 Ma when magnetic reversals are absent. This megachron and the following decrease in magnetic reversals correlate also with the intensity of mantle magmatism. Thus mantle plumes coming from the mantle/core boundary controlled the magnetic reversal frequency [25]. This conclusion is supported by plume generated maximum magmatism being developed during the megachron period of 80-150 Ma (Fig. 8). Moreover, a similar periodicity for the frequency of magnetic reversals which have a period of about 30 Ma (half-period in 14-17 Ma) [29] whose boundaries coincide with peaks of blueschist data as noted above: 30, 60, 90 Ma. In turn these periods, Cretaceous megachron (with 30 Ma displacement) and maximum mantle magmatism can be joined into a longer periodical dependence with 180 Ma period (90 Ma half period) that coincide with the duration of the greatest tectonic epochs : Alpine (0-200 Ma), Hercynian (220-400 Ma), Caledonian (400-580 Ma and Balkalian (600-800 Ma).

All the correlations noted above are not casual and are explained by the global regularities of multi-layer convection in the upper and lower mantle and the Earth's core. These are complicated by short periods (about 30 Ma) of mantle plumes generated at the core/mantle boundary [25, 16]. The two layer model of mantle convection shows a strong correlation with the stable subduction zones with descending flows in the lower mantle convective cell. The transitional or unstable relations between the lower and upper mantle convective cells and mantle plume pulses are reflected in the cyclic implusing of the tectonic processes. We hope by these explanations to show the main spatial-time regularities of glaucophane schist origin and the preservation of blueschists and eclogites.

REFERENCES

1. E. Abbate, V. Bortolotti, P. Passerini, and G. Princiipi. The rhythm of Phanerozoic ophiolites, *Ophioliti.* **10**, 109-139.(1985).

2. W.F. Brace, W.G. Ernst, and R.W. Kallberg. An experimental study of tectonic overpressure in Franciscan Rocks, *Geol. Soc. Amer. Bull.* **81**, 3125-1338.(1970).

3. W.D. Carlson and J.L. Rosenfeld. Optical determination of topotactic aragonite-calcite growth kinetics: metamorphic implications, *Jour. Geol.* **89**, 615-618.(1981).

4. B.V. Chesnokov and V.A. Popov. Change of volume of quartz in eclogites of the South Urals, *Rep. Acad Sci. USSR.* **162**, 176-178.(1965) (*in Russian*).

5. C. Chopin. Coesite and pure pyrope in high-grade blueschists of the Western Alps. A first record and some consequences, *Contrib. Mineral. Petrol.* **86**, 107-118.(1984).

6. M. Cloos. Flow melanges: numerical modelling and geologic constraints on the origin in the Franciscan subduction complex, California, *Geol. Soc Amer. Mem.* **93**, 330-345.(1982).

7. M. Cloos. Blueschists in the Franciscan complex of California: petrotectonic constraints on uplift mechanisms, *Geol. Soc. Amer. Mem.* **164**, 77-93.(1986).

8. R.G. Coleman. Plate tectonic emplacement of upper mantle peridotites along continental edges, *Jour. Geophys. Res.* **76**, 1212-1222.(1970).

9. N.L. Dobretsov, *Glaucophane schists and eclogite-glaucophane-schist complexes in the USSR.* Nauka, Novosibirsk. (1974). (*in Russian*).

10. N.L. Dobretsov. The new overthrusting model for the blueschist metamorphism with references to Franciscan-Great Valley problems, *Ofioliti.* **4**, 132-141.(1979).

11. N.L. Dobretsov. Blueschists and eclogites: a possible plate tectonic mechanism for the emplacement from the upper mantle, *Tectonophysics.* **186**, 253-268.(1991).

12. N.L. Dobretsov, M.M. Buslov, and V.A. Simonov. Associated ophiolites, blueschists and eclogites of Altai mountains, *Doklady Acad. Nauk SSSR.* **318**, 413-417.(1991) (*in Russian*).

13. N.L. Dobretsov, R.G. Coleman, J.G. Liou, and S. Maruyama. Blueschist belts in Asia and possible periodicity of blueschist facies metamorphism, *Ofioliti.* **12**, 445-456 (1987).

14. N.L. Dobretsov and A.G. Kirdyashkin. Dynamics of subduction zones: models of accretion wedge formation and transportation of blueschists and eclogites, *Geolog. & Geoph.* **N3**, 5-9 (1991) (*in Russian*).

15. N.L. Dobretsov and A.G. Kirdyashkin. Subduction zone dynamics: models of accretionary wedge, *Ofioliti.* **17**(1), 155-164.(1992).

16. N.L. Dobretsov and A.G. Kirdyaskin. Experimental modelling of two-layer convectioons in the Earth, *Tectonophysics* (1993) (in press).

17. N.L. Dobretsov, N.V. Sobolev, and V.S. Shatsky, *Eclogites and blueschists in folded belts.* Nauka, Novosibirsk (1989) (*in Russian*).

18. N.L. Dobretsov and T. Watanabe. Comparison of ophiolites and blueschists of Sakhalin and Hokkaido, *Ophioliti* (1992) (in press).

19. S. Dong, Q. Shen, D. Sun, and L. L., *Metamorphic map of China, 1:4,000,000. Map and explanatory text.* Geol. Publ. House, Beijing. (1986).

20. W.G. Ernst. California blueschists, subduction and significance of tectonostratigraphic terrains, *Geology.* **12**, 436-440.(1984).

21. W.G. Ernst. Tectonic history of subduction zones inferred from retrograde blueschist P-T paths, *Geology.* **16**, 1081-1084.(1988).

22. K.J. Hsu. Exhumation of high-pressure metamorphic rocks, *Geology.* **19**, 107-110.(1991).

23. E. Jagoutz, V.S. Shatsky, N.V. Sobolev, and N,P, Pokhilenko. *Ph-Nd-Sr isotopic study of the Kokchetav massif, the outcrop of the lower lithosphere.* in *Inter, Geol. Congress 28th.* Washington, D.C.: (1989). (extended abstract)

24. A.G. Kirdyashkin, *Heat gravitational flows and heat-exchange in asthenosphere.* Nauka, Novosibirsk. (1989).(*in Russian*).

25. R.L. Larson and P. Olson. Mantle plumes control magnetic reversal frequency, *Earth Planet. Sci. Lett.* **107**, 437-447.(1991).

26. J.C. Liou, S. Maruyama, X. Wang, and S. Graham. Pre-Cambrian blueschist terrains of the world, *Tectonophysics.* **181**, 97-111.(1990).

27. J.G. Liou, S.A. Graham, S. Maruyama, X. Wang, and X. Xiao. Proterozoic blueschist belt in W. China: best documented Precambrian blueshict in the world., *Geology.* **17**, 1127-1131.(1989).

28. S. Maruyama, J.C. Liou, R.G. Coleman, and N.L. Dobretsov. Blueschists in Asia and their significance in evolving Asia, *Jour. Geol.* (199?) (in press).

29. A. Mazuad and C. Laj. The 15 Ma geomagnetic reversal periodicity: a quantitative test, *Earth Planet. Sci. Let.* **107**, 689-697.(1991).

30. A. Nur and Z. Ben-Avraham, Break up and accretion tectonics, In: *Accretionary Tectonics in the Circum-Pacific Regions,* M. Hashimoto and S. Uyeda (Ed.), pp. 3-18, Terra Sci. Pub., Tokyo. (1983).

31. J.P. Platt. Dynamics of orogenic wedges and the uplift of high pressure metamorphic rocks, *Geol. Soc. Amer. Bull.* **97**, 1037-1053.(1986).

32. H.F. Ryan and D.W. Scholl. The evolution of fore-arc structures along an oblique convergent margin, Central Aleutian arc, *Tectonics.* **8**, 65-71.(1989).

33. T.J. Schopf. Permian-Triassic extinctions: relation to sea-floor spreading, *Jour. Geol.* **82**, 129-143.(1974).

34. V.S. Shatsky, E.J. Jagoutz, O.A. Kozmenko, and N.V. Sobolev. The geochemistry of ultra high pressure rocks from Kokchetav massif (Kazahstan), *Geochim. Cosmochimica Acta.* (1992) (in press).

35. N.V. Sobolev, N.L. Dobretsov, A.B. Bakirov, and V.S. Shatsky, Eclogites from various types of metamorphic complexes in the USSR and problems of their origin, In: *Blueschists and Eclogites (Memoir),* B.W. Evans and E.H. Brown (Ed.), Vol. 164, pp. 349-363, Geol. Soc. Amer., Boulder, Colo. (1986).

36. N.V. Sobolev and V.S. Shatsky. Diamond inclusions in garnets from metamorphic rocks, *Nature.* **343**, 742-746 (1990).

37. P.R. Vail, R.M. Mitochum, and S. Thompson, Seismic stratigraphy and global changes of sea level. IV Global cycles of relative changes of sea level, In: *Seismic stratigraphy (memoir)*(Ed.), Vol. 26, pp. 83-116, Amer. Assoc. Pet. Geol., Tulsa (1977).

38. Z. Wang, X. Xiao, and Y. Tang, On blueschists in south Tian-Shan, China (Report #1). International Geological Correlation Program (Project 283): 1990) (Novosibirsk)

39. L.P. Zonenshain and M.I. Kuzmin. Absolute reconstructions of the Paleozoic oceans, *Earth Planet. Sci. Lett.* **74**, 103-116 (1985).

Blueschists discovered in Kumishi, South Tianshan and their tectonic significance.

GAO JUN[1], TANG YAOQING[1], ZHAO MIN[1], WANG JUN[1], and WU HANQUAN[2]

[1]Institute of Geology, CAGS, Beijing, China
[2]Institute of Geology and Mineral Resources, CAGS, Xi'an, China.

Abstract - Blueschists have been recently discovered in Kumishi, South Tianshan. They occur as one kind of blocks in the matrices of ophiolite melange and contain glaucophane and crossite. They are the result of subduction event which happened along a back-arc ocean basin along the southern margin of the Paleo-Asian Ocean in the early Paleozoic.

Key Words - Kumishi blueschist, glaucophane, crossite, Paleo-Asian Ocean.

OUTLINE OF GEOLOGY AND OCCURRENCE OF BLUESCHIST

Kumishi area is located in South Tianshan, just near the north margin of the Tarim Platform (Fig. 1). An early Paleozoic ophiolitic melange exists in south Kumishi town, its about 20 km wide from N to S [2]. A metamorphic terrane occurs in south Kumishi town [3]. Kumishi metamorphic terrane is a Precambrian continental block spilt from the Tarim platform. It consists of a series of amphibolite facies metamorphic rock containing schists, gneisses and gneissic migmatitic granite. The average U-Pb age of sixteen single grains of zircon in the gneiss is 2600-2800 Ma. The Nd model age of the augen biotite gneiss is 1709-2076 Ma. These two separate age determinations indicate that the Kumishi metamorphic terrane is a Precambrian block. However, the Rb-Sr whole rock isochron age of the augen biotite gneiss is 451 ± 18.7 Ma. This age probably represents the tectono-thermal event which produced the Kumishi metamorphic terrane.

It is thought that a divergence from the Tarim platform took place as the Paleo-Asian Ocean crust was subducted towards the north margin of Tarim platform. The K/Ar age of the biotite from the augen biotite gneiss is 350 Ma. this age indicates the collision and closure events that took place between the Kumishi metamorphic terrane and the Tarim platform. An early Paleozoic ophiolitic melange exists in the southern part of the metamorphic terrane. The exotic blocks of the melange are mainly ophiolites with just a few gneiss blocks. The size of exotic block occurrence changes from several cm^2 to several hundred m^2. The matrix of the melange is mainly Silurian greenschist (S). A lot of thrusts occur within the ophiolitic melange and they dip SSW (200-210°). The ophiolitic suite consists of sedimentary rock, volcanic rock, cumulates and peridotite. Chemical discrimination diagrams [2] indicated that the ophiolite suite could be oceanic crust that formed in a back-arc basin. This crust is considered to have formed in the Early Paleozoic (O-D).

A blueschist band, several meters wide, was found to be intercalated with a greenschist layer as a matrix material in the melange near the major ophiolite suite (Fig. 1). Both glaucophane and crossite are present in the blueschist band.

IDENTIFICATION OF GLAUCOPHANE AND CROSSITE

The types of blueschist are epiodote-chlorite-schist and epidote-chlorite-quartz-schist. The typical mineral association is glaucophane (or crossite) +chlorite+zoisite+epidote+quartz. This is a typical lower greenschist facies mineral association. The glaucophane and crosstie

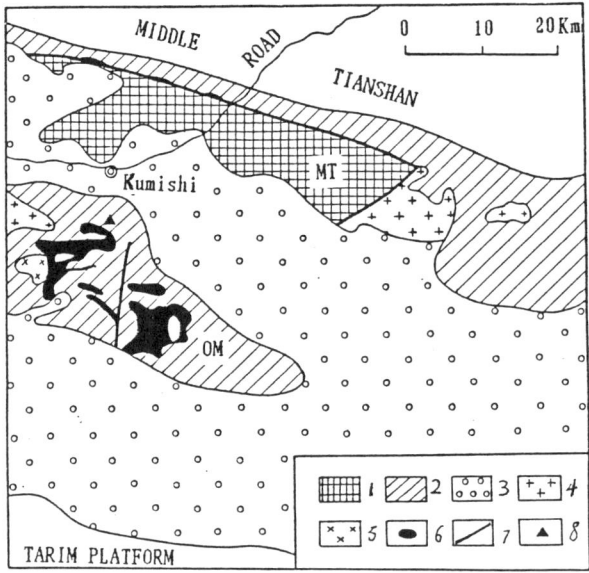

Figure 1. The geologic sketch map of the Kumishi region. MT- Metamorphic terrane, OM - Ophiolitic melange. 1. Precambrian metamorphic rocks., 2. Paleozoic strata, 3. Quaternary, 4. Granite, 5. Diorite, 6. Mafic and ultramafic rocks, 7. Fault, 8. The outcrop of blueschists.

are closely intergrown with the chlorite. The modal volume of the blue amphibole is minimal usually less than 5%. The blue amphiboles show macroprismatic and acicular grain habits. Their grain size is very fine varying from about 0.02 x 0.1 to 0.04 x 0.5mm. Their pleochrism is strong, changing from dark blue, light blue to colorless. We chose seven grains with distinct pleochrosim and established their chemical composition by the electron microprobe. The results are shown in Table 1.

The electron microprobe data were re-calculated into the amphibole formula: $[A_{0-1} B_2 C^{IV}_5 T^{IV} O_{22} (OH,F,Cl)_2]$ and Fe^{3+} is estimated using Cameron's (1974) formula ($Fe^{3+}= Na_B+Al^{IV}- [(Na+K)_A+2Ti+Cr]$) and Fe^{2+} was calculated using B. Reynard's (1988) formula ($Fe^{2+}=$ Total Fe - Fe^{3+}x 1.1) (These formulas were given to us by letter from Mr. Wu Hanquan). We calculated the formulas for the seven separate amphibole grains and plotted these on an amphibole compositional diagram taken from [1] . (Fig. 2) Four of the seven grains analyzed by the electron microprobe are glaucophanes and three of them are crossite.

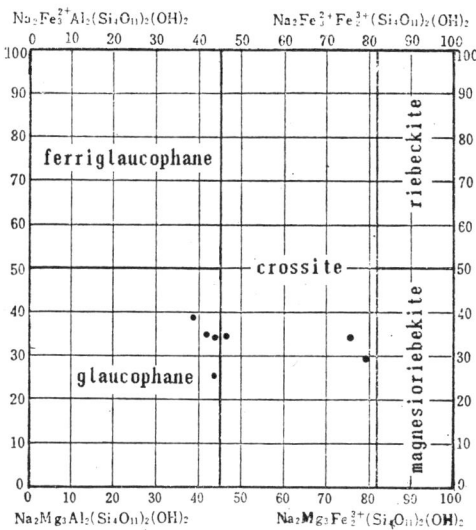

Figure 2. Compositons of glaucophane and crossite from Kumishi blueschists.

TECTONIC SIGNIFICANCE

During the development of plate tectonics many geologists have paid close attention to the glaucophane group because it is an indicator mineral for high pressure and low temperature conditions. The presence of a high pressure and lower temperature blueschist belt is used as important evidence to establish paleo-subduction zones. The Tianshan orogen is an important tectonic region during the evolution of the Paleo-Asian Ocean. The Kumishi area is just located within a contiguous zone between South Tianshan and the Tarim platform. Does the discovery of glaucophane and crossite in the Kumishi ophiolitic melange indicate that there existed a subduction of the Paleo-Asian Ocean in this zone? The answer is no according to recently compiled data. First, the grain size of glaucophane and crossite is very fine. They only intergrow with epidote/chlorite and quartz. Phengite is not present, indicating that the pressure was not very high when glaucophane and crossite formed. Secondly, Kumishi ophiolitic melange has been proved to have originated within a back-arc basin [2].

The glaucophane and crossite in the Kumishi ophiolitic melange probably only indicate a collision and closure event of back-arc small ocean basin. Up to now there are no radiometric ages on the blueschist. The authors consider that the age of blueschist is probably 350 Ma and believe that it is related to the tectono-thermal event that produced the larger metamorphic terrane. The 350 Ma also probably represents the time of closure of the back-arc basin along the north margin of the Tarim platform as part of the Paleo-Asian Ocean tectonic province.

This article is a preliminary report on blueschist discovered in Kumishi ophiolitic melange. More detailed studies are continuing. The results of the current studies will be published in the future.

Table 1. Electron microprobe data of the glaucophane and crossite in blueschists in Kumishi, South Tianshan

components	1	2	3	4	5	6	7
SiO_2	55.00	55.73	55.57	56.45	54.33	54.11	52.91
TiO_2	0.13	0.00	0.08	0.05	0.07	0.11	0.09
Al_2O_3	5.76	5.15	6.43	5.69	5.42	2.63	2.59
Cr_2O_3	0.12	0.12	0.03	0.04	0.06	0.00	0.02
FeO(Total)	17.35	17.39	16.71	16.46	16.94	21.90	21.04
NiO	0.11	0.11	0.00	0.00	0.16	0.03	0.00
MgO	9.93	10.30	10.69	10.84	11.78	10.31	11.31
MnO	0.16	0.17	0.13	0.14	0.17	0.28	0.15
CaO	1.34	3.19	0.78	1.98	2.92	1.11	2.83
K_2O	0.00	0.00	0.00	0.00	0.00	0.00	0.00
Na_2O	6.27	5.21	6.67	6.24	5.09	6.02	5.91
Total	96.71	97.33	97.05	97.82	96.83	97.29	96.86

Number of cation

		1	2	3	4	5	6	7
T	Si	7.9992	8.0081	7.9917	8.0532	7.8898	8.0396	7.9070
	Al^{iv}	0.0008	0.0000	0.0083	0.0000	0.1112	0.0000	0.0930
C	Al^{vi}	0.9847	0.8716	1.0753	0.9527	0.8172	0.4685	0.3637
	Ti	0.0194	0.0000	0.0082	0.0053	0.0079	0.0126	0.0101
	Fe^{3+}	0.7526	0.5669	0.7731	0.7548	0.7043	1.4717	1.4189
	Fe^{2+}	0.7773	1.2001	0.3267	0.0000	0.3912	0.7307	0.6668
	Mg	2.1528	2.5146	2.2916	2.3055	2.5333	2.2829	2.5191
	Mn	0.0199	0.0212	0.0158	0.0168	0.0207	0.0253	0.0194
	Ni	0.0125	0.0125	0.0000	0.0000	0.0136	0.0033	0.0000
	Cr	0.0133	0.0133	0.0038	0.0047	0.0063	0.0000	0.0020
	Ca	0.2370	0.0000	0.0000	0.0000	0.0000	0.0000	0.0000
B	Ca	0.0000	0.4916	0.1204	0.2901	0.4544	0.1771	0.4532
	Na	1.4744	1.2420	1.5521	1.5345	1.1550	1.4441	1.1523
	Fe^{2+}	0.5256	0.2664	0.3275	0.1754	0.3926	0.3708	0.3945
A	Na	0.2922	0.2038	0.3069	0.1923	0.2807	0.5033	0.5595
	K	0.0000	0.0000	0.0000	0.0000	0.0000	0.0000	0.0000

Acknowledgements
This study was supported by grants from the National Natural Science Foundation and Project 305 of Xinjiang, Chinese Academy of Geological Sciences.

REFERENCES

1. W.R. Phillips and D.T. Griffen, *Optical mineralogy*. Freeman and Com., San Francisco. (1981).

2. W. Wu, C. Jiang, L. Li, and F. Yang. *Kumishi tectonic melange, South Tianshan and tectonic significance*. in *Second Symposium on Geology and Mineral Resources, Tianshan, Xianjiang*. (1991). (*in Chinese*).

3. W. Zhu, C. Wang, and R. Ma. The determination of Kumishi metamorphic terrane in Xinjiang and the tectonic significance of its isotopic age, *Bull. Nanjing Univ. Geoscience*. 3, 367-372.(1991) (*in Chinese with English abstract*).

EARLY ARC PLUTONIC ROCKS IN THE OLYUTOR RANGE, NORTHEASTERN KAMCHATKA, RUSSIA

H. TANAKA[1], P. K. KEPEZHINSKAS[2], S. MIYASHITA[3] and I. REUBER[4]

[1] *Department of Earth Sciences, Faculty of Science, Yamagata University, Koshirakawamachi 1-4-12, Yamagata 990, JAPAN*
[2] *Institute of the Lithosphere, Academy of Sciences of Russia, Staromonetny per., 22, Moscow, RUSSIA*
[3] *Department of Geology and Mineralogy, Faculty of Science, Niigata University, Igarashi Ninomachi 8050, Niigata 950-21, JAPAN*
[4] *Institut Dolomieu, 15, rue Maurice Gignoux, 38031 Grenoble Cedex, FRANCE*

Abstract. The geology, petrography and geochemistry of the Epilchik ultramafic pluton and the North Machevna gabbroic pluton in the southern Olyutor Range of Northeastern Kamchatka are presented and discussed together with the geodynamic framework. Upper Cretaceous volcanogenic-sedimentary rocks in the range, intruded by many ultramafic and mafic to intermediate plutons of Paleogene age, constitute a prolongation of the submarine Shirshov Ridge, which separates the Aleutian basin from the Komandorsky basin. The Epilchik pluton is an Alaskan-type complex, concentrically zoned from dunitic core to gabbroic margin, and was formed by early olivine- and subsequent clinopyroxene-dominated fractionation from a primary magma of olivine tholeiite or high-alumina basalt composition. The North Machevna pluton, composed of gabbroic rim and quartz dioritic to quartz monzodioritic core, exhibits geochemical and mineralogical features of calc-alkaline rock series, originated from a primary magma of high-alumina basalt composition. The Epilchik and North Machevna plutons were formed in an intraoceanic (primitive) island arc of Late Cretaceous age, which was situated above subduction zone where two oceanic plates were converged.

INTRODUCTION

Studies of early arc plutons and arc-roof ultramafic-mafic complexes provide much information on the geodynamic evolution of the continental crust at convergent plate margins. Though subduction-related plutons in intraoceanic island arcs are exposed infrequently, they may represent the early stages of the development of the continental crust (ex.[1-7]). While the Kamchatka peninsula is characterized by active volcanoes in continuation of the Kurile arc, its northeastern part has recorded a long history of variable events related to island arc settings at the junction between the Kurile and Aleutian arcs [8,9]. In the Olyutor Range, Northeastern Kamchatka, many ultramafic and mafic to intermediate plutons of Paleogene age are distributed in Upper Cretaceous volcanogenic-sedimentary rocks, which formed in oceanic and intraoceanic island arc environments. These plutons provide clues as to unravelling the character and history of parental materials of primary arc magmas, and record the history of differentiation of the arc magmas and that of the crustal growth in the intraoceanic island arc. In this paper, the geology, petrography and geochemistry of an ultramafic pluton and a gabbroic pluton are presented and discussed together with the geodynamic framework.

GENERAL GEODYNAMICS AND GEOLOGICAL OUTLINE OF THE OLYUTOR RANGE

The Northeastern Kamchatka region is the southern extension of the North Kamchatka-Koryak fold belt, composed of oceanic and arc-derived terranes of Paleozoic to Cenozoic age, which were accreted to the Eurasian margin during Cretaceous and Tertiary times [8,10]. The Northeastern Kamchatka terranes frame the northern and western parts of the Komandorsky basin, which is separated from the Aleutian basin by the submarine Shirshov Ridge (Fig. 1). While the Aleutian basin consists of a remnant of the Kula plate of Mid-Cretaceous age, trapped behind the Aleutian arc [11], the Komandorsky basin is considered to be newly formed oceanic crust as a result of spreading during Late Cretaceous and Tertiary times, as indicated by the high heat flow (approximately 2.5 times higher than in the Aleutian basin) [12]. Upper Cretaceous volcanogenic-sedimentary rocks are widely distributed in the Olyutor Range, located on the continental extension of the Shirshov Ridge. The Upper Cretaceous rocks show a close spatial and genetic relation with the submarine Shirshov Ridge [12-14]. The Olyutor Range is thought to be a remnant intraoceanic island arc of Late Cretaceous age, which was formed above the subduction zone where two oceanic plates were converged [12].

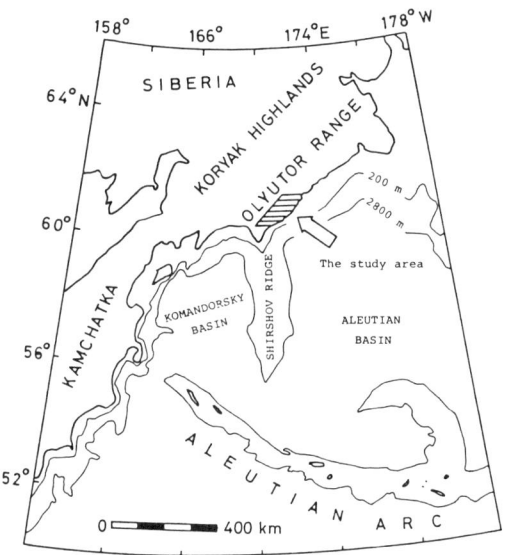

Figure 1. Index and tectonic map of Northeastern Kamchatka, Russia.

Figure 2 shows the generalized geological map of the southern Olyutor Range, Northeastern Kamchatka. This region is mainly underlain by the Vatyna Group which is composed of oceanic volcanogenic-sedimentary rocks of Albian to Campanian age, and the Achayvayam Group which consists of intraoceanic island arc volcanogenic-sedimentary rocks of Maastrichtian to Danian age [13-16]. The oceanic and arc terranes have been thrust northward over the Upper Cretaceous-Paleogene Koryak flysch sequences[17], and are covered by Late Cenozoic volcanic rocks on the west side [18].

The oldest rocks exposed in the Olyutor Range are Albian-Turonian pillow lavas of MORB affinity associated with cherts and pelagic limestones [19]. They are overlain by chert, alkali basalt and ferrobasalt of Coniacian to early Santonian age. The rocks pass gradually to late Santonian-Maastrichtian oceanic sequences of basalt interbedded with chert [16]. These oceanic rocks of the Vatyna Group are partially overlapped in time with the activity of the Achayvayam Group, which is composed of basalt, basaltic andesite, pyroclastics and tuffite [14,15]. The Achayvayam rocks are arc tholeiites and low-K calc-alkaline lavas, typical of intraoceanic island arcs [20]. The widespread distribiution of basaltic rocks is a characteristic feature recognized in intraoceanic island arcs [21], such as the Tonga and Kermadec [22.23], the Mariana [24] and the Aleutian [7,25].

Numerous ultramafic to intermediate plutons of Paleogene age intruded into the Upper Cretaceous volcanogenic-sedimentary rocks in the southern Olyutor Range. Many of these plutons are located within faulted zones and elongated in the E-W direction. Two plutons of

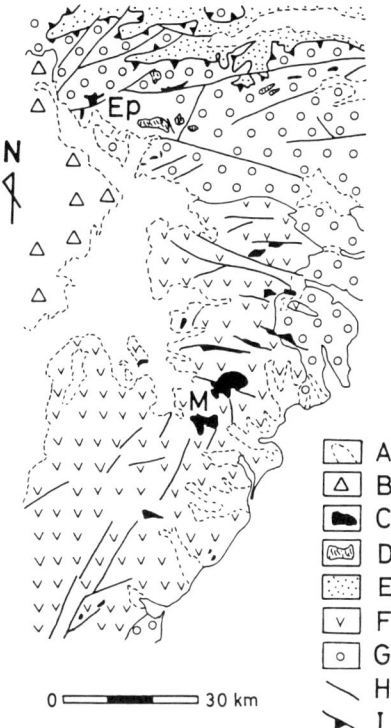

Figure 2. Generalized geological map of the southern Olyutor Range, Northeastern Kamchatka, Russia. A, Alluvium; B, Pliocene to Recent volcanic rocks; C, Mafic to intermediate plutons; D, Ultramafic plutons; E, Flysh of presumably Senonian age; F, The Achayvayam Group of Maastrichtian to Danian age formed in an oceanic island arc; G, The Vatyna Group of Albian to Campanian age formed in an ocean; H, Fault; I, Thrust fault; Ep, The Epilchik pluton; M, The Machevna pluton.

different characters are described in the following chapters; the Epilchik ultramafic pluton and the Machevna gabbroic pluton. The Epilchik pluton intruded into the oceanic Vatyna Group, and the Machevna pluton into the Achayvayam Group arc. In the southern Koryak highlands, K/Ar ages of about 65 to 40 Ma are obtained on the mafic to intermediate plutonic rocks, and pebbles of ultramafic rocks and gabbroic rocks are found in Oligocene deposits [15,26].

THE EPILCHIK ULTRAMAFIC PLUTON

Many ophiolites and ultramafic complexes are distributed in the Koryak highlands, where they are known to be within accretionary terranes of oceanic crust of Paleozoic to Late Cretaceous age [17]. The ages of the oceanic crust which were subducted or obducted in the region become young oceanward, and four distinct emplacement ages of the ophiolites and ultramafic complexes, from Late Jurassic to Eocene, are recognized [10]. The Epilchik ultramafic pluton occurs in the northrn part of the southern Olyutor Range, the most oceanward in the Koryak highlands.

The Epilchik ultramafic pluton comprises East and West plutons. Both plutons are Alaskan-type ultramafic complexes, showing a concentric structure: a dunitic core passes progressively into wehrlite and clinopyroxenite of the rim, and into a discontinuous margin of gabbro (Fig. 3). The dunite cores of the two plutons are distributed in topographically higher parts. They contain wehrlitic to clinopyroxenitic veins of 1 cm to 30 cm wide, whose volume

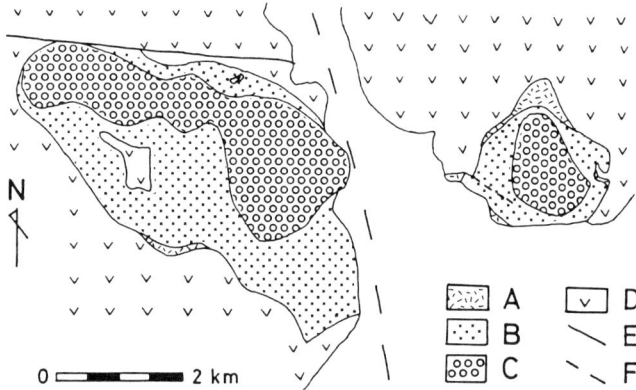

Figure 3. Geological map of the Epilchik pluton. A, Gabbro; B, Clinopyroxenite and wehrlite; C, Dunite; D, The Vatyna Group composed mainly of basalt, dolerite, basaltic tuff and chert; E, Fault; F, Inferred fault.

decreases toward the centers of the plutons. Several gabbroic dikes up to 20 m in wide are observed in the rim wehrlite and clinopyroxenite. The field evidence indicates that the dunite intruded earlier than the wehrlite and clinopyroxenite, which was followed by the emplacement of gabbro. The clinopyroxenite of the southeastern rim of the West pluton is intruded by pervasive pegmatitic hornblende-plagioclase veins of a few cm to 1 m wide and by several dikes of leucocratic trondhjemite, from a few m to 10 m in width.

Wall rocks of the Epilchik pluton are the Vatyna Group composed mainly of basalt, dolerite, basaltic tuff and chert. All the rocks are metamorphosed at least to the greenschist facies, and contain such metamorphic minerals as actinolite, epidote, chlorite, albite, sericite and rarely alkali amphibole or pumpellyite. Three metamorphosed basalts near the Epilchik pluton, reported in Bogdanov and Fedorchuk [16], straddle fields of oceanic-floor basalts and island arc basalts on diagrams Ti/100 - Zr - Y·3 and Ti/100 - Zr - Sr/2 [27] (not shown) and on a diagram Zr/Y - Zr [28] (not shown).

Structures are only weakly developed in the Epilchik pluton. Layering caused by variation in the proportions of olivine and clinopyroxene is locally developed in the ultramafic rocks. The marginal gabbros and the wehrlite parts of the rims, which occupy topographically lower parts of the East and West plutons, show vertical to near-vertical foliations and layering, respectively. Layering recognized within the dunite cores flattens toward higher parts. The internal structures of the rocks, like a upheaved plug, may indicate that the Epilchik pluton was emplaced as a hot crystal mush with profound mobility [2,29]. However, evidence of contact metamorphism caused by the intrusion of the Epilchik pluton is not present, because the neighboring host rocks have undergone regional (?) metamorphism of a similar grade to that of the marginal gabbros of the pluton, at least up to the greenschist facies. The reverse zoning from the dunitic cores to the gabbroic margins recognized in the Epilchik pluton is the most conspicuous feature of the Alaskan-type ultramafic complexes. The reverse zoning in some Alaskan-type complexes such as the Emigrant Gap area [29] and northern Venezuela [30] have been explained by flow differentiation of mafic magmas. However, Barriere [31] revealed that flow differentiation is unlikely to be an important process in intrusions with a diameter or width over 100 m. The constituent rocks of the Epilchik pluton had intruded at least three times, as mentioned before. This evidence suggests that the reverse zoning of the Eplichin pluton is probably attributed to successive emplacements of a roughly differentiated, deeper level magma chamber, as exemplified in the Duke Island ultramafic complex, Alaska [32] and the Smartville complex, California [33].

In contrast to our magmatic emplacement model, Aleksandrov et al. [13] pointed that the Epilchik East and West plutons were a single layer-like nappe, dissected by erosion and young faults, and that the peculiar structure of the plutons are due to recumbent fold of an ultramafic complex, which was originally layered into a dunite and a clinopyroxenite with a

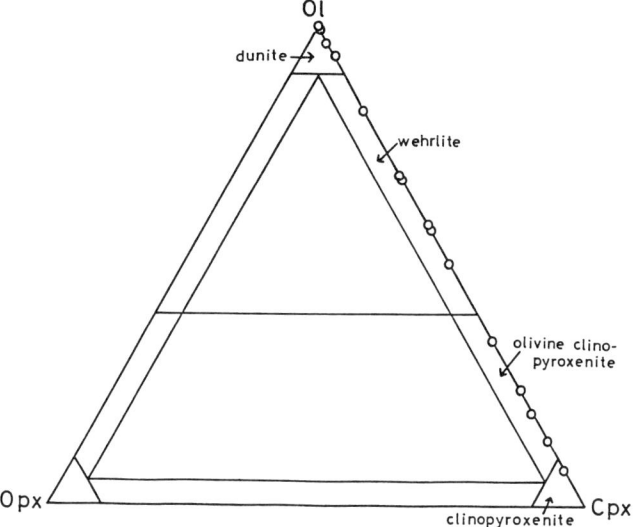

Figure 4. Modal compositions of the Epilchik ultramafic rocks plotted in a olivine (Ol) - clinopyroxene (Cpx) - orthopyroxene (Opx) diagram. After the nomenclature of Streckeisen [34].

narrow wehrlite zone. Following Aleksandrov et al.'s [13] idea, Fujita and Newberry [10] and Stavsky et al. [8] considered that the Late Cretaceous oceanic crust of the Olyutor terrane, including the the Epilchik pluton, has been thrust over the Upper Cretaceous Koryka flysch sequences during Paleocene or Eocene time.

Modal compositions of the Epilchik ultramafic rocks are shown in an olivine - clinopyroxene - orthopyroxene diagram (Fig. 4). The ultramafic rocks are essentially composed of olivine and clinopyroxene in varying amounts and completely devoid of orthopyroxene, which is a characteristic feature recognized universally in Alaskan-type ultramafic complexes [2,32,35,36].

The massive dunites are characterized by the dominance of remarkably homogeneous olivine up to 5 mm, minor chromian spinel and trace amounts of diopside and pargasitic hornblende. Olivines in two dunites from the East and West plutons exhibit the identical and restricted Fo content (Fo_{89}), typical of olivine from cumulate rocks [37]. The olivines, serpentinized in varying degrees, are slightly poorer in Fo content than olivine (Fo_{90-92}) for source peridotite of oceanic-ridge basalts [38,39], but positively comparable to that (Fo_{89}) in mantle peridotites beneath island arcs [40]. Cr/(Cr+Al) ratios of the spinel, ranging from 0.53 to 0.83, are comparable to that of spinels in southeastern Alaskan intrusives [41].

The wehrlite and clinopyroxenite are composed of various proportions of olivine (Fo_{81-87}) and clinopyroxene, accompanied sometimes by pargasitic hornblende and chromian magnetite in trace amounts. The clinopyroxenes in the ultramafic rocks are diopside characterized by high Ca and Mg, and their compositions (not shown) closely correspond to those from Alaskan-type ultramafic complexes [2,30,36,37]. The clinopyroxenes contain variable Al_2O_3 (1.5 - 4.0 wt.%), low TiO_2 (0.05 - 0.5 wt.%) and have high Cr_2O_3 up to 1.0 wt.%. The TiO_2-poor nature of the clinopyroxenes suggests that a primary magma of the Epilchik pluton does not have alkali basalt chemistry [42].

The marginal gabbros generally contain hornblende and plagioclase, and some gabbros have clinopyroxene as major igneous minerals. They are metamorphosed at least up to the greenschist facies, and commonly include epidote, clinozoisite, chlorite, actinolite, sericite, albite and rarely alkali amphibole as metamorphic minerals. The frequent presence of igneous hornblende in the marginal gabbros and the existence of pegmatitic hornblende-plagioclase veins in the rim clinopyroxenite suggest that later magmas had high water contents, probably more than 3 wt.% H_2O [43].

Table 1.
Bulk rock chemical compositions of plutonic rocks of the Epilchik pluton. Rock types of the analysed samples are as follows; E139 & E107 - dunite; E09 & E85 - wehrlite; E133 & E103 - clinopyroxenite; E02 & E102 - gabbro. Rock analyses were made by means of XRF.

	East pluton				West pluton			
Name	E139	E09	E133	E02	E107	E85	E103	E102
SiO_2	38.75	41.55	48.59	44.88	36.45	44.96	48.08	45.56
TiO_2	0.01	0.19	0.32	1.20	0.01	0.11	0.30	0.88
Al_2O_3	0.48	2.26	3.05	18.00	0.37	1.54	3.19	21.20
Fe_2O_3	2.68	4.80	3.09	5.83	4.22	3.35	2.83	4.45
FeO	7.84	8.16	4.47	5.81	6.22	5.72	4.17	5.01
MnO	0.20	0.21	0.16	0.16	0.18	0.18	0.14	0.21
MgO	46.34	30.56	19.47	5.57	43.91	30.23	20.79	3.96
CaO	0.36	8.92	19.18	9.13	0.21	11.35	18.20	10.62
Na_2O	n.d.	0.02	0.07	3.09	n.d.	0.05	0.09	3.72
K_2O	n.d.	0.01	n.d.	1.75	n.d.	n.d.	n.d.	0.79
H_2O+	3.20	3.21	1.52	3.13	7.89	2.41	2.17	2.77
H_2O-	0.14	0.11	0.08	0.13	0.54	0.09	0.03	0.03
P_2O_5	0.01	0.01	n.d.	1.33	n.d.	n.d.	0.01	0.80
Total	100.01	100.01	100.00	100.01	100.00	99.99	100.00	100.00
ppm								
Nb	n.d.	0.4	1.0	2.0	n.d.	0.5	0.8	1.6
Zr	8.5	11.7	13.4	75.9	8.8	10.8	14.4	48.6
Y	8.1	8.2	8.1	29.7	8.2	8.2	8.3	27.0
Sr	10.0	35.1	52.4	886	4.5	46.9	54.7	1425
Rb	0.6	0.3	0.7	24.7	0.4	0.1	n.d.	4.0
Th	0.1	n.d.	0.1	2.0	0.7	0.7	0.8	2.4
Pb	n.d.	0.8	1.0	0.9	n.d.	1.0	0.9	2.5
Ga	1.5	4.7	5.1	21.4	1.4	3.8	4.5	22.6
Zn	66.6	80.7	37.5	76.4	62.0	49.5	37.5	88.6
Cu	7.4	11.4	5.8	306	4.1	4.7	6.2	41.4
Ni	895	586	147	1.9	1028	396	194	n.d.
Co	140	112	57.3	47.2	134	98.0	55.5	32.9
Cr	994	1653	1027	n.d.	1349	2266	1514	4.3
V	8.7	93.3	136	420	6.9	42.8	126	325
Ba	13.8	17.3	21.7	271	13.5	15.5	22.3	277
Ce	n.d.	n.d.	n.d.	29.6	n.d.	n.d.	1.0	31.6

Eight chemical analyses of ultramafic rocks and gabbros are presented in Table 1. Major and trace elements were analysed by means of XRF. The major elements chemistry of the ultramafic rocks is intimately related to their modal mineralogy. SiO_2, Al_2O_3 and CaO increase, but MgO decreases greatly from the MgO-rich rock (dunite) to the CaO-rich rock (clinopyroxenite), which directly reflects an increase of clinopyroxene abundance and a decrease of olivine abundance toward the clinopyroxenite. The variation trend of major elements may indicate that the ultramafic rocks were formed by magmatic accumulation of early olivine and subsequent clinopyroxene from a mafic magma. The marginal gabbros have significantly lower $Mg/(Mg+Fe^{2+}+Fe^{3+})$ ratios and are remarkably depleted in Ni and Cr compared with the ultramafic rocks. Trace element abundances vary with the modal mineralogy, mineral chemistry and $Mg/(Mg+Fe^{2+}+Fe^{3+})$ ratios of the rocks. Conspicuous variations of trace elements are marked decreases of Ni and Cr, and general increases of Sr, Ba, Zr, Y, Ga, V and Cu with decreasing $Mg/(Mg+Fe^{2+}+Fe^{3+})$ ratio, as shown in Fig. 5. The variation trend of trace elements is consistent with the early olivine- and subsequent clinopyroxene-dominated fractionation.

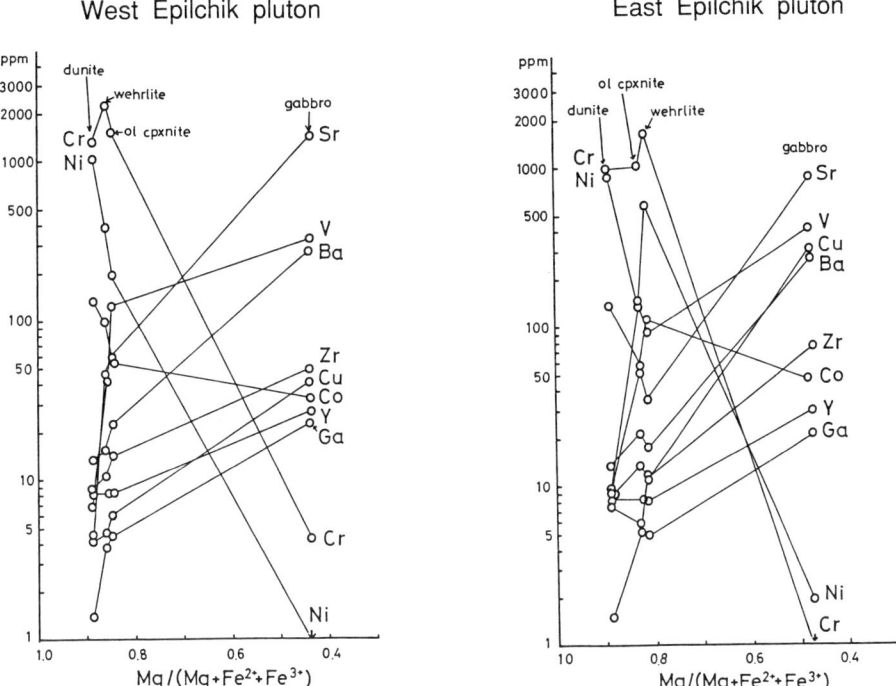

Figure 5. Variations of trace elements contents with $Mg/(Mg+Fe^{2+}+Fe^{3+})$ ratios in the Epilchik rocks.

THE MACHEVNA GABBROIC PLUTON

The Machevna gabbroic pluton comprises the North and South intrusions. The North Machevna pluton is roughly concentrically zoned, with fine-grained and cumulate textured gabbroic rocks on the rim and quartz dioritic to quartz monzodioritic rocks toward the core. A small amount of hornblende-plagioclase-olivine clinopyroxenite, intruded by many dikes of leucocratic granodiorite, occurs in the northern rim (Fig. 6). The northeastern contact of the pluton is characterized by a complicated zone of gabbro and basalt, both the rocks have been metamorphosed to the amphibolite facies, presumably by synshear, hydrothermal fluid infiltration [44]. All the constituent rocks of the pluton have undergone pervasive low-grade metamorphism up to the greenschist facies, and commonly contain epidote, chlorite, actinolite, sericite, albite and prehnite as metamorphic minerals.

Country rocks of the Machevna pluton are the Achayvayam Group, composed mainly of basalt and basaltic andesite. Country rocks cropping out near the pluton have been metamorphosed up to the amphibolite facies and consist chiefly of green hornblende, actinolite, epidote, biotite, chlorite and sodic plagioclase. Chemical compositions of two amphibolites are shown in Table 2. The amphibolites have compositions of island arc basalts with moderate MgO and low Nb, Zr, Ti and Y [45,46], and are clearly distinguished from boninitic lavas reported from some intraoceanic island arcs of the Western Pacific [47].

Modal compositions of the North Machevna rocks are plotted on a quartz - plagioclase - potash feldspar diagram (Fig. 7). The rim gabbroic rocks are fairly heterogeneous, and some of them contain quartz and potash feldspar in minor amounts. The gabbroic rocks have clinopyroxene, hornblende and magnetite in varying amounts as essential mafic minerals, accompanied locally by lesser amounts of olivine and biotite. The clinopyroxenite, sporadically occurring in the northern rim, is characterized by calcic plagioclase (An_{85-90}) and moderately Fe-rich olivine (Fo_{77-79}). Calcic plagioclase (An_{85-100}) and moderately

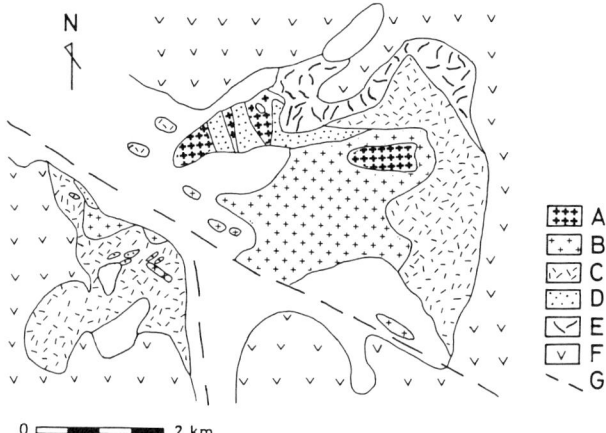

Figure 6. Geological map of the North Machevna pluton. A, Leuococratic granodiorite; B, Quartz diorite and quartz monzodiorite; C, Gabbroic rocks; D, Clinopyroxenite; F, The Achayvayam Group composed mainly of basalt and basaltic andesite; G, Inferred fault.

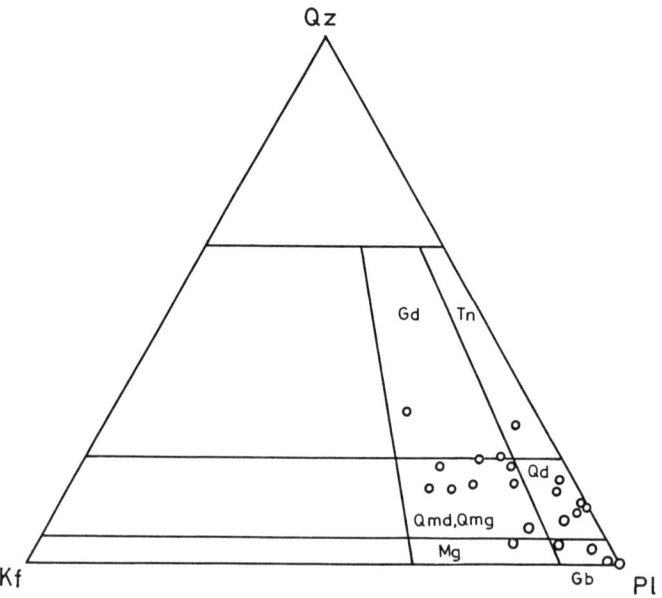

Figure 7. Modal compositions of the North Machevna rocks plotted in a quartz (Qz) - plagioclase (Pl) - potash feldspar (Kf) diagram. Gb, Gabbro; Qd, Quartz diorite; Tn, Tonalite; Mg, Monzogabbro; Qmd, Quartz monzodiorite; Qmg, Quartz monzogabbro; Gd, Granodiorite. After the nomenclature of Streckeisen [34].

Fe-rich olivine (Fo$_{60-80}$) are similar to clinopyroxenites that commonly occur together in arc cumulate gabbros [3,33,36,48], but are distinct from cumulate gabbros from mid-oceanic ridges, oceanic islands, or tholeiitic layered intrusions [45,50]. Clinopyroxene in the clinopyroxenites display compositions straddling fields of diopside, salite and augite, slightly richer in Fe* (total Fe as Fe^{2+}) than those from the Epilchik ultramafic pluton (not shown). They have variable Al$_2$O$_3$ (1.2 - 3.7 wt.%) and low TiO$_2$ (0.05 - 0.50 wt.%), typical of clinopyroxenes from non-alkali basalts [42].

Quartz diorite and quartz monzodiorite are main constituents of the core of the North Machevna pluton. Main mafic minerals of the quartz diorite and quartz monzodiorite are commonly hornblende, biotite and magnetite, and minor clinopyroxene is contained in some rocks. The hornblende and biotite in the rocks have fairly high Mg/(Mg+Fe*) (Fe* signifies total Fe as Fe^{+2}) ratios ranging from 0.60 to 0.74, reflecting high abundance of magnetite [51].

The Machevna rocks are fairly abundant in magnetite, and modal % of magnetite reaches 8 vol. % in the rim gabbroic rocks and up to 4 vol.% in the core granitic rocks. Magnetic susceptibility of the rocks measured at the outcrops using Kappameter KT-5 are high. The gabbroic rocks commonly show magnetite susceptibility of 50 ~ 200×10^{-3} SI unit, and the granitic rocks those of 30 ~ 50×10^{-3} SI unit (Fig. 8). The Machevna rocks are classified into the magnetite-series of Ishihara [52], because the boundary value of magnetic susceptibility between the magnetite-series and ilmenite-series granitic rocks is 3 ~ 4×10^{-3} SI unit. Plutonic rocks with high magnetic susceptibility similar to the Machevna rocks are distributed in the Tanzawa pluton, central Japan, formed in an immature island arc (paleo-Izu-Ogasawara arc) in Early Miocene time [53].

Chemical compositions of 11 Machevna rocks are presented in Table 2. They exhibit a wide range of SiO$_2$ from 42.7 to 66.1 wt.%. FeO*(total Fe as FeO)/MgO ratios of the rocks remain approximately constant with increasing SiO$_2$, which is a characteristic of calc-alkaline rock series [21]. On a variation diagram of Na$_2$O+K$_2$O - FeO* - MgO (not shown), the Machevna rocks delineate a calc-alkaline differentiation trend comparable to calc-alkaline plutonic rocks from the Unalaska Island in the Aleutian rc [54] and the New Guinea Mobile belt [1,55]. The samples with greater than 45 wt.% SiO$_2$ have moderate K$_2$O, and plot within the range of the medium- to high-K andesites of Gill [56] on a SiO$_2$ - K$_2$O

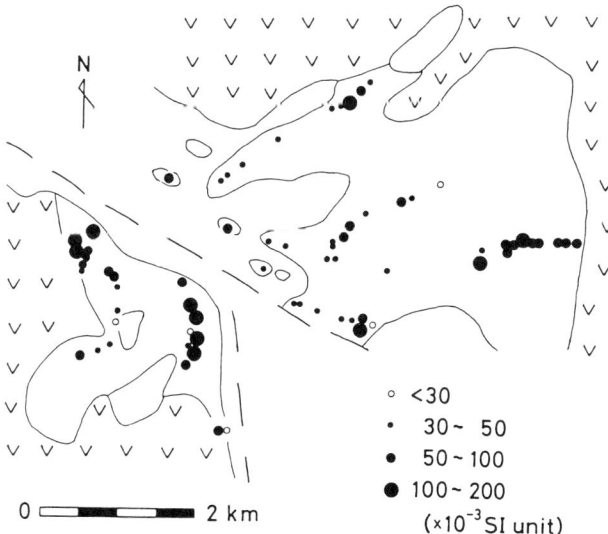

Figure 8. Magenetic susceptibility of the plutonic rocks in the North Machevna pluton.

Table 2.
Bulk rock chemical compositions of plutonic and metamorphic rocks in and near the North Machevna pluton. Rock types of the analysed samples are as follows; Ma23 & Ma24 - clinopyroxenite; Ma14, Ma11 & Ma17 - gabbro; Ma18 - quartz monzogabbro; Ma110 - monzogabbro; Ma48 - quartz diorite; Ma67 - quartz monzodiorite; Ma90 & Ma16 - granodiorite; Ma69 & Ma19 - amphibolite. Rock analyses were made by means of XRF.

Name	Ma23	Ma24	Ma14	Ma11	Ma17	Ma18	Ma110	Ma48	Ma67	Ma90	Ma16	Ma69	Ma19
SiO_2	44.54	44.75	42.70	42.90	47.92	51.98	53.40	53.54	58.52	59.56	66.11	49.11	49.77
TiO_2	0.24	0.39	0.52	0.65	0.68	0.64	0.59	0.51	0.54	0.51	0.38	0.76	0.78
Al_2O_3	4.79	7.74	16.59	15.99	15.22	16.17	16.81	18.12	17.24	16.95	15.69	15.26	15.63
Fe_2O_3	3.85	4.61	7.07	7.33	5.54	4.10	4.74	3.53	3.49	3.09	2.06	5.19	3.25
FeO	6.68	6.65	6.18	5.80	6.06	5.15	3.98	3.85	2.64	2.92	1.81	5.45	6.11
MnO	0.21	0.21	0.18	0.22	0.22	0.18	0.17	0.17	0.15	0.13	0.05	0.18	0.16
MgO	22.79	17.24	9.52	9.39	7.74	5.39	4.54	5.32	2.65	3.03	1.68	7.04	7.73
CaO	13.19	14.93	15.51	13.92	11.95	8.23	8.47	9.27	4.72	6.23	4.51	9.59	9.86
Na_2O	0.12	0.24	0.56	0.86	2.20	2.83	3.36	2.10	4.48	3.27	3.23	3.32	3.04
K_2O	0.06	0.34	0.06	0.20	0.73	2.35	1.96	1.33	3.21	2.35	2.92	1.11	1.01
H_2O+	3.42	2.83	1.05	2.07	1.41	2.48	1.53	2.07	1.97	1.65	1.30	2.56	2.39
H_2O-	0.09	0.06	0.05	0.27	0.00	0.18	0.17	0.08	0.08	0.09	0.13	0.23	0.12
P_2O_5	0.01	0.01	0.01	0.40	0.33	0.31	0.28	0.12	0.30	0.21	0.13	0.19	0.16
Total	99.99	100.00	100.00	100.00	100.00	99.99	100.01	100.01	99.99	99.99	100.00	99.99	100.01
ppm													
Nb	0.9	0.9	1.0	1.1	1.2	1.6	1.9	1.7	3.8	3.1	2.9	2.0	1.5
Zr	15.2	15.8	22.4	26.0	31.0	58.1	47.9	41.6	72.0	67.5	84.5	49.1	63.3
Y	10.1	9.9	8.0	16.6	16.9	19.8	19.9	17.3	23.4	19.5	16.2	18.0	18.5
Sr	139	287	812	845	1032	910	877	598	812	782	609	710	663
Rb	1.2	6.1	n.d.	1.9	8.1	34.9	16.9	15.3	42.0	27.8	31.1	18.3	15.7
Th	1.0	0.9	1.9	1.9	1.7	1.8	2.1	1.3	3.2	2.5	2.3	2.0	1.7
Ga	6.4	9.6	1.8	17.0	17.8	17.2	3.8	16.8	3.3	2.8	15.3	3.3	3.0
Pb	2.2	3.1	17.2	2.4	4.0	5.8	17.3	3.7	18.4	16.8	1.0	17.1	15.7
Zn	70.0	71.9	66.1	99.3	98.5	90.5	60.6	75.1	63.4	50.8	19.1	70.9	82.7
Cu	21.0	58.4	16.5	478	182	98.0	151	106	45.1	22.1	13.9	32.2	63.4
Ni	384	272	47.7	52.8	49.7	31.0	12.5	15.2	2.4	8.8	5.0	36.1	63.9
Co	86.2	71.0	58.7	54.8	47.5	33.1	38.1	26.2	18.2	17.6	14.1	46.5	47.7
Cr	1037	922	95.5	233	160	86.4	26.6	68.0	n.d.	28.5	5.8	93.0	236
V	148	236	428	466	414	304	300	335	204	219	189	376	277
Ba	18.6	56.4	33.3	93.1	254	577	489	181	586	430	507	255	152
Ce	n.d.	n.d.	n.d.	2.4	10.7	13.2	16.5	11.6	21.5	12.6	17.8	12.5	14.4

diagram (not shown). They are, however, depleted in Rb (less than 42 ppm) and have very high K/Rb ratios from 560 to 960, which is a feature of plutonic rocks from intraoceanic island arcs such as the Aleutian arc [7] and Papua New Guinea [5]. On an average tholeiitic MORB-normalized Spiderdiagram of Pearce [57], the Machevna rocks display a selective enrichment of large ion lithophile (LIL) elements such as Sr, K, Rb, Ba and Th, and a depletion of high field strength (HFS) elements such as Nb, Zr, Ti and Y (Fig. 9), which is a characteristic feature of rocks occurring typically in subduction-related island arc settings [45,47,58,59]. The Machevna rocks have fairly low Nb, Y and Rb/Zr among subduction-related igneous suites, suggesting their primitive island arc origin [60].

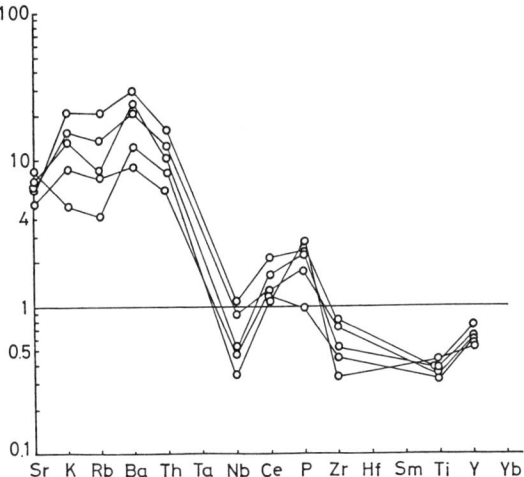

Figure 9. MORB-normalized geochemical patterns of Pearce [57] for the North Machevna gabbroic to granodioritic rocks.

GEODYNAMIC AND PETROGENETIC IMPLICATIONS OF THE PLUTONS

The Olyutor Range is situated on the continental extension of the Shirshov Ridge, which separates the Komandorsky basin from the Aleutian basin (Fig. 1). The Komandorsky basin was formed as a result of local spreading in the west side of the Kula plate during Late Cretaceous to Late Miocene time. The Olyutor Range is a remnant island arc of Late Cretaceous age, which was formed above the subduction zone where two oceanic plates were merged [12]. The Olyutor island arc began its accretion to the Eurasian plate as early as Eocene [8,10]. In this way, the Paleogene plutons (65 - 40 Ma) of the Olyutor Range appear to be connected with the late stage of igneous activity, which occurred in the intraoceanic island arc during subduction and accretion of oceanic crust. Their E-W elongation and internal structures may suggest vertical magma intrusions into E-W extensional zones related to the westward subduction of the trapped Kula plate [9].

The Epilchik ultramafic pluton shows some charcteristic features of Alaskan-type ultramafic complexes; the concentric zoning from the dunitic cores to the gabbroic margins, the common presence of clinopyroxene and chromian spinel, and the complete absence of orthopyroxene. The major and trace elements chemistry of the Epilchik ultramafic rocks is in harmony with their formation by the early olivine- and subsequent clinopyroxene dominated fractionation. The Alaskan-type complexes are commonly believed to have formed in magma reservoirs beneath volcanoes overlying subduction zones in island arcs [30,32,36,37]. In the Epilchik pluton, olivine crystallized before clinopyroxene, which was succeeded by plagioclase. This order of crystallization is experimentally obtained for high-alumina basalt and alkali olivine basalt at pressures less than 17 kbar under anhydrous conditions and at pressures less 23 kbar under water-undersaturated conditions [40], and commonly recognized in island arc tholeiites [58]. The Fo content of olivine (Fo89) in the Epilchik dunite is the same as the olivine assumed to be in equilibrium with primary arc basalt magmas in the Quaternary Northeast Japan [40]. The marginal gabbros in the Epilchik ultramafic pluton have siginificantly lower $Mg/(Mg+Fe^{2+}+Fe^{3+})$ ratios and are remarkably depleted in Cr and Ni (less than 5 ppm) compared to the core ultramafic rocks (Fig. 5). This fact may indicate that the marginal gabbros do not represent a primary magma (chilled margins) of the Epilchik pluton, but correspond to residual liquids formed by olivine- and clinopyroxene-dominated fractionation from the primary magma. The clinopyroxene chemistry reveals its non-alkali basalt lineage. The primary magma of the Epilchik pluton is thought to be an olivine tholeiite or a high-alumina basalt.

The Machevna pluton displays geochemical and mineralogical features of subduction-related island arc calc-alkaline rock series; calc-alkaline variation trends on diagrams of FeO^*/MgO vs. SiO_2 and of Na_2O+K_2O - FeO^* - MgO, the selective enrichment of LIL elements with constant depletion of HFS elements compared with the average tholeiitic MORB, and the presence of calcic plagioclase (An_{85-90}) and moderately Fe-rich olivine (Fo_{77-79}) in the clinopyroxenite. The Machevna rocks are depleted in Rb, Nb, and Y among island arc igneous suites, and may represent plutonic rocks formed in intraoceanic (primitive) island arcs such as the Aleutian arc [7,54], Papua New Guinea [5] and Jamaica [61]. The clinopyroxene in the pluton has compositions typical of that from non-alkali basalts. One Machevna gabbro with 48 wt.% SiO_2 (Ma17 in Table 2) is relatively rich in Sr and Ba, and its major and trace elements chemistry is more closely comparable to that of the primary high-alumina basalt, estimated by Tatsumi et al. [40] and Sakuyama and Nesbitt [59], than that of the primary olivine tholeiite. The primary magma of the Machevna pluton is supposed to be a high-alumina basalt, derived from partial melting of the HFS element-depleted mantle source that has undergone the selective enrichment of LIL elements by aqueous and silicious fluids originated from the down-going slab [46,47,58,59,62]. Because the olivine and clinopyroxene from the Machevna clinopyroxenite are slightly depleted in Mg compared with those from the Epilchik pluton, the primary magma of the Machevna pluton seems to be less magnesian than that of the Epilchik pluton.

The Epilchik pluton is located about 50 km north of the Machevna pluton. The marginal gabbros of the Epilchik pluton are apparently poor in Ni and Cr compared with the rim gabbros of the Machevna pluton. The Epilchik ultramafic rocks are completely devoid of plagioclase, but the Machevna clinopyroxenite contains a small amout of plagioclase. These facts may imply that there is no direct petrogenetic link between the Epilchik pluton and the Machevna pluton. However, the two plutons have an important common feature; they are composed of plutonic rocks formed in the subduction-related intraoceanic island arc. The plutonic rocks of the two plutons, along with the pre-existing oceanic crust and some early extrusive lavas, may have contributed to the growth of the lower crust of the Olyutor intraoceanic island arc, which was situated above the subduction zone where two oceanic plates converged [6,12,63,64]. In the central intraoceanic Aleutian arc, ultramafic cumulates are formed in magma reservoirs from mantle-derived olivine tholeiite and high-alumina basalt at the crust-mantle boundary. These cumulates are overlain by gabbroic rocks of the lower crust, and differentiated volcanic and plutonic rocks that constitute the upper crust [7,37,65]. The Epilchik pluton may be comparable to the arc-roof ultramafic cumulates of the Aleutian arc, and the Machevna pluton to the overlying gabbroic rocks.

Acknowledgements

We express sincere thanks to S.F. Sobolev, M.Y. Krasotov, G.V. Ledneva. A.D. Platonov, N.A. Polyakova and I.V. Kepezhinskas for accompanying us in the field trip in 1990. We are also grateful to N.A. Bogdanov for organizing the field trip, to K. Aoki and N. Tsuchiya for access to and help on the XRF analysis, and to R.G. Coleman for critical reading of the manuscript.

REFERENCES

1. G.C. Brown. Calc-alkaline intrusive rocks : their diversity, evolution, and relation to volcanic arcs. In: *Andesites: Orogenic Andesites and related Rocks*. R.S. Thorpcd (Ed.). pp.437-461. John Wiley & Sons, Chichester (1982).
2. A.W. Snoke, W.D. Sharp, J.E. Wright and J.B. Salleby. Significance of mid-Mesozoic peridotitic to dioritic intrusive complexes, Klamath Mountains - western Sierra Nevada, California. Geology. **10**, 160-166 (1982).
3. A.R. Chivas, A.S. Andrew, A.K. Sinha and J.R. O'Neil. Geochemistry of a Pliocene-Pleistocene oceanic-arc plutonic complex, Guadalcanal. Nature. **300**, 139-143 (1982).
4. L.E. Burns. The Border Ranges ultramafic and mafic complex, south-central Alaska : cumulate fractionates of island-arc volcanics. Can. J. Earth Sci. **22**, 1020-1038 (1985).
5. J.B. Whalen. Geochemistry of an island-arc plutonic suite: the Uasilau-Yau Yau intrusive complex, New Britain, P.N.G. J. Petrol. **26**, 603-632 (1985).
6. L.P. Gromet and L.T. Silver. REE variations across the Peninsular Ranges batholith : implications for batholithic petrogenesis and crustal growth in magmatic arcs. J. Petrol. **28**, 75-125 (1987).

7. S.M. Kay, R.W. Kay, G.P. Citron and M.R. Perfit. Calc-alkaline plutonism in the intra-oceanic Aleutian arc, Alaska. In: *Plutonism from Antarctica to Alaska*. S.M. Kay and C.W. Rapela (Eds). Geol. Soc. Am. Spec. Pap. **241**, 233-255 (1990).
8. A.P. Stavsky, V.D. Chekhovitch, M.V. Kononov and L.P. Zonenshain. Plate tectonics and palinspastic reconstructions of the Anadyr-Koryak region, northeast USSR. Tectonics. **9**, 81-101 (1990).
9. I. Reuber, P.K. Kepezhinskas, N.A. Bodganov, H. Tanaka, S. Miyashita, S.F. Sobolev, M.Y. Krasotov and G.V. Ledneva. Geometry and structure of early arc plutons in NE Kamchatka (U.S.S.R). C. R. Acad. Sci. Paris t.312, Serie II, 289-294 (1991). *(in French with abridged English version)*.
10. K. Fujita and J.T. Newberry. Accretionary terranes and tectonic evolution of northeast Siberia. In: *Accretion Tectonics in the Circum-Pacific Regions*. M. Hashimoto and S. Uyeda (Eds). pp. 43-57. TERRAPUB, Tokyo (1983).
11. A.K. Cooper, M.S. Marlow and D.W. Scholl. Mesozoic magnetic lineation in the Bering Sea marginal basin. J. Geophys. Res. **81**, 1916-1934 (1976).
12. N.A. Bogdanov. Geology of the Komandorsky deep basin. J. Phys. Earth. **36**, S65-S71 (1988).
13. A.A. Aleksandrov, N.A. Bogdanov, S.A. Palandzhyan and V.D. Chekhovich. Tectonics of the northern part of the Olyutorsk zone of the Koryak Highlands. Geotectonics. **14**, 241-249 (1980).
14. N.A. Bogdanov, V.S. Vishnevskaya, A.N. Sukhov, A.V. Fedorchuk and V.D. Chekhovich. Oceanic olistostromes of the western shores of the Aleutian basin (Bering Sea). Geotectonics. **16**, 397-402 (1982)
15. E.S. Alekseyev. Fundamental features of the evolution and structure of the southern part of the Koryak Highlands. Geotectonics. **13**, 57-64 (1979).
16. N.A. Bogdanov and A.V. Fedorchuk. Geochemistry of Cretaceous oceanic basalts of the Olyutorski Range (Bering Sea). Ofioliti. **12**, 113-124 (1987).
17. S.A. Palandzhian. Ophiolite belts in the Koryak Upland, Northeast Asia. Tectnophysics. **127**, 341-360 (1986).
18. P.K. Kepezhinskas. Origin of Cenozoic volcanic series of Komandorsky basin framing according to geochemical and experimental data. Geologicky Zbronik-Geologica Carpathica. **38**, 71-81 (1987).
19. V.S. Vishnevskaya. Middle Late Cretaceous radiolarian zonation of the Bering region, U.S.S.R. Mar. Micropaleontol. **11**, 139-149 (1986).
20. A.V. Fedorchuk. The Machevna volcanoclastic complex of the Campanian-Danian. In: *Geology of the Southern Koryak Highland*. N.A. Bogdanov, V.S. Vishnevskaya, P.K. Kepenzhinskas, A.N. Sukhov and A.V. Fedorchuk. pp. 108-122. Nauka, Moscow (1987). *(in Russian)*.
21. A. Miyashiro. Volcanic rock series in island arcs and active continental margins. Am J. Sci. **274**, 321-355 (1974).
22. A. Ewart, W.B. Bryan and J.B. Gill. Mineralogy and geochemistry of the younger volcanic islands of Tonga, S.W. Pacific. J. Petrol. **14**, 429-465 (1973).
23. J.W. Cole. Regional distribution and character of active andesite volcanism : Tonga-Kermadec-New Zealand. In: *Andesites: Orogenic Andesites and Related Rocks*. R.S. Thorpe (Ed.). pp.245-258. John Wiley & Sons, Chichester (1982).
24. A. Meijer. Regional distribution and character of active andesite volcanism : Mariana-Volcano Islands. In : *Andesites : Orogenic Andesites and Related Rocks*. R.S. Thorpe (Ed.). pp. 293-306. John Wiley & Sons, Chichester (1982).
25. S.M. Kay, R.W. Kay and G.P. Citron, Tectonic controls on tholeiitic and calc-alkaline magmatism in the Aleutian arc. J. Geophys. Res. **87**, 4051-4072 (1982).
26. N. P. Mitrofanov, A.M. Podolskiy, N.Y. Kostin, M.A. Talalay and S.D. Sheludchenko. The Koryak volcano-plutonic complex. Intern. Geol. Rev. **22**, 1335-1345 (1980).
27. J.A. Pearce and J.R. Cann. Tectonic setting of basic volcanic rocks determined using trace elements analyses. Earth Planet. Sci. Lett. **19**, 290-300 (1973).
28. J.A. Pearce and M.J. Norry. Petrogenetic implications of Ti, Zr, Y and Nb variations in volcanic rocks. Contrib. Mineral. Petrol. **69**, 33-47 (1979).
29. O.B. James. Origin and emplacement of the ultramafic rocks of the Emigrant Gap area, California. J. Petrol. **12**, 523-560 (1971).
30. C.G. Murray. Zoned ultramafic complexes of the Alaskan type : Feeder pipes of andesitic volcanoes. In: *Studies in Earth and Space Sciences (Hess Volume)* R.E. Shagan and others (Eds). Geol. Soc. Am. Mem. **132**, 313-335 (1972).
31. M. Barriere. Flow differentiation : limitation of the Bagnold effect to narrow intrusions. Contrib. Mineral. Petrol. **55**, 139-145 (1976).
32. T.N. Irvine. Petrology of the Duke Island ultramafic complex, southeastern Alaska. Geol. Soc. Amer. Mem. **138**, 240p. (1974).
33. J.S. Beard and H.W. Day. Petrology and emplacement of reversely zoned gabbro-diorite plutons in the Smartville complex, northern California. J. Petrol. **29**, 965-995 (1988).
34. A. Streckeisen. To each plutonic rocks of its proper name. Earth Sci. Rev. **12**, 1-33 (1976).
35. H.P. Jr. Taylor. The zoned ultramafic complexes of south-eastern Alaska. In; *Ultramafic and Related Rocks*. P. J. Wyllie (Ed.). pp. 97-121, John Wiley & Sons, New York (1967).
36. A.W. Snoke, J.E. Quick and H.R. Bowman. Bear Mountain igneous complex, Klamath Mountains, California ; an ultrabasic to silicic calc-alkaline suite. J. Petrol. **22**, 501-552 (1981).

37. W.K. Conrad and R.W. Kay. Ultramafic and mafic inclusions from Adak Island : crystallization history, and implications for the nature of primary magmas and crustal evolution in the Aleutian arc. J. Petrol. **25**, 88-125 (1984).
38. D. Muir and C.E. Tilley. Basalts from the northern part of the rift zone of the Mid-Atlantic Ridge. J. Petrol. **5**, 409-434 (1964)
39. A.P. Le Roex, A.J. Erlank and H.D. Needham. Geochemical and mineralogical evidence for the occurrence of at least three distinct magma types in the "FAMOUS" Region. Contrib. Mineral. Petrol. **77**, 24-37 (1981).
40. Y. Tatsumi, M. Sakuyama, H. Fukuyama and I. Kushiro. Generation of arc basalt magmas and thermal structure of the mantle wedge in subduction zones. J. Geophys. Res. **88**, 5815-5825 (1983).
41. H.J.B. Dick and T. Bullen. Chromian spinel as a petrogenetic indicator in abyssal and alpine-type peridotites and spatially associated lavas. Contrib. Mineral. Petrol. **86**, 54-76 (1984).
42. J. Leterrier, R.C. Maury, P. Thonon, D. Girard and M. Marchal. Clinopyroxene compositions as a method of identification of the magmatic affinities of paleo-volcanic series. Earth Planet. Sci. Lett. **59**, 139-154 (1982).
43. C.W. Burnham. Magmas and hydrothermal fluids. In: *Geochemistry of Hydrothermal Ore Deposits, 2nd Edition.* H.L. Barnes (Ed.). pp. 71-136. John Wiley & Sons, New York (1979).
44. P.A. Flagler and J.G. Spray. Generation of plagiogranite by amphibolite anatexis in oceanic shear zones. Geology. **19**, 70-73 (1991).
45. M.R. Perfit, D.A. Gust, A.E. Bence, R.J. Arculus and S.R. Taylor. Chemical characteristics of island-arc basalts : implications for mantle sources. Chem. Geol. **30**, 227-256 (1980).
46. J.A. Pearce. Trace element characteristics of lavas from destructive plate boundaries. In : *Andesites : Orogenic Andesites and Related Rocks.* R.S. Thorpe (Ed.). pp.525-548. John Wiley & Sons, Chichester (1982).
47. L. Beccaluva and G. Serri. Boninitic and low-Ti subduction-related lavas from intraoceanic arc-backarc systems and low-Ti ophiolites : a reappraisal of their petrogenesis and original setting. Tectonophysics. **146**, 291-315 (1988).
48. T.E. Smith, C.H. Huang, M.J. Walawender, P. Cheung and C. Wheeler. The gabbroic rocks of the Peninsular Ranges batholith, southern California : cumulate rocks associated with calc-alkaline basalts and andesites. J. Volcanol. Geotherm. Res. **18**, 249-278 (1983).
49. R.J. Arculus and K.J.A. Wills. The petrology of plutonic blocks and inclusions from the Lesser Antilles island arc. J. Petrol. **21**, 743-799 (1980).
50. J. S. Beard. Characteristic mineralogy of arc-related cumulate gabbros : implications for the tectonic setting of gabbroic plutons and andesite genesis. Geology. **14**, 848-851 (1986).
51. S. Ishihara. The granitoid series and mineralization. Econ. Geol. 75th Anniversary Volume. 458-484 (1981).
52. S. Ishihara. The magnetite-series and ilmenite-series granitic rocks. Mining Geol. **27**, 293-305 (1977).
53. K. Sato. Miocene granitoid magmatism at the island-arc junction, central Japan. Modern Geol. **15**, 367-399 (1991).
54. M.R. Perfit, H. Brueckner, J.R. Lawrence and R.W. Kay. Trace element and isotopic variations in a zoned pluton and associated volcanic rocks, Unalaska island, Alaska : a model for fractionation in the Aleutian calcalkaline suite. Contrib. Mineral. Petrol. **73**, 69-87 (1980).
55. D.R. Mason and J.A. McDonald. Intrusive rocks and porphyry copper occurrences of the Papua New Guinea-Solomon Islands region : a reconnaissance study. Econ. Geol. **73**, 857-877 (1978).
56. J. Gill. *Orogenic Andesites and Plate Tectonics.* Springer-Verlag, Berlin · Heidelberg (1981).
57. J.A. Pearce. Role of the sub-continental lithosphere in magma genesis at active continental margins. In: *Continental Basalts and Mantle Xenoliths.* C.J. Hawkesworth and M.J. Norry (Eds). pp.230-249. Shiva Publishing Limited, Cheshire (1983).
58. J.A. Pearce, S.J. Lippard and S. Roberts. Characteristic and tectonic significance of supra-subduction zone ophiolites. In: *Marginal Basin Geology.* B.P. Kokelaar and M.F. Howells (Eds). Geol. Soc. London Spec. Publ. **16**, 77-94 (1984).
59. M. Sakuyama and R.W. Nesbitt. Geochemistry of the Quaternary volcanic rocks of the Northeast Japan arc. J. Volcanol. Geotherm. Res. **29**, 413-450 (1986).
60. G.C. Brown, R.S. Thorpe and P.C. Webb. The geochemical characteristics of granitoids in contrasting arcs and comments on magma sources. J. Geol. Soc. London. **141**, 413-426 (1984).
61. J.A. Pearce, N.B.W. Harris and A.G. Tindle. Trace element discrimination diagrams for the tectonic interpretation of granitic rocks. J. Petrol. **25**, 956-983 (1984)..
62. A.J. Crawford, L. Beccaluva, G. Serri and J. Dostal. Petrology, geochemistry and tectonic implications of volcanics dredged from the intersection of the Yap and Mariana trenches. Earth Planet. Sci. Lett. **80**, 265-280 (1986).
63. R.W. Kay. Island arc processes relevant to crustal and mantle evolution. Tectonophysics. **112**, 1-15 (1985).
64. R.W. Kay and S.M. Kay. Petrology and geochemistry of the lower continental crust : an overview. In: *The Nature of the Lower Continental Crust.* J.B. Dawson, D.A. Carswell, J. Hall and K.H. Wedepohl (Eds). Geol. Soc. Spec. Publ. **24**, 147-159 (1986).
65. S.M. Kay and R.W. Kay. Role of crystal cumulates and the oceanic crust in the formation of the lower crust of the Aleutian arc. Geology. **13**, 461-464 (1985).

Geodynamic Evolution of Continental Margins in Eastern Asia and Tectonic Setting of East China Sea

EDMUND Z. CHANG, XUDONG YING, DA ZHOU & LIBING WANG
Department of Geology and Applied Earth Sciences, Stanford University, CA 94305, USA

Abstract - A huge continent including Yangtze craton, the southeast coastal area, western Taiwan and continental shelves of South China sea and East China sea existed since the beginning of Phanerozoic and is called Yangtze-Cathaysian. There is no considerable continental growth afterward and Paleozoic ocean in South China is not significant. The so-called Huanan Calidonides is an aulacogen on continental crust.
The relative movement between North China block (Sino-Korean craton) and South China block (Yangtze-Cathaysian craton) had been right-lateral along a transform fault during the Late Paleozoic. Devonian deep water deposits on oceanic crust existed only in East Qinling and Imjingang of Korean peninsula. Subduction of a small ocean evidenced by metamorphism, magmatism and deformation is seen only in East Qinling. Carboniferous and, probably, Permian clastics and minor carbonates of shallow marine and terrestrial facies were deposited in an E-W trending narrow basin called Central Latitudinal basin which is controlled by the transform fault and the biota of the basin was communicated with the contemporary sedimentary basins in South China, North China and West China. There has been no wide Mesozoic oceanic crust between North China and South China blocks. Tectonic framework of central China during Late Paleozoic is quite similar to that of western North America during Cenozoic.
Amalgamation of Sino-Korean craton with Siberian craton at the end of Paleozoic changed the moving direction of the former to the southeast and contraction was prevailing between the already juxtaposed North China and South China blocks and the intervening Dabie-Su-Lu microcontinent. The southeast corner of North China block was slipped into the concave of the microcontinent, and enabled the latter to be underthrusted to a depth of more than 90 km to form ultrahigh pressure metamorphic complex in Early and Middle Triassic time. Down-going oceanic crust is not a prerequisite of A-type subduction such as the cases in Dabie-Su-Lu and Pamir. Tan-Lu fault, primarily a hinge fault, took place at the climax of contraction between North China and South China blocks in Late Triassic time. Differential uplifting and the consequent erosion of Dabie and Su-Lu terranes, which are constituted by piles of low-angle thrust sheets, led to an apparent left-lateral offset of Dabie and Su-Lu. Su-Lu was uplifted higher and unroofed deeper than Dabie. An ancient river system along the E-W trending suture had been drained off into Songpan-Ganzi ocean to the west. Collision of Indian plate shifted the relative movement between North China and South China blocks to left-lateral and initiated the Fen-Wei graben.
The continental margin along the eastern Asian converted from transform to active since Late Jurassic. Accretionary complexes in Ryukyus, Taiwan, Mindoro, Palawan and northern Borneo demonstrate subduction-related magmatism, melange and metamorphism.

Key Words - Geodynamics, tectonics, continent-continent collision, subduction, late Mz melange, ultrahigh pressure metamorphism, transform plate boundary, Tan-Lu, Dabie, Su-Lu, Eastern Asian, East China Sea

INTRODUCTION

The East China Sea, marginal to the western Pacific, makes up one section of the continental margin along eastern Asia. It is difficult to understand the tectonic setting of the East China Sea without a broad picture of the Mesozoic continental margin of eastern Asia. It is also difficult to know the Mesozoic continental margin without a broad picture of the continental dynamics. Ideas about the tectonic evolution of eastern Asia during the Late Paleozoic and Mesozoic are diverse and controversial. For example, when and how did North China and South China collide? Is there a Precambrian continent southeast of the Yangtze carton or, instead, is there a Caledonian Huanan (South China) fold belt ? When did the subduction process along eastern Asia start during the Mesozoic? At what time was there a change from contraction to extension producing sedimentary basins in eastern China ? The senior author has published a few papers [86, 100, 101, 102], addressing some of the above questions. Most of the ideas are still valid, but some of them need to be revised. This paper provides some new ideas based on recently available data in order to foster more discussion of those problems.

I. TECTONIC FEATURES OF SOUTH CHINA

There is no problem for most geologists to accept that a Pre-Sinian craton underlies the upper, middle and lower reaches of the Yangtze River. The oldest overlying strata on this craton is the so-called Sinian, whose type section is the Yangtze gorge. Biostratigraphic study, especially the discovery of the Ediacaran fauna in Yunnan indicates that the Yangtze area has been continental since 850 Ma [75]. Grabau (1924) proposed that the coastal region of southeastern China is also an ancient continent called Cathaysia. Since Grabau's suggestion, Early Paleozoic fossils have been discovered from low-grade metamorphic assemblages south of Jiangnan (South of Yangtze) oldland. Geologists started to abandon Grabau's idea and put the Caledonides and Variscanides into these provinces. In the last ten years, geochronologic studies provide evidence of Pre-Sinian and Precambrian metamorphic and igneous rocks within a vast area in the Yangtze craton and areas southeast of it (Fig. 1 and Table 1). These facts enable geologists to challenge the existence of post-Sinian orogenic belts in the southeastern China. Shui and others [69, 70] suggested that the Cathaysian continent collided with the Yangtze continent in late Proterozoic time and that post-collisional rifting formed an early Paleozoic trough between the two continents. They proposed that before Devonian, the trough gradually closed from the northeast to the southwest, producing the folded and metamorphosed lower Paleozoic rocks of the so-called Huanan Caledonides. Ren and others [63] believed that there should be a Precambrian continent south of Yangtze craton but it is located offshore the mainland. They called it the Indochina-South China Sea continent and maintained that Grabau's Cathaysian is Caledonian. They based their argument on three points: 1) There is no unconformity between the Precambrian and Cambrian, whereas a major unconformity is found beneath Devonian mature sandstone; 2) Sinian volcanic rocks and Ordovician and Silurian turbidites reveal an unstable environment; and 3) The Lower Paleozoic rocks are metamorphosed and folded.

Within the last few years, isotopic studies support the Precambrian age of Cathaysia. Although Mesozoic intrusive and extrusive rocks, as well as sedimentary rocks, now cover most of the Cathaysian basement, many of the Yanshanian granitic intrusions and extrusions in Zhejiang, Jiangxi, and Guangdong Provinces are proved to have originated from anatexis of Precambrian basement by isotope studies (no. 17, 18, 28 & 32 in Table 1). The early Paleozoic Zhuguangshan batholith, for example, is a remobilized product of Precambrian basement (no. 29 of Table 1). Zhang and Sun [98] indicated that almost all the Phanerozoic granitic intrusions in south China are of crustal origin, with high $^{87}Sr/^{86}Sr$ initial values. Jahn and others [36] pointed out that: "available Sm-Nd isotopic data for Phanerozoic granitoids and metasediments from the South China foldbelts and Taiwan invariably show Proterozoic model ages (T_{DM}) ranging from 1 to 2.5 Ga, with a mean of 1.54 ± 0.30 Ga". Li and others [44] concluded from a comparative study of Sm-Nd isotopic and U-Pb zircon data that southern China crust was formed before 1.2-1.4 Ga and mainly before 1.8 Ga, with no apparent additional growth of crust during late Proterozoic and Paleozoic time.

Paleozoic or Mesozoic ophiolites, associated pelagic sediments, subduction-related volcanoplutonic rocks or paired metamorphic belts, indicatives of an active continental margin are not present in South China. In other words, there is no evidence to confirm the existence of Early Paleozoic or Mesozoic oceanic crust, such as the "Banxi Ocean" of Hsu and others [24], south of the Yangtze River. Instead, ophiolite and blueschist that are present along the southern margin of the Yangtze craton yield late Proterozoic ages [9, 68, 69]. Ren and other [63] suggested that the Huanan Caledonides might be an aulacogen developed on a continental crust or a "miogeosyncline". Jahn and others [36] concluded that "the influence of oceanic plate subduction was relatively limited" and "the paleo-Tethys seaway might have been opened by fracturing and the small seaway then closed shortly afterwards". We suggest that a single Precambrian continent including the Yangtze craton and the area southeast of it was already in existence at the very beginning of Phanerozoic. This continent would also include the

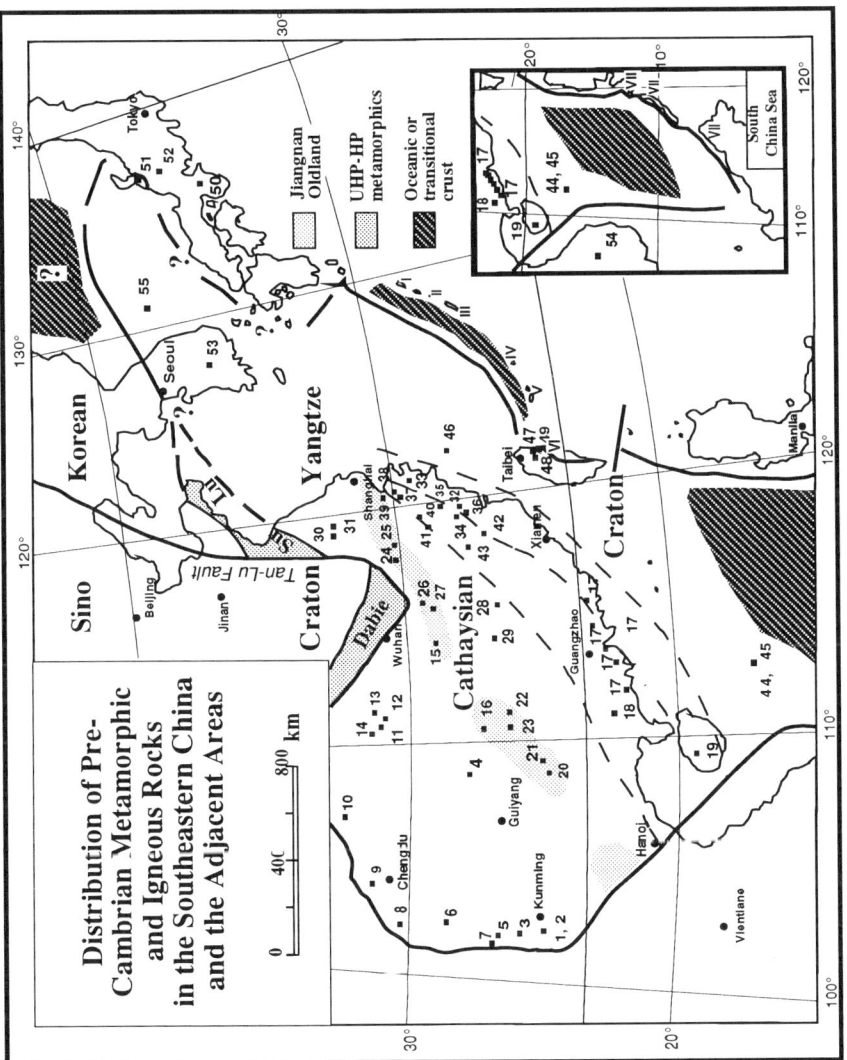

Figure. 1 Distribution of Precambrian metamorphic and igneous rocks in the southeastern China and the adjacent areas. Heavy line (except for Tan-Lu fault) is the boundary of Yangtze-Cathaysian Craton (or South China block). Dotted areas mark the Jiangnan (South Yangtze) Oldland. For chronologic data of the numbered sample locations, please see Table 1. Numbers I-VII are locations of stratigraphic columns in Fig. 11.

Table 1. Pre-Cambrian Dating in the SE China and Adjacent Areas (See Figure 1 for locations)

No	Lithology, Formation and Location	Age (m.y.) and Method	Reference
1	Dahongshan Group, East Yunnan	1720 (Rb-Sr)	Dong et al., 87
2	Dahongshan Group, East Yunnan	1900 (U-Pb, zircon)	Dong et al., 87
3	Yuanmou Granite, Yuanmou, Yunnan	619 (U-Pb, zircon), 719 (K-Ar)	Zhang & Sun, 88
4	alaskite in Fanjingshan Group, Guizhou	905-966 (K-Ar)	Dong et al., 86
5	Tongde gneiss, Dukou, Sichuan	2957 (Pb isotope)	Zhang, 87
6	Kangding Group, Shaba, Mianning, Sichuan	2404 (Rb-Sr)	Zhang, 87
7	Dukou Granite, Dukou, Sichuan	833-867 (K-Ar)	Zhang & Sun, 88
8	Huangcaoshan Granite, Luding-Shimian, Sichuan	649-808 (K-Ar)	Zhang & Sun, 88
9	Peng-Guan Granite, Guanxian, Sichuan	818 (K-Ar), 1043 (U-Pb, zircon)	Zhang & Sun, 88
10	Hannan-Zhongba Granite, Nanjiang, Sichuan	800 (Rb-Sr), 1036 (U-Pb, zircon)	Zhang & Sun, 88
11	Huangling Granite in Kongling Group, Hubei	2375 (U-Pb, zircon)	Hubei Geol. Bureau, 89
12	Huangling Granite, Yichang, Hubei	846 (K-Ar)	Zhang & Sun, 88
13	Zircon in Kongling Group, Hubei	2850 (U-Pb, zircon)	Liu, 86
14	Sandouping Granite, Yichang, Hubei	879 (K-Ar)	Zhang & Sun, 88
15	Jiuling Granite, Getengling, Hunan	844 (U-Pb, zircon), 937 (39Ar/40Ar)	Zhang & Sun, 88
16	agglomerate in Banxi Group, Hunan	900-1100 (Rb-Sr)	Ren et al., 90
17	Yenshan granite in Haifeng, Yangjiang, Dianbai, Taishan, Xinyi, Guangdong	1430-2000 (Inherited age)	Zhu, 75
18	Yunkai Group, West Guangdong	1248 (U-Pb, zircon)	Shui, 86
19	Shilu Group, Hainan	1288-1325 (Pb isotope)	Jiang & Ma, 85
20	Bendong Granite in Sibao Group, Guangxi	1063 (Rb-Sr)	Guangxi Geol. Bureau, 85
21	Sanfang-Yuanbao Granite, Rongshui, Guangxi	730 (Rb-Sr), 712 (U-Pb, zircon)	Zhang & Sun, 88
22	Motienling Granite, North Guangxi	2860 (U-Pb, zircon)	Guangxi Geol. Bureau, 85
23	Huashan Granite, Guangxi	1900-2000 (Inherited age)	Tang, 1988
24	Langsi Granite in Shangsi Group, Anhui	953 (K-Ar)	Anhui Geol. Bureau, 87
25	Xucun Granite in Shangsi Group, Anhui	877-913 (K-Ar)	Anhui Geol. Bureau, 87
26	Jiuling Granite, Xiaguan, Jingan, Jiangxi	838 (K-Ar)	Zhang & Sun, 88
27	Jiuling Granite, Changling, Nanchang, Jiangxi	765 (K-Ar)	Zhang & Sun, 88
28	Taoshan Granite, South Jiangxi	1918 (Inherited age)	Ren et al., 90
29	Zhuguangshan Granite, Jiangxi	1162, 1382, 1900 (Inherited age)	Deng, 87
30	East Ningzhenshan, Jiangsu	1771 (K-Ar)	Jiangsu Geol. Bureau, 90
31	East Ningzhenshan, Jiangsu	1123 (Rb-Sr)	Jiangsu Geol. Bureau, 90
32	source of volcanics, South Zhejiang	1180-1320 (Nd isotope))	Chen et al., 87
33	Yenshan granite, East Zhejiang	2300 (Inherited age)	Yu, 88
34	Zircon in conglomerate, Pre-Cambrian, S. Zhejiang	1436 (U-Pb, zircon)	Shui, 88
35	metamorphics in basement, S. Zhejiang	977 (Rb-Sr)	Zhejiang Geol. Bureau, 88
36	gneiss, South Zhejiang	1400-1800 (Rb-Sr, U-Pb, zircon)	Shui, 88
37	Chencal Group, East Zhejiang	1569, 1813 (Rb-Sr)	He et al., 88
38	Chencal Group, East Zhejiang	1438 (U-Pb, zircon)	Shui et al., 86
39	peridotite in Chencal Group, E. Zhejiang	892 (K-Ar, cpx)	Shui, 86
40	gneiss, Suichang, Zhejiang	1569-1983 (Rb-Sr)	Shui, 86
41	Wu'ao gneiss, Longquan, Zhejiang	2004 (U-Pb)	Shui, 86
42	Dikou Gneiss, Fujian	1822-1851 (U-Pb, Pb-Pb, zircon)	Shui, 86
43	Qinghua, West Fujian	2412 (U-Pb, zircon)	Shui, 86
44	gneiss in Xiyong-1 well, Xisha, South China Sea	627 (Rb-Sr, whole rock)	Nanhai Pet. Headqu., 85
45	gneiss in Xiyong-1 well, Xisha, South China Sea	1465 (Rb-Sr)	Nanhai Pet. Headqu., 85
46	gneiss, Lingfeng-1 well, East China Sea	1680 (Rb-Sr)	Liu, 90
47	Tananao Group, Taiwan	1000-1700 (U-Pb, zircon)	Jiang, 86
48	Nanao Granite, Taiwan	506-637 (Sm-Nd)	Jiang et al., 84
49	Tailuko Granite, Taiwan	1668 (U-Pb, zircon)	Jiang et al., 84
50	Reyka Gneiss, Nara, Japan	1782 (U-Pb, zircon)	Geol. Sury. Japan, 82
51	Hida Gneiss, Japan	1493 (U-Pb, zircon)	Geol. Sury. Japan, 82
52	gneiss pebble, Mino, Japan	1640-1680 (U-Pb, zircon)	Geol. Sury. Japan, 82
53	Xiaobai Group, South Korea	1822 (?)	Korean Geol. Survey, 86
54	gneiss, Kunshong massif, Cambodia	2300 (U-Pb)	Ren et al., 90
55	gneiss and migmatite, Ullung Island, Sea of Japan	2729 (gneiss), 1983-2231 (migmatite)	Minato et al., 85

basement of western Taiwan, the East China Sea except for Okinawa Trough, the southern part of the Korean Peninsula, part of the Sea of Japan, and the Hida belt of the Southwest Japan. In the south, this continent includes most of the South China Sea except for younger oceanic crust developed here, the Hainan Island and the Cambodian and Shan blocks. The Cambodian block was later involved in the Late Paleozoic accretionary fold belt and the Shan block was involved in the Indosinian (or Triassic) and younger fold belt in western Yunnan, Burma, Thailand and Malaysia. This huge Precambrian continent is called the Yangtze-Cathaysian craton in this paper (Fig. 1).

II. COLLISION BETWEEN THE SINO-KOREAN CRATON AND THE YANGTZE-CATHAYSIAN CRATON

One of the most controversial problems concerns the actual mechanics and timing of the collision between the North China and South China blocks. In most papers [e.g., 13, 56, 61, 95], the South China block has been equated with the Yangtze craton. We include Cathaysia within the South China block. In this case, the South China block is almost of the same size as the North China block, i.e., the Sino-Korean craton. This has important implications that will be discussed below.

1. Wudang-Tongbuo-Dabie-Su-Lu microcontinent consisting of Precambrian basement and probably Paleozoic cover strata

Between the North China block and South China block in East Qinling, the Wudang, Tongbuo and Dabie Mountains as well as Su-Lu area, Precambrian is widely distributed. The Foziling and Sujiahe Groups in North Huaiyang area, the Erlangping, Qinling, and Kuanping Groups in East Qinling area, and Dabie Groups in Wudang-Tongbuo-Dabie area all yield typical Precambrian flora [1, 19, 27, 65] (see Table 2 & Fig. 2). Dating of the Precambrian rocks of the above Groups gave 2500 to 600 Ma and some younger ages. Although the age data should be screened carefully before using them, Precambrian protolith exists and occupies a major part of the region based on Precambrian fossils anyhow.

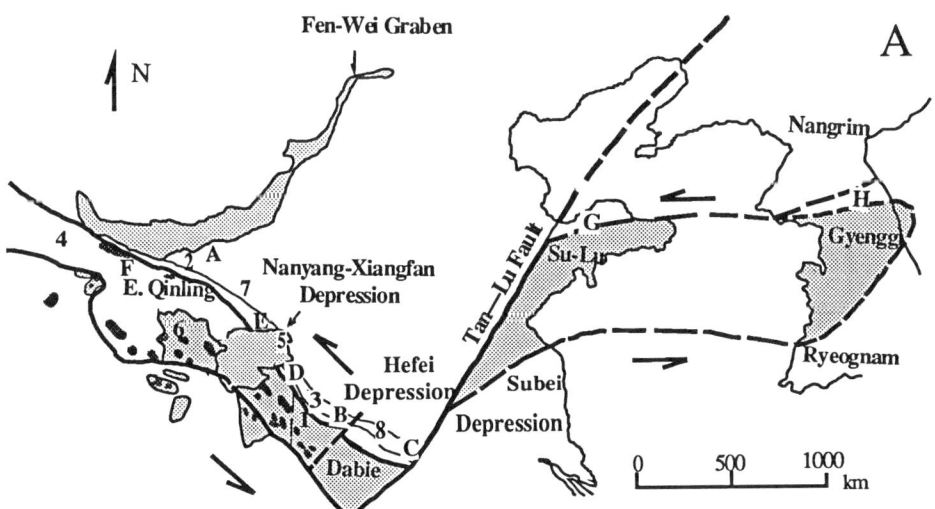

Figure 2a Location of Precambrian (shown by numbers) and Paleozoic (shown by letters) fossils in East Qinling, Wudang, Tongbuo, Dabie, Su-Lu and Imjingang areas. For fossil lists, please see Table 2. The relative movement between North China block and South China block has been left-lateral since Cenozoic. It was, however, right-lateral during Late Paleozoic and Mesozoic. See text for details

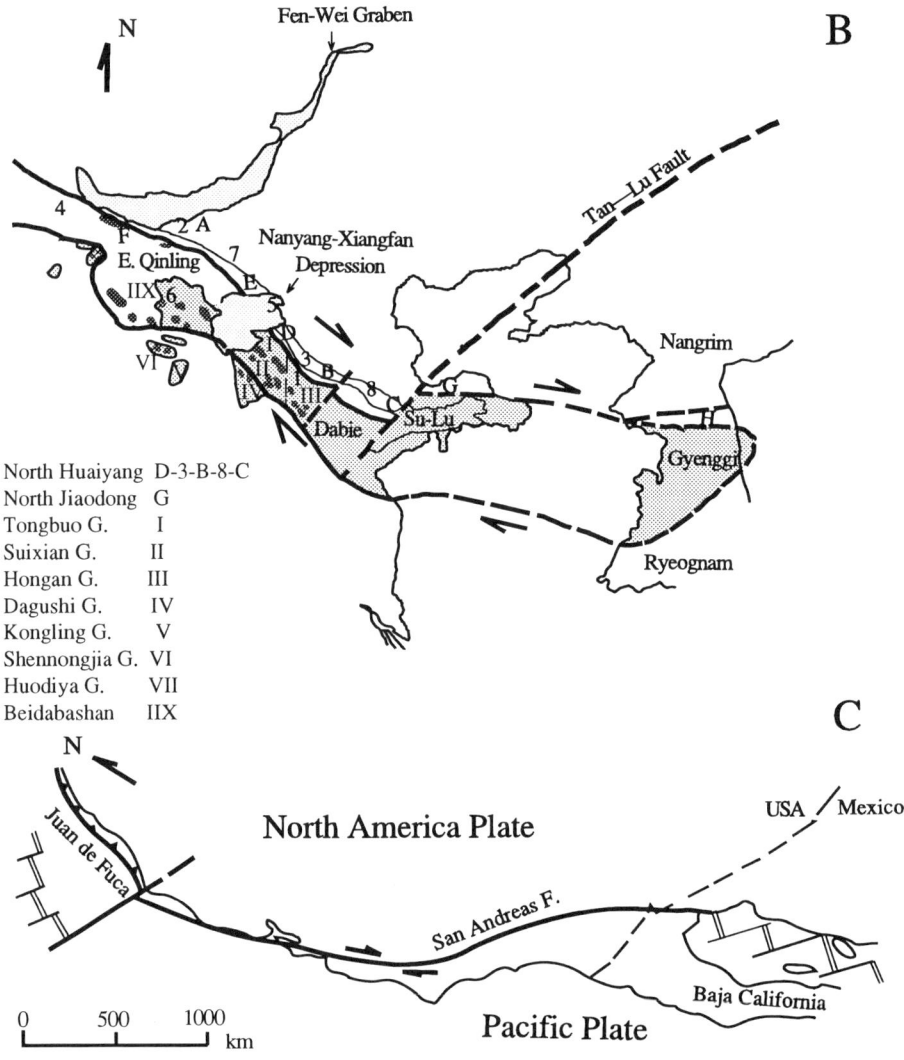

North Huaiyang D-3-B-8-C
North Jiaodong G
Tongbuo G.   I
Suixian G.   II
Hongan G.   III
Dagushi G.   IV
Kongling G.   V
Shennongjia G. VI
Huodiya G.   VII
Beidabashan   IIX

Figure 2b Before the onset of Tan- Lu fault in Late Triassic, the transform fault along the suture between North China and South China blocks can be correlated with San Andreas fault.

Figure 2c San Andreas fault links two mid-ocean ridge systems in Gulf of California and offshore Washington-Oregen. Subduction is active currently in the latter area

Figure 2a, **2b** and **2c** are of the same scale.

Whether Phanerozoic rocks are present in the aforementioned metamorphic assemblages is still a problem. The age data younger than 600 Ma could represent the age of protolith or the age of metamorphism and, therefore, should be carefully examined. However, Phanerozoic fossils have been discovered in the previously-designated Precambriam low-grade metamorphic rocks of many places (see Table 2 & Fig. 2), for example, Lower Paleozoic fossils in Erlangping, Taowan and Qinling Groups and Lower Silurian-Devonian fossil in Liuling Group of East Qinling, and Upper Paleozoic spores in Sujiahe, Xinyang and Luzhenguan Groups in North Huaiyang. The areas with Paleozoic fossils are either set apart from Precambrian units on the provincial geologic maps or left without explanation. This contradiction of biostratigraphic observations, however, is compatible with the conclusion that East Qinling and Wudang-Tongbuo-Dabie region consists of a series of imbricated structural sheets made of Precambrian, Early and Late Paleozoic rocks as well as tectonic mixtures of those different ages of rocks [50, 57, 90].

We are not sure whether the identified Paleozoic rocks are cover strata on Precambrian basement or bioherms capping seamounts and their associated rocks from a down-going ocean slab. But the lithology, fauna and metamorphic grade of the fossil-bearing rocks are not active continental margin assemblages. One study might enlighten us on this point. On the northern slope of the Dabie Mountains, the Yangshan Formation is proved to be Lower Carboniferous by its contained plant fossils. But in the limestone gravels of conglomerate in the lower part of the same formation, conodonts, ostracods, cyanophyts and heliolitids have been found [38]. Three conodonts are Ordovician and Silurian in age, and *Heliolites* cf. *anhuiensis* Ye et Deng is a typical member of Lower Silurian in lower Yangtze region. It is well known that there are no Upper Ordovician and Silurian within the Sino-Korean craton, the most probable source of such Early Paleozoic limestone gravels is the Dabie Mountains. As this limestone and its biota are the same as those strata on Yangtze craton, it is quite possible that Yangtze-type cover strata were widespread also in Dabie Mountains in Early Carboniferous time and their clasts were shed to the north to form the conglomerate of the Yangshan Formation. It is significant that no high-grade metamorphic clasts, representing the gneiss, amphibole, eclogite and granulite of the present Dabie Mountains, have been found in this Carboniferous limestone conglomerate. This suggests that the deep unroofing down to the Precambrian basement took place some time after Early Carboniferous.

Our conclusion is that Wudang-Tongbuo-Dabie as well as Su-Lu areas constitute a microcontinent (called WTDSL microcontinent later) between North China and South China blocks with Precambrian basement. The overlying Yangtze-type Early Paleozoic cover strata were subsequently removed by Late Paleozoic erosion. There is no lithologic evidence to show that oceanic crust existed between WTDSL microcontinent and South China block prior to collision. Thousands of mafic dikes and bi-model volcanics in Wudang, Suixian and Hongan Groups of the western part of WTDSL microcontinent provide evidence of failed rifting extending between South China block and WTDSL microcontinent during Proterozoic and Early Paleozoic time [27, 65]. The WTDSL microcontinent is considered as part of South China block, because there is no evidence of relative motion between them and their basements are similar.

2. Paleozoic ocean between North China and South China blocks diachronously closed before and after Devonian time

Early Paleozoic sedimentary sequences in East Qinling are similar in lithology and facies to the platform sediments of Yangtze craton and North Dabashan area south and west of the Wudang Mountains. But Devonian sediments in East Qinling show apparent differences in lithofacies and thickness than those of the Yangtze and North Dabashan areas. Devonian deep water turbidites in East Qinling are very thick (>12,000 m) [65] and show a provenance to the north, indicating that Devonian sedimentary basin was located immediately to the south of Sino-Korean craton. The Devonian fauna, however, is generally correlated to the benthonic fauna of South China, and called Xiangchuan type [2, 88]. Along the west margin of the Nanyang-Xiangfan Cenozoic basin in Xichuan County, there are Middle Devonian beach facies clastic rocks with ripple marks and cross-bedding and the Upper Devonian consists of platform margin bioherm facies based on sedimentary and biota features [19]. To the east of this Cenozoic basin, no Devonian and older Paleozoic rocks are present. The deep water sedimentary basin of Devonian time might not extend eastward beyond East Qinling and, consequently, Devonian sedimentary basin might be just a bay of considerable size but not an open ocean of several thousand kilometers wide.

Table 2a. Precambrian Fossils found in East Qinling North Huaiyang, North Jiaodong, and Imjingang areas

(1) Pt_1 Dabie Group (Regional Geologic Report of Anhui Province, 1987)
Microflora (blue algae and bacteria): (north Hubei) *Leiominuscula minuta, L. incrassata, Margominuscula rugosa, M. antiaua, Trachminuscula sp., Zonosphaeridium minutum, Dictyosphaera sp., Leiopsophaosphaera sp., Trematosphaeidium minutum, Polyporata obsoleta, Lignum punctulosum* and *Eosynechococcus mooreii*

(2) Pt_1 Qinling Group (Regional Geologic Report of Henan and Shaanxi Provinces, 1989)
    1. Yanlinggou Formation
        Microflora: *Protosphaeridium sp., Trachysphaeridium planum, T. sp., Lophosphaeridium sp., Trimatosphaeridium sp., Synsphaeridium coglutinatum (Tim.)?, Leiosphaeridium? sp.* and *Taeniatum sp.*
        Stromatolite: *Kussiella kussiensis Krylov*
    2. Taowan Formation
        Microflora: *Lignum nematoideum Sin, L. cf. punctulosum Sin et Liu, Polyporata microporosa Sin et Liu, P. obsoleta, ?Triangumorpha sp., Trematosphaeridium holtedahlii Tim., Taeniatum crassum Sin et Liu, Laminerites sp., Leiopsophosphaera sp., Margominuscula sp.* and *Trachysphaeridium sp.*

(3) Pt_1 Sujiahe Group (Regional Geologic Report of Henan Province, 1989)
Microflora: *Pseudozonosphaera asperella Sin et Liu, P. rugosa Sin et Liu, P. verrucosa Sin et Liu, P. nucleolata Sin et Liu, P. longilicata, Trachysphaeridium rude Sin et Liu, T. Incrassatum Sin, T. cultum (Andr.), T. rugosum Sin, Hubeisphaera sp., Asperatopsophosphaera umishanensis Sin et Liu, A. bavliensis Schep., A. partialis Schep., Trematosphaeridium holtedahlii Tim, Polyporata obsoleta Sin et Liu* and *Turuchania sp.*

(4) Pt_2 Kuanping Group (Regional Geologic Report of Shaanxi Province, 1989)
Stromatolite: *Cryptozoon giganteum, Straticonophyton* and *Stratifera*

(5) Pt_2 Xinyang Group (Regional Geologic Report of Henan Province, 1989)
Microflora: *Pseudozonosphaera asperella, P. sinica, Trachysphaeridium rude, T. planum, T. regosum, T. minor, T. sp., Asperatopsophosphaera umishanensis, Paleamorpha figurata, Trematosphaeridium holtedahlii, Polyporata obsoleta, Leiominuscula sp., Lignum nematoideum* and *Taeniatum sp.*
Spore: *Azonomonoletes sp.* and *Stenozonoletes sp.*

(6) Pt_2 Wudang Group and Suixian Group (Regional Geologic Report of Hubei Province, 1990)
Microflora: *Leiominuscula sp., Margominuscula sp., M. errucosa, M. rugosa, Lophominuscula sp., Quadratimorpha sp., Q. simplisis, Dictyovsphaera sp., Leiofusa cf. digitata, Taeniatum crassum, Polyporata sp., Trematosphaeridium cf. minutum, Asperatopsophosphaera sp., A. umishansis var minor, Leiopsophosphaera sp., Protonucellosphaeridium sp., Lophosphaeridium sp.* and *Lignum sp.*

(7) Pt_3 Erlangping Group (Regional Geologic Report of Henan Province, 1989)
Microflora: *Stenomarginata sp., Zonosphaeridium sp., Pseudozonosphaeridium sp., Protoleiosphaeridium sp., Laminarites sp., Polyporata obsoleta, Taeniatum sp., Protosphaeridium densum?, Trematosphaeridium sp.* and *Asperatopsophosphaera bavlensis*

(8) Pt_3 Foziling Group (Regional Geologic Report of Anhui Province, 1987)
Microflora: *Laminarites antiquissimus Eichw., Stenomarginata pussila Naum, Protosphaeridium sp., Trimatosphaeridium sp., Taeniatum sp.* and *Polyporata obsoleta Sin et Liu.*

Carboniferous is a critical time to test our tectonic model, because pelagic or semi-pelagic sediments of Carboniferous time would have existed there if an open ocean had been there between North China and South China blocks. On the contrary, if only shallow marine or terrestrial sediments can be found all over this area, the story should be different. In East Qinling, the Carboniferous from north to south consists of clastic rocks and a coal assemblage of fluvial and swamp facies; clastic rocks and sparse coal of littoral and tidal flat facies; and clastic rocks and carbonate of littoral and shallow marine facies [19]. Benthonic brachiopods and corals are the predominant fossils of marine Carboniferous in East Qinling and can be correlated to those in Yangtze craton. Plant fossils occur in littoral and terrestrial facies

Table 2b. Phanerozoic Fossils Discovered from "Precambrian" Assemblages in East Qinling, North Huaiyang, North Jiaodong and Imjingang areas

(A) Qinling Group (Pt$_1$) (both in Henan and Shaanxi): Early Paleozoic fossils
 Microflora: *Archaeohystrichosphaeridium? sp. (appearing since Cambrain), Micrhystridium? sp.* and *Ooidium? sp.*
 Insect tooth: *Scolecodont*
 Chitinozoa?

(B) Sujiahe Group (Pt$_1$) (Xinxian County, Henan): Late Paleozoic spores
 Spore: *Acanthotriletes cf. impolitus* and *Lophotriletes sp.*

(C) Luzhenguan Group (Pt$_1$) (Huoshan County, Anhui): Late Paleozoic spores
 Spore: *Verrucosisporites sp., Reticulatasporites sp.* and *Cyclogranisporites sp.*

(D) Xinyang Group (Pt$_2$) (Nanwan reservoir, Xinyang County, Henan): Late Paleozoic, probably Carboniferous spores
 Spore: *Lophotriletes sp., Periplectotriletes sp.* and *Leiotriletes sp.*

(E) Erlangping Group (Pt$_3$) (Tongbuo County, Henan): Probably Early Paleozoic
 Microflora: *?Micrhystridium sp.* and *Lophosphaeridium sp.*
 Needle-bone of sponge
 Radiolarian: *Liosphaeridae, Stylosphaeridae* and *Cyrtoidae* (Xixia County, Henan)

[The above data are from Regional Geology of Henan (1989), Hubei (1990), Anhui (1987) and Shaanxi (1989) Provinces]

(F) Taowan Group (Pt$_1$) (Luonan, Shaanxi): Cambrian fossils (Wang et al., 1989)
 Small shelly fossils: *Auriculatespira andunca, Orgzoconcha prisca, Actinotheca mirus, A. brevituba, Sulcavitus sp., Allatheca sp., Chacelloria cf. altaica, Archiasterella antinque, Allonnia sp., Actinoites diclinatus, Niphadus xihaopingensis Duan, Parakorilithes mammillatus, Circotheca sp., Actinoites diclinatus, Korilithes sp., Actinotheca manicae, Halkieria (Sachites) sp., Coleolus sp., ?Sfenotheca sp., Helcionella sp., ?Camena sp., Pojitaia runnegari, Pelagiella sp.* and *Barskoria Simplex.*

(G) Penglai Group (Pt) (Qixia County and Huangxian County, Shandong): Carboniferous and probably Devonian fossils (Yang, 1992)
 Brachiopod: *?Gigantoproductus sp.* (C1), *Marginiferids sp.* (C), and *Spiriferids sp.*
 Plant: *Lepidodendrales*

(H) Imjingang Group (Rimjin River area of Korean peninsula): Middle Devonian and Lower Carboniferous fossils, including brachiopods, gastropods, zoarium, crinoids, ostrocads, Lepidodendrales and charophyta have been found (Yang, 1992).

sediments of Carboniferous in East Qinling are also commonly seen in South China. Further to the east in North Huaiyang area, Carboniferous sediments are proved to exist in many places, although some of them are slightly metamorphosed. In Gushi-Luoshan-Xinyang of North Huaiyang area (see Fig. 2-A), the well-studied Carboniferous section consists of 4500-7500 meters of clastic rocks with coal and minor limestone of marine and terrestrial facies. Plant fossils of Lower Carboniferous are the same as those in the Yangtze craton. Marine fossils of Middle Carboniferous such as brachiopod, coral, gastropod, fusulinid, bivalve, ostracod and conchostraca include common members from North, South and West China. Upper Carboniferous flora shows affinity of both North China and South China. These facts reveal that North Huaiyang had been an intermediate sedimentary region which contains no pelagic or semi-pelagic sediments and its fauna and flora could communicate between North, South and West China since Middle Carboniferous time. This observation is a negative evidence against a big and deep ocean between North China and South China blocks in Carboniferous time.

In Qixia and Huangxian County (G in Fig. 2-A) of Shandong Province, Early Carboniferous and probably Late Devonian brachiopods of Yangtze affinity have been found recently [91] (see Fig. 2 and Table 2). Further to the east, in Imjingang area of Korean Peninsula, terrestrial clastics and shallow marine limestones with Devonian and probably Carboniferous fossils crop out in a 20 km wide E-W trending belt. Apparently, a narrow

Carboniferous sedimentary basin extending from East Qinling through North Huaiyang, North Jiaodong (Qixia-Huangxian) to Imjingang area was infilled by shallow marine and terrestrial sediments along the suture zone between the North China and South China blocks and is called Central Latitudinal basin (Fig. 3).

Collision between North China and South China blocks probably took place in Caledonian (before Devonian) and started in the middle section, i.e., the North Huaiyang and North Jiaodong areas (see Fig. 2-A), where Devonian sediments are absent, whereas in East Qinling and Imjingang areas in the western and eastern sections, thick Devonian turbidite was still accumulated on oceanic crusts. Oceanic crust consumption might happen in East Qinling and North Tongbuo areas only in Devonian time as supported by isotopic study (from 399 to 348 Ma) and structural analyses [50, 57]. Prior to the deposition of Carboniferous strata, the intermediate ocean was totally closed. The Carboniferous Central Latitudinal basin occupied the suture zone between North China and South China blocks which was a transform fault with right-lateral strike-slip movement at that time (Fig. 3). We are not sure whether any small stripes of oceanic crust were trapped between North China and South China blocks and covered by those Carboniferous sediments. Detailed geophysical investigations might solve this problem.

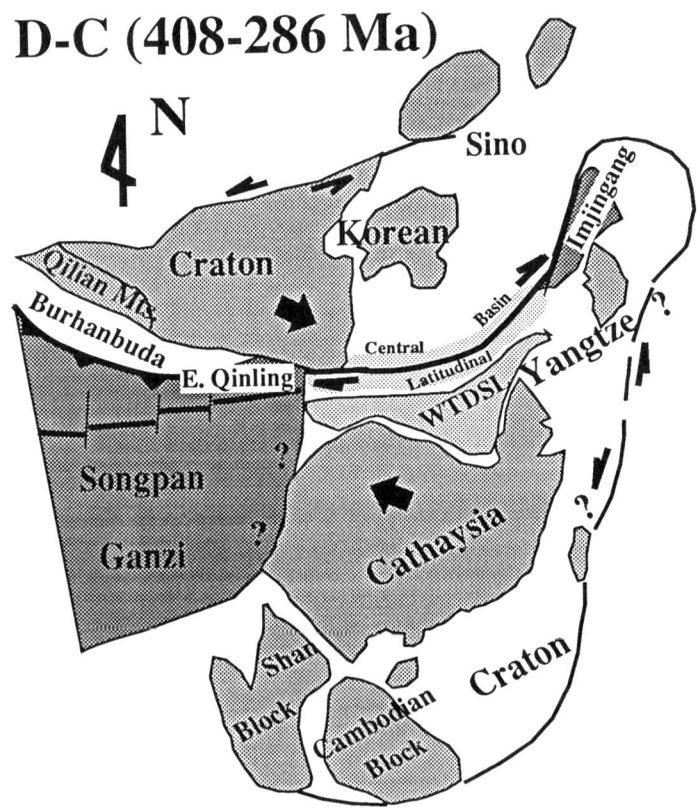

Figure 3 Tectonic reconstruction of eastern China in Devonian and Carboniferous. Note that oceanic crust existed in East Qinling and Imjingang areas in Devonian time, but not between North China and South China blocks, where Carboniferous is shallow marine and terrestrial deposits and the sedimentary basin is called Central Latitudinal basin (dotted area).

3. North margin of Wudang-Tongbuo-Dabie-Su-Lu microcontinent subducted beneath Sino-Korean craton in Early-Middle Triassic time

The important discovery of coesite- and diamond-bearing eclogite [77, 78, 79, 89] in the Dabie Mountains and its offset part from Donghai in northern Jiangsu to Rongcheng and Weihai in eastern Shandong, i.e., Su-Lu area, [12, 99] has attracted great attention in the world, because it means that ultrahigh pressure conditions (28-40 kb and 700-900 °C) (Bohlen and Boettcher, 1982) existed when the mineral assemblage formed. Ultrahigh pressure metamorphis occur in 90-120 km depth and the geothermal gradient is extremely low at about 7-9 °C / km. It is important to note that petrochemical and Nd isotope studies [77, 79, 89] suggest the protolith of the ultrahigh pressure metamorphic rocks derived mainly from upper continental crust with some mafic enclaves.

Recent dating of the coesite- and diamond-bearing eclogite as well as its country rocks ranges about 221-244 Ma [e.g., 44, 77, 78, 99]. This age requires that the ultrahigh pressure metamorphic event was in Early and Middle Triassic time. This event is about 120 to 140 Ma later than the disappear of once the deep ocean environment between North China and South China blocks at the end of Devonian. As detailed above, the contact between these two blocks is considered to be a right-lateral transform fault unrelated to oceanic crust subduction. The present-day analog is the San Andreas fault connecting two mid-ocean ridges namely Gulf of California in the south and Juan de Fuca in the north (Fig. 2). This configuraion is why subduction-related magmatic arcs can hardly be expected on either side of the boundary between North China and South China blocks. During Permian time, a narrow sedimentary trough, i.e., the Central Latitudinal Basin, within the transform fault disppeared as the whole region began to uplift.

At the end of Paleozoic, Siberian block collided with Sino-Korean block in the north, and the two blocks moved together toward the southeast (Fig. 4). South China block is of the similar size of North China block as mentioned above, and the inertia allowed it keep moving to the west as before, carrying WTDSL microcontinent with it. The primitive concave shape of WTDSL microcontinent fitted in the southeastern corner of North China block, consequently, the northern margin of the microcontinent has to go down beneath the southern margin of North China block (Fig. 5). Considering the big inertia of North China and South China blocks moving toward each other and the small size of the subducted area of WTDSL microcontinent compared with the total area of the two blocks (about 1: 1,000), it is not necessary to call upon any special mechanics to push one piece of continent crust beneath another. Obviously, the underthrusting continent crust is not necessarily intact and may be fragmental, its frontal zone must be severely distorted and imbricated. Attachment to down-going oceanic crust is not a pre-requisite to subducting and doubling the crust. A current analog is seen in continent-continent collision at the Pamir syntaxis (Molnar, 1992, personal comm.), where there is no evidence of down-going oceanic crust preceded the doubling crust into the outer mountain arcs. Collision appears to be continental shortening of the upper part of the crust by means of large-scale and severe folding, thrusting, decollement as well as ductile shearing and to be continental thickening of the lower crust by means of underplating.

Another problem is how continent-continent collision without preceding down-going oceanic crust can produce a geothermal gradient low enough to form ultrahigh pressure metamorphism. The key point is that the involved terrane should be old, cold and dry supracrust such as Dabie and Su-Lu areas at the very beginning. From the recent study by Coleman and Wang [10], the lower part of an old Precambrian continent which has been dehydrated can be in such a condition (point "bottom" in Fig. 6). If such old, cold and dry continental is push to 90 km deep, quartz and graphite will be converted to their allomorphs, coesite and diamond (points A and B in Fig. 6).

How fast can a piece of continental crust be transported to depths of 90 km or more? In the Himalayan Mountains where the maximum thickness of earth's crust (including sialic upper crust and mafic lower crust) is 75 km, the measured maximum exhumation rate is 10-20 mm / year (see Fig. 18 of Ma and others, [55]). If an area is completely in isostatic compensation condition, no uplifting or subsidence should be detected provided that structural arching-up and down-warping are negligible. As the Himalayas have been strongly uplifting in the recent geologic time, the area should be under isostatic incompensation condition and the growth rate of the mountain root would be bigger than the rate of exhumation (10-20 mm / year), assuming that the density is consistent from the top to the bottom of the continental

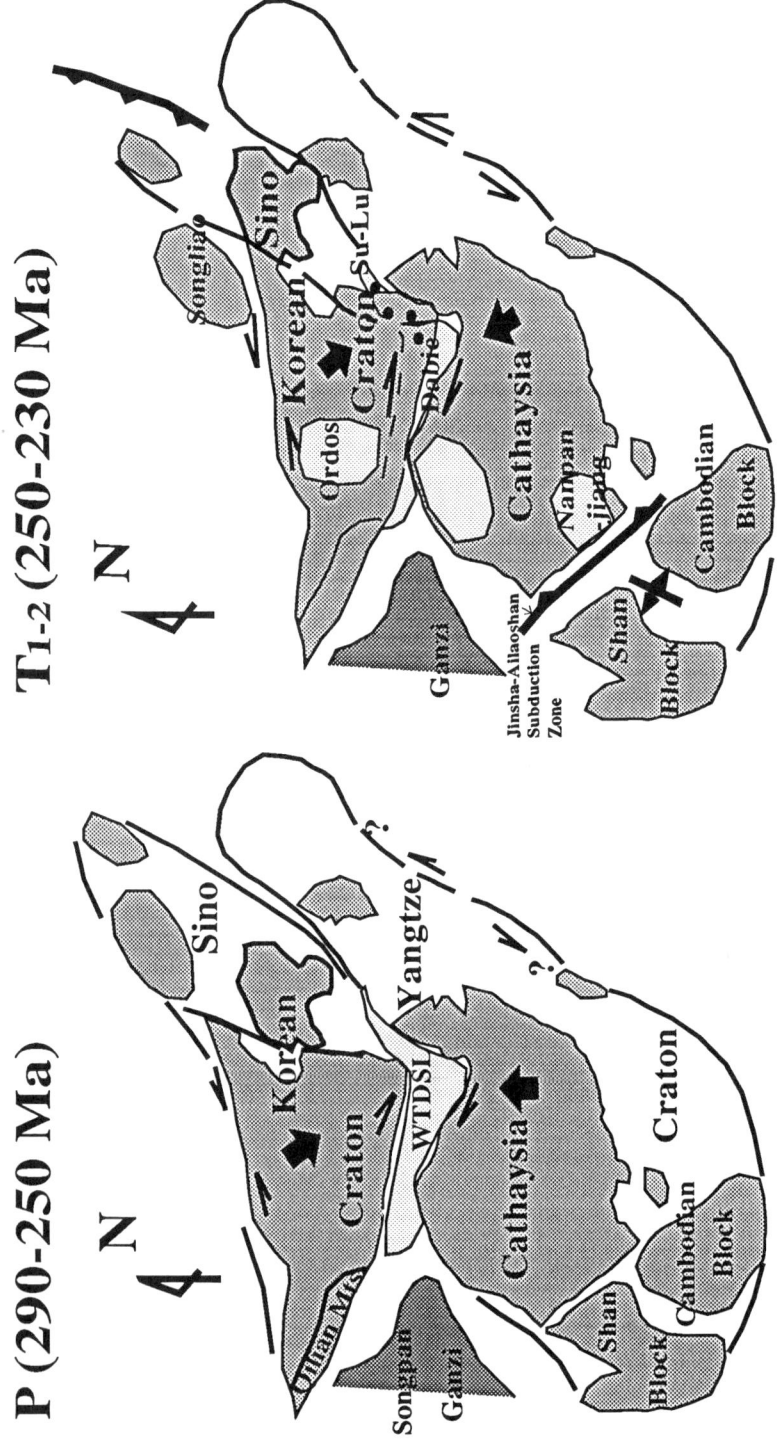

Figure 5 Tectonic reconstruction of eastern China in Early and Middle Triassic. Note that the northern margin of Wudang-Tongbuo-Dabie-Su-Lu (WTDSL) microcontinent was underthrusted beneath the southern margin of Sino-Korean craton and ultra-high pressure metamorphism occurred in the deeply buried Dabie and Su-Lu areas. Songpan-Ganzi area was a turbidite basin probably on a trapped oceanic crust. Ordos, Sichuan and Nanpanjiang basins were initiated. Black circles are coesite- and diamond-bearing eclogites.

Figure 4 Tectonic reconstruction of eastern China at the end of Permian. Note that North China shifted its moving direction from east to southeast as a result of amalgamation between North China block and Siberian block, and that the right-lateral movement to the north and south of Wudang-Tongbuo-Dabie-Su-Lu (WTDSL) microcontinent.

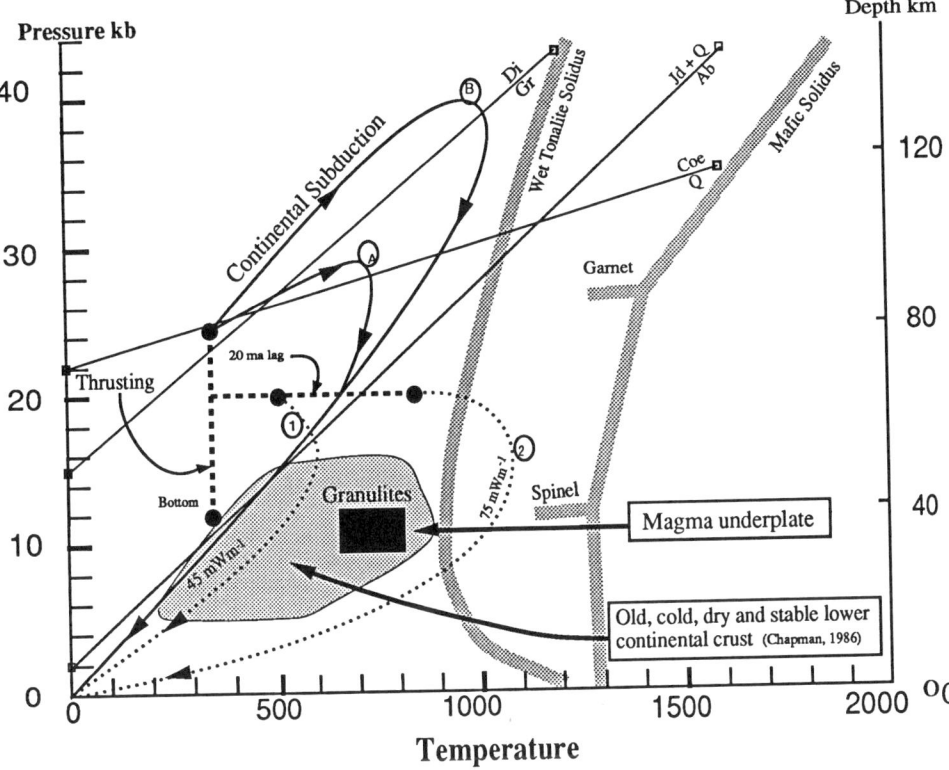

Figure 6 Pressure vs temperature diagram showing P-T paths (solid lines with arrows) in case of continent-continent collision (Coleman & Wang, in press), assuming a crustal thickness of ~40 km and nearly instantaneous thickening of the crust by low angle underthrusting. The P-T realm of old, cold, dry and stable lower continental crust is highlighted according to Chapman's data (1986). The base of the ~40 km thick continental crust is with a heat flow smaller than 45 mWm^{-2}. As a result of anatexis (or dehydrated) and radioactive decay of the heat-generating elements of the crust, the old crust become colder and drier through geologic time, and eventually, the lower continental crust at ~ 40 km depth is of low thermal conductivity of 1.5 Wm^{1}K^{-1} and in thermal equilibrium at about 365 °C (Chapman, 1986). At points A and B, coesite and diamond could form respectively. If the uplifting and unroofing are quick enough, then the ultrahigh pressure metamorphic complex can be preserved. Di = diamond, Gr = graphite, Jd = jadeite, Q = quartz, Ab = albite, Coe = coesite. 1) and 2) represent the case of fast exhumation and the case of 20 m.y. delay of exhumation respectively.

crust. Bear in mind that the protolith of coesite- and diamond-bearing eclogites are supracrustal (upper crust) material. In reality, it should be more difficult to bring down lighter sialic material to a depth greater than the root of the current Himalayas, i.e., from 75 to 100 km. It is reasonable to assume that the average down-going rate of upper crust from surface to 100 km deep is less than 5 mm / year. Such a journey will take 20 Ma. In other words, if a piece of crustal material of WTDSL microcontinent started to be underthrusted beneath the Sino-Korean craton at the end of Permian (248 Ma), that piece of sialic material could reach 100 km deep in 228 Ma. Therefore, it is not surprising to find a Early and Middle Triassic ultrahigh pressure metamorphic age in a collision zone where severe contraction between the two blocks started in Permian time when Siberian and Sino-Korean blocks jointed and moved together southeastward.

But why coesite-bearing rocks are exposed only in the eastern section of the suture? The shape of collided continents are very important. As mentioned before, the northern side of

WTDSL microcontinent is considered concave toward the north, whereas the southeastern front of Sino-Korean craton is a sharp corner, usually called "spur". When the "spur" of one continent invades the concave part of another continent or microcontinent, the point-source could cause double crust by increased stress. Areas where coesite- and diamond-bearing eclogite have been discovered such as Alps, Norway and Khazakstan might also have had similar conditions. The two syntaxes at the two corners of the India subcontinents, Jerum in the west and Assam in the east are similar and might provide conditions for ultrahigh pressure metamorphism at depth. Tiny grains of diamond has been found in ophiolite melange along Yarlung Zangpu suture in Tibet (Internal report from Chinese Academy of Geologic Sciences). This is a clue to the above suggested prediction. Collided continents without producing a point-source or syntaxis and without involving huge continents which have great total momentum might not provide enough stress to form double crust and produce ultrahigh pressure condition, and coesite and diamond will not form.

4. Initiation of Tan-Lu fault in Late Triassic time

The Tan-Lu fault formed right at the tip of the "spur" in the Late Triassic when the continental shortening reached its climax and ultrahigh pressure metamorphic complex started to exhume quickly from the depth (Fig. 7). The normal erosion and uplift rates of continental crust is 1-5 mm / yr, i.e., 1-5 km / m.y. [18]. Based on radiometric data, Hacker and Peacock mentioned that the average long-term exhumation rate of the Dora Maria coesite-bearing eclogite in Alps is about 1.5 km / m.y. [17]. According to the recent study by Coleman and Wang [10], a much more rapid uplift by either erosion or low angle faulting at rates near 13-25 mm / yr, i.e., 13-25 km / m.y. are required to allow preservation of the ultrahigh pressure metamorphic assemblage before thermal relaxation oversteps the solidus of the supracrustal rocks and granitic melts begin to form. How big an uplifting and unroofing rate is enough to preserve ultrahigh pressure metamorphic complex is still open to question. A tectonic model of the geologic history of ultrahigh pressure metamorphism of Dabie and Su-Lu areas is shown in Fig. 8. The apparent left-lateral offset of the southern boundary of WTDSL microcontinent, or the offset between coesite-bearing Dabie and Su-Lu areas, is about 450 km, but no similar brittle deformation can be found beyond the south end of the fault in Yangtze craton. In the north, apparent offset along Tan-Lu fault decreases northward and diminishes to zero in Lower Liaohe graben. Tan-Lu fault becomes "one of most enigmatic tectonic elements in eastern Asia" as mentioned by Yin and Nie [95].

The enigma will be lessened if the Tan-Lu fault is not regarded as a "pure" left-lateral strike-slip fault but primarily as a hinge fault instead. Dabie and Su-Lu areas are parts of the collision complex and consist of a stack of very low-angle thrust sheets with southern vergence. The southern boundary of WTDSL microcontinent, or the contact between WTDSL microcontinent and South China block, is probably along subhorizontal thrust surfaces. The Tan-Lu hinge fault made the eastern block, i.e., the Su-Lu area, being uplifted more than ten kilometers higher than the western block, i.e., Dabie area, the apparent horizontal offset of the southern boundaries of them could be a few hundreds kilometers after deep erosion. It should be considered that movement along the Tan-Lu fault might include a strike-slip component besides differential vertical movement and that younger sediments cover the northern part of western block (Hefei depression) and southern part of eastern block (Subei depression). This geometric reconstruction could allow the suprisingly 450 km apparent displacement on geologic map (Fig. 9). This model is consistent with the observation that coesite-bearing eclogite is more popular in Su-Lu area than Dabie area, because the former area was uplifted higher and exposed deeper than the latter area.

The widespread ten-kilometer thick Triassic turbidite in Songpan-Ganzi and West Qinling is the detritus influx carried by a big river from east to west along the weak zone, i.e., the suture between North China and South China blocks. The Dabie Mountains were uplifted higher than the western section and the closest ocean was in Songpan-Ganzi and not the proto-Pacific to the east. From the paleogeographic maps [74], it is clear to find out that Triassic deltaic and turbiditic deposits shed westward with the lithofacies shifted from shallow epicontinental sea to bathyal sea in the same direction. From the available data, a Triassic turbidite basin did not exist in the East China sea as suggested by Yin and Nie [95].

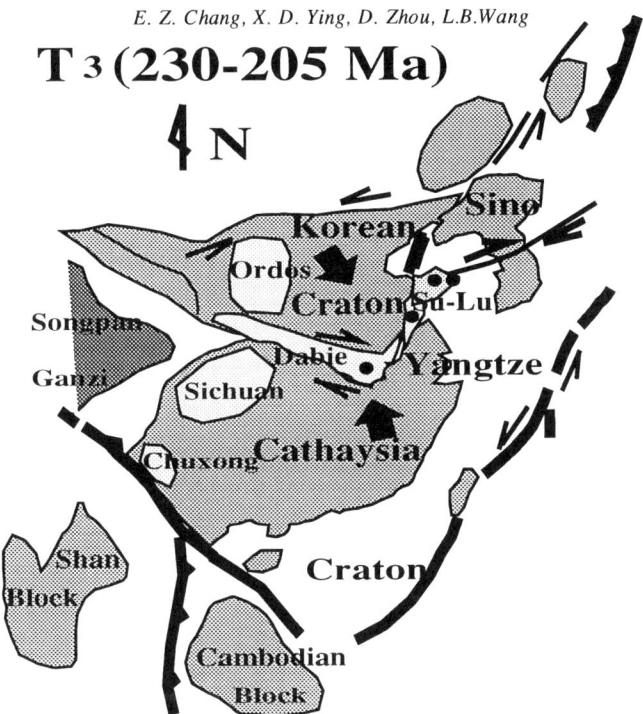

Figure 7 Tectonic reconstruction of eastern China in Late Triassic. Low-angle thrust sheets in the suture zone between North China and South China blocks in Dabie and Su-Lu areas were cut by Tan-Lu fault which is primarily a hinge fault with subsequential left-lateral strike-slip component. Dabie and Su-Lu areas were uplifted differentially. Black circles show coesite- and diamont-bearing eclogites. Note that depositional center in the southwestern Yangtze craton shifted from Nanpanjiang to Chuxong.

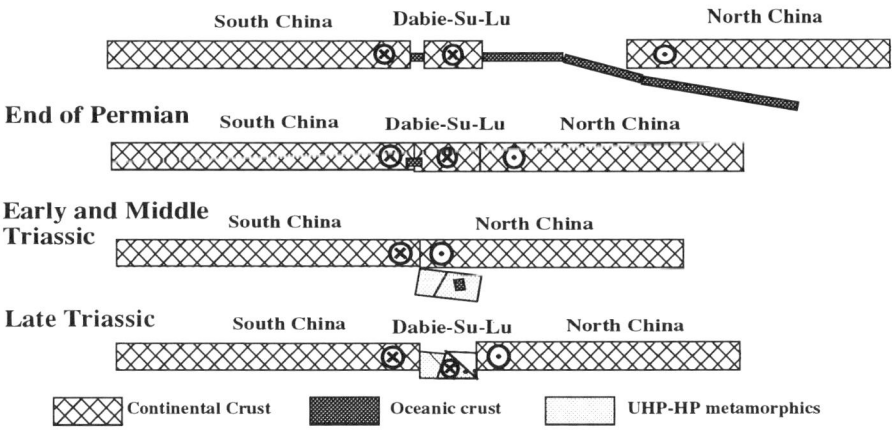

Figure 8 Tectonic model of ultra-high pressure metamorphism in Dabie and Su-Lu areas. Subduction of oceanic crust along the southern active continental margin of Sino-Korean craton was going on in Devonian and earlier geologic time and can be identified in East Qinling area. At the end of Paleozoic, the boundary between North China and South China blocks was primarily a right-lateral transform fault. Amalgamation of Sino-Korean craton and Siberian craton changed the moving direction of Sino-Korean craton and Dabie-Su-Lu area was push to underthrust beneath the former. Protolith of continental material of Dabie-Su-Lu was subjected to ultra-high pressure metamorphism at about 90-100 km depth. Subsequential extention combined with the buoyance of the stacked and wedged mountain-root enabled Dabie-Su-Lu terranes came to the surface quickly to have ultra-high pressure metamorphic complexes preserved.

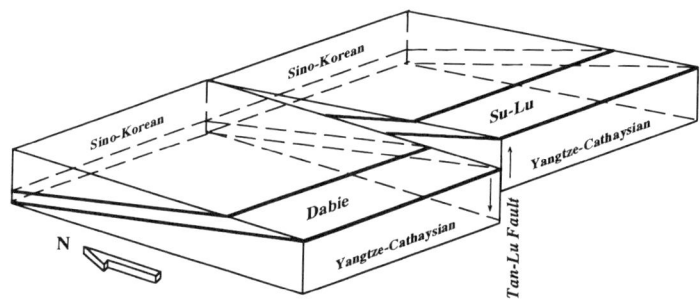

A. Differential uplifting along a hinge fault

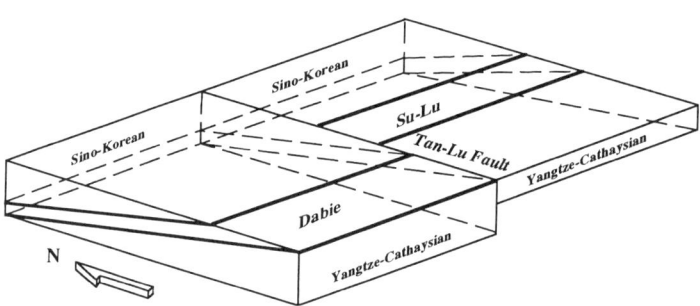

B. Apparent offset after erosion

Figure 9 Kinetic model of Tan-Lu fault. The fault is primarily a hinge fault with an axis far in the north. The eastern block was uplifted higher than the western block, therefore, the apparent offset of the ultra-high pressure (UHP) metamorphic terranes, Dabie and Su-Lu, is very prominent because the two areas consist of very low angle thrust sheets and the boundary between Dabie-Su-Lu and South China block is subhorizontal.

III. TECTONIC NATURE OF THE CONTINENTAL MARGIN OF THE YANGTZE-CATHAYSIAN CRATON DURING THE MESOZOIC

1. A transform-fault type continental margin during the early Mesozoic

The first observed Mesozoic tectonism in the Yangtze-Cathaysian craton is the Indosinian which climaxed in the late Triassic. Indosinian tectonism is evident only in folds of the cover strata in the lower reach of Yangtze River. There are no Indosinian (Triassic time) active continental margin rocks or subduction-related volcanics in southeastern China. Triassic granitic rocks of crustal origin with high $^{87}Sr / ^{86}Sr$ have been reported from Hunan, Jiangxi, Guangdong and Guangxi Provinces [63, 98]. However, no ophiolite or blueschist of this same age has yet been found. The Triassic age of Banxi melange of Hsü and others [24] has not found supporting evidences from palaeontologic and isotopic studies as well as the intensive mapping of the area. In Nanpanjiang area of Guangxi Province, a several kilometer thick Triassic turbidites sequence is widespread. It is a basin related to an east-facing subduction zone [84, 101]. But the volcanic arc itself was largely removed by later strike-slip fault in Ailaoshan area (see Fig. 5).

Jurassic and Cretaceous tectonism are the well-known Yanshanian orogeny. During the first phase of the Yanshanian (or the early Yanshanian) in the Early and Middle Jurassic, a series of left-lateral NNE-trending faults, such as Changle-Xiamen, Lishui-Haifeng, Wuchuan-Sihui, and many other small faults formed (Fig. 10). These faults are temporally and genetically related to the Tan-Lu fault which is one of the most important consequences of the collision between Sino-Korean craton and Yangtze-Cathaysian craton. The widespread S-type granitic intrusions in Zhejiang, Fujian, Jiangxi, Hunan, Guangdong and Guangxi Provinces, together with hydrothermal and skarn-type W-Sn-Mo-Bi mineralization, originated from anatexis of the continental crust [36, 98] and are spatially related to those NNE-trending faults (Fig. 10). For example, the granitic belt in eastern Jiangxi stretches in NNE direction. The individual granitic bodies, however, are arranged in a N-S direction, revealing that the intrusions are controlled by the secondary extension induced by NNE left-lateral shearing [37]. The same situation is seen in northern Fujian [63].

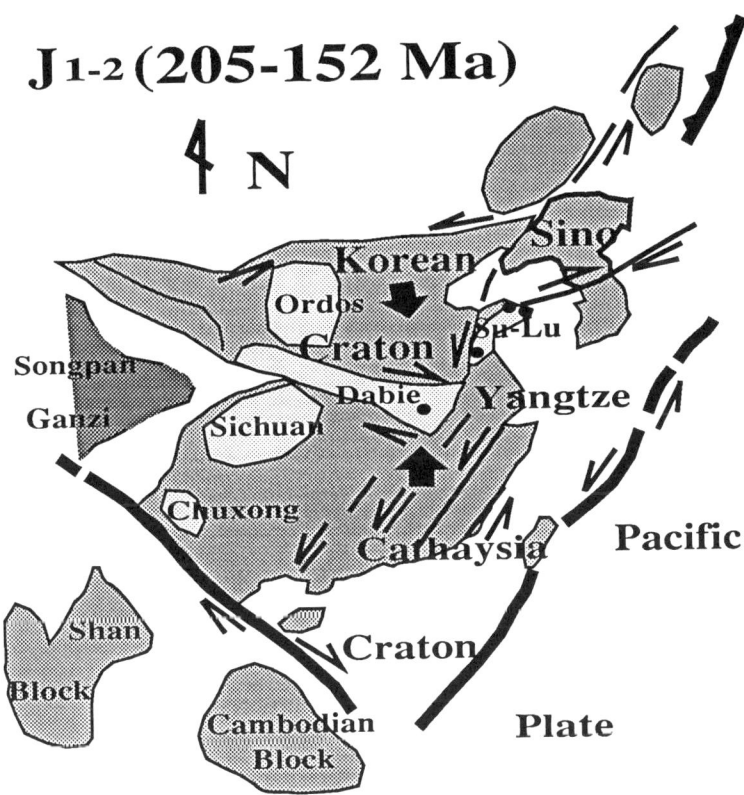

Figure 10 Tectonic reconstruction of eastern China in Early and Middle Jurassic. Left-lateral strike-slip faults were prevailing in southeastern China. From the coast inland, the major faults are Changle-Xiamen, Lishui-Haifeng, Wuchuang-Sihui and Yongxiu-Yulin. Note that the eastern continental margin of Asia was a transform fault.

It was believed by many authors [e.g., 43] that an active continental margin of Indosinian-Yanshanian ages existed along the southeast coast of mainland China, but no subduction complex has been found. Prediction of such a subduction complex located offshore has not been substantiated either, because the recent offshore drilling indicates that the continental shelves are underlain by Precambrian continental crust (no. 46 in Table 1).

The widespread Mesozoic intrusions and extrusions are thought to be evidence of subduction-related calc-alkaline magmatism. However, recent isotope and rare-earth element studies [63, 85, 98] of the granitoid intrusions indicate that Mesozoic magmatism in southeastern China is complicated genetically. The granitoid intrusions derived from upper and lower crust but not belong to calc-alkaline series. The copper-iron bearing magmatism in the lower Yangtze area might be partially from the mantle but have been strongly contaminated by crustal material [98]. Actually, the eastern continental margin of the Yangtze-Cathaysian craton was a transform fault during the Early Triassic through the Middle Jurassic time. The two pieces of the oldest (or Jurassic) oceanic crust in the western Pacific shown on the map of "Bedrock Geology of the World" by Larson and Pitman [42], i.e., the piece to the east of the Mariana Trench and the piece to the west of the Timor Sea between the northwestern Australia and the Nusa Tenggara (or Lesser Sunda) Archipelago can be matched nicely. The left-lateral offset of the two pieces of oceanic crust might give a hint of the existence of a long-distance left-lateral displacement along the big transform fault at the eastern continental margin of Yangtze-Cathaysian craton and its southern extension.

2. An active continental margin during the late Mesozoic

The eastern continental margin of the Yangtze-Cathaysian craton experienced a remarkable change in the Middle Jurassic time. The transform margin was converted into an active continental margin. Westward subduction was then active from the Late Jurassic to the Early Cretaceous (the second phase of the Yanshanian or the middle Yanshanian), and continued into Late Cretaceous (the third phase of the Yanshanian or the late Yanshanian) and Paleogene.

Based on geophysical studies of the last fifteen years, the easternmost structural belt of the East China Sea, i.e., the Eastern Marginal Folded Uplift was the same tectonic unit of the Ryukyu island arc before the opening of the Okinawa Trough [49]. This is the only accretionary belt along the eastern margin of the Precambrian Yangtze-Cathaysian craton during Mesozoic. This accretionary belt extends southward into the Central Range of Taiwan, Philippine and Borneo Island, and to the north into the Japanese Islands.

a) The Late Jurassic and Early Cretaceous melange in the northern and central Ryukyus

Ryukyu Islands consist of inner and outer tectonic belts. The two belts parallel each other and are divided by the Butsuzo Tectonic Line. This Tectonic Line is a left lateral strike-slip fault with a NW-SE contractional component and extends into Japan as the boundary between the Shimanto terrane and the Jurassic to earliest Cretaceous Chichibu + Kurosegawa + Sambagawa terranes [28]. The southern end of the Butsuzo Tectonic Line is offset by a right-lateral strike-slip fault and connects through offshore area with the Longitudinal Valley fault of eastern Taiwan. The Ryukyu Islands, together with the Butsuzo Tectonic Line, are offset by two major NW-trending left-lateral strike-slip faults, Tokara in the north and Miyako in the south, as revealed by detailed submarine and geophysical investigations [40].

The inner belt of the Ryukyu Island arc is made of Late Jurassic to Early Cretaceous melange. It is represented by the Yuwan Formation in Amami-Oshima of the northern part of the central Ryukyus [40] (I in Fig. 11). The Yuwan Formation consists of siliceous shale, sandstone, chert and basalt and contains abundant exotic blocks, from several meters to 3,000 meters across, of bedded chert, basalt and limestone. Most of the sedimentary rocks show intraformational folds with SE vergence. Carboniferous, Permian and Triassic fossils are found in the exotic blocks whereas the matrix contains Late Jurassic and Early Cretaceous radiolarians. It means that the sedimentary process are Late Jurassic and Early Cretaceous, as Carboniferous, Permian and Triassic rocks were incorporated into the debris flows. The bedrock could have come from the Yangtze-Cathaysian craton in the west, but it might also have come from seamounts in the east. One of the seamounts carried by the downgoing slab has been studied. The K-Ar age of a titanaugite-kaersutite camptonite sheet associated with the basalt of the seamount is measured to be 103.8 Ma [62].

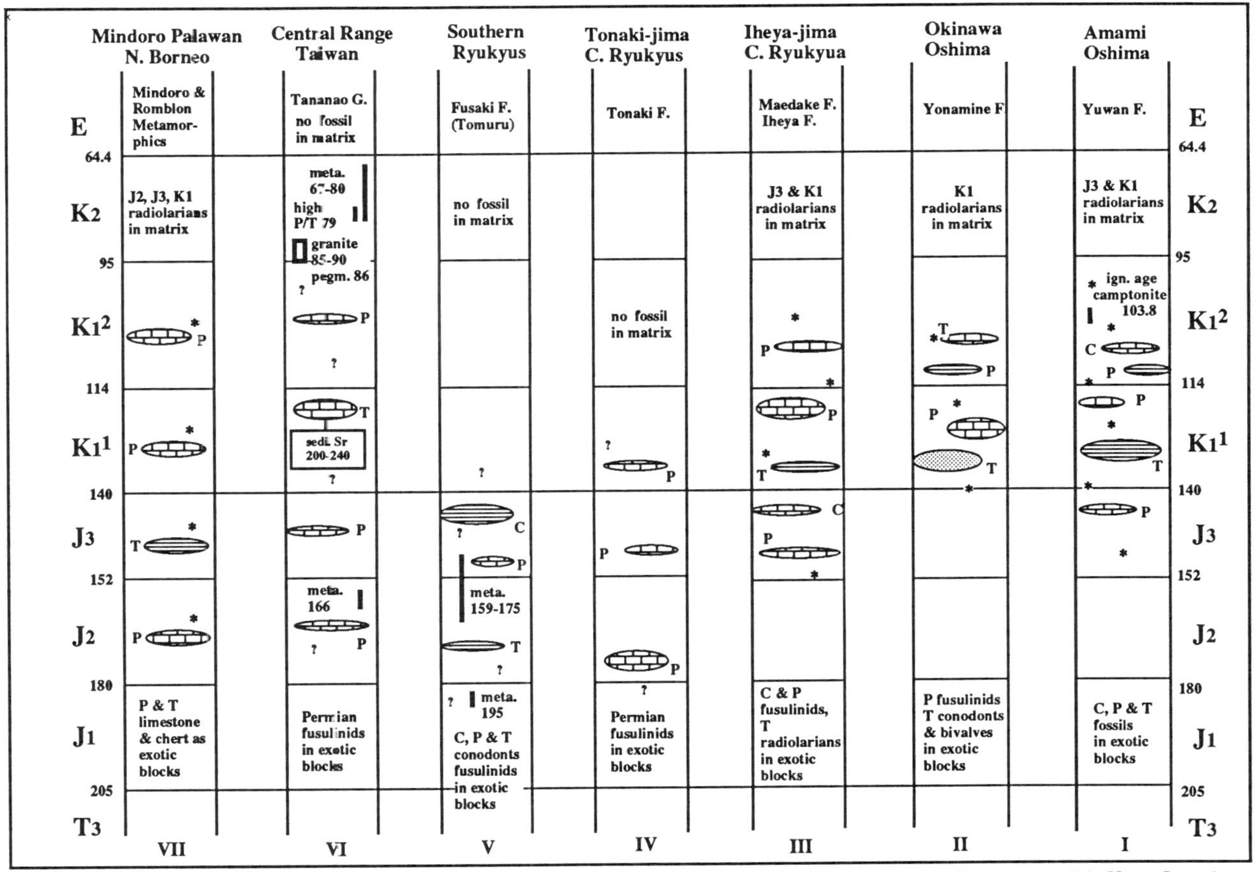

Figure 11 Stratigraphic columns of Ryukyus, Central Range of Taiwan and Mindoro-Palawan-N. Borneo areas. Note that the sediments are mainly Upper Jurassic and Lower Cretaceous in ages as proved primarily by radiolarian whereas the exotic blocks yield Carboniferous, Permian and Triassic fossils. The melange was subjected to different grades of metamorphism from place to place. For locations of columns I-VII, see Fig. 1.

Similar melange of the same age is also found in Okinawa-jima of the southern part of the central Ryukyus (II in Fig. 11). In Motobu Peninsula, the Yonamine Formation is chaotic melange consisting of mudstone, sandstone, conglomerate, tuff, greenstone and siliceous rocks, and exotic blocks of various size and lithology. Middle Permian fusulinids and Middle and Late Triassic conodonts are found from the exotic blocks and Early Cretaceous radiolarians are found in the matrix [40]. The so-called Nakijin "Formation" also in the Motobu Peninsula of Okinawa-jima, consists of limestone and mudstone intercalated with andesitic lava and yields Halobia styriaca Mojsisovics and Triassic ammonites [29]. This "Formation" is probably a huge piece of exotic block, because similar limestone containing Halobia and ammonite fossils [30], were reported in the Cretaceous melange at the Hedo Cape of northernmost Okinawa-jima.

The Iheya Formation and Maedake Formation of Iheya-jima (III in Fig. 11), offshore west of Okinawa, contain Late Jurassic and Early Cretaceous radiolarians in the matrix [73]. Carboniferous and Permian fusulinids are obtained from exotic blocks of limestone [29]. The Tonaki Formation of the Tonaki-jima of central Ryukyus (IV in Fig. 11), where Permian fusulinids such as *Yabeina* sp., *Kahlerina* sp., *Schwagerina* sp., etc [41] are found in the exotic limestones, is also melange of probably the same age, although so far no fossils evidence has been found in the matrix.

Fujita [16] mentioned that the Chichibu belt of Japanese Islands is characterized by melange that consists of Carboniferous, Permian and Triassic limestone, chert, greenstone and mudstone as exotic blocks within Upper Jurassic and Early Cretaceous sedimentary matrix. Similar melange is also reported from the Cretaceous Shimanto Supergroup of the Amami-Oshima Island [62]. Thus, the older accretionary belt in the inner part of Ryukyus is transitional with the younger and outer accretionary belt.

b) The Late Jurassic and Early Cretaceous melange in the southern Ryukyus

In the southern Ryukyus (V in Fig. 11), the Fusaki Formation in the inner belt is a melange complex. Exotic blocks of chert in the formation yield Late Carboniferous, Permian as well as Triassic conodonts and radiolarians, but no fossils have been found in the muddy matrix so far [31]. It was suggested that the age of the Fusaki Formation is probably Middle to Late Jurassic, because correlative complexes are found in the central and northern Ryukyus as well as in the Mino-Tamba Terrane and the Northern Chichibu Terrane of Southwest Japan, where the age of the matrix is more reliably constrained.

In the southern Ryukyus, another stratigraphic unit is called the Tomuru Formation, i.e., the Yaiyama metamorphic rocks. This Formation is thought to be the oldest formation of the Ryukyu islands and is composed of greenschist and blueschist, whose protolith is made up of pelite, psammite, siliceous and hyaloclastic rocks, pillow lava, gabbro and others. The Tomuru Formation is actually a mixture of various lithology of different metamorphic grade, together with some serpentinite bodies. Radiometric ages of the metamorphism are Jurassic, i.e., 159-175 Ma by the K-Ar method [60, 66] and 195 Ma by the Rb-Sr method [67]. Nishimura and Itaya [59] reported that K-Ar ages of phengite in blueschist of Ishigaki Island of the southern Ryukyu range from 174 to 194 Ma The age range is correlated to the older episode of the Nagasake metamorphism and the younger episode of the Sangun metamorphism in the Inner Belt of Southwest Japan [60]. The metamorphic age reveals that the subduction process had been active during the Middle and Late Jurassic. The Carboniferous, Permian and Triassic fossils in the exotic blocks, however, can only tell us that the subduction was more recent.

c) The Tananao complex of the Central Range of Taiwan

The oldest geologic-tectonic element of Taiwan is the pre-Tertiary metamorphic complex cropping out in the eastern part of the Central Range, called the Tananao Group (VI in Fig. 11 & Table 3). The Tananao Group is mainly black schist, greenschist, siliceous schist and minor limestone. A stratigraphic sequence of the Group has been proposed, however, the basis of classification is mainly lithologic and stratigraphic relations of the subdivided units can not be accurately established [20]. U-Pb zircon dating of two granitic plutons within the

Tananao schist reveals that these bodies intruded about 85-90 Ma ago. The inherited zircon-Pb component of the Mesozoic intrusive, however, suggests that Precambrian continental crust (ca. 1,000-1,700 Ma old) was involved in the generation of the granitic magma. The Sm-Nd model ages of Mesozoic and Cenozoic sedimentary rocks of Taiwan range from 1.38 to 1.99 Ga [36] and the model ages of late Mesozoic (ca. 90 Ma) granitic rocks are in the range 1.22-1.33 Ga [35], reinforcing the idea that Precambrian basement underlies the central and western Taiwan. But the nature of this continental margin in Mesozoic is unknown. Faure and Charvet (1987) tried to extend the Sangun belt in central Japan through Yaeyama Islands of Ryukyu arc to the Central Range of Taiwan and attributed the belt to Permian orogeny. But the Tananao schist has long been considered to be the result of the late Mesozoic subduction process along the Asian continent by other authors [4, 21, 22].

The western part of the Tananao Group, the Tailuko belt, is characterized by chloritic and biotitic greenschist facies low-pressure assemblages except in the north where amphibolite facies assemblages are associated with emplacement of remobilized granitic rocks, metamorphic grade increases gradually eastward [13]. In the Tailuko belt, pegmatite associated with earlier biotite-muscovite S-type granite (White & Chappel, 1977) yields K-Ar and Rb-Sr radiometric ages of 86 Ma [94]. The eastern part of the Tananao Group is the Yuli belt. It lacks marble layers and granitic intrusions and instead contains serpentinized mafic-ultramafic igneous bodies and rare tectonic lenses of high-pressure epidote-bearing barrositic amphibolite. Some authors consider that the Tailuko and Yuli belts form a paired metamorphic belt [93], but Lo and others [52] reported that glaucophane schists also exist in the Tailuko belt in the Hoping-Chipan area of Hualien County. Based on the above data, Yui and others [96] advocate that the Tailuko belt consists of a subduction/collision complex formed from the Late Permian to Middle Jurassic, whereas the Yuli belt represents a younger subduction complex formed from the Late Jurassic to Cretaceous. This conclusion is based on the facts that fusulinids and coral fossils of Permian ages have been found in the Tailuko belt [92] and that some of the marbles in the Central Range of Taiwan are Tatarian age (230-243 Ma) according to the study of oxygen isotopes [97]. Nevertheless, those marbles and limestones could be exotic blocks. Study by Jahn and others [33, 34, 35] reported that the $^{87}Sr/^{86}Sr$ initial ratio of marble samples of the Tananao group best fits the seawater Sr isotope evolution curve for 200-240 Ma, suggesting that the deposition of limestone (converted to marble later on) was during the time range from Early Triassic to Early Jurassic, but age ranges of 190-215 Ma; 235-245 Ma and 280-310 Ma are all possible based on the data by Jahn and others (see Fig. 3 and Table 1 of [34]). Regardless of the uncertainty of the timing, the reported ages can only tell the time of deposition of the limestone which was converted to marble later, but not the time of subduction if the limestone is solely exotic blocks within melange.

The problem is whether or not the marbles behave as exotic blocks. Lu [53] and Hsü [23] believed that both the Tailuko and Yuli belts are melange. The marble might have come from a Permian-Triassic reef-capped seamount, ocean plateau [75, 97] or from the Permian-Triassic cover strata of the Yangtze-Cathaysian craton, because the fossils are exactly the same as those in the upper Permian Changxing Limestone on the mainland. Nevertheless, the timing of the formation of the melange and, therefore, the subduction process which produced the melange must be younger than the exotic block. The metamorphic age of the complex might reveal the time of subduction. A Pb isotope study of the marble in Tananao Group gave an older (166 Ma) metamorphic age [35]. The barrositic amphibolites of the Yuli belt crystallized at 79 Ma [32, 33]. In the same belt, the $^{40}Ar/^{39}Ar$ age of nephrite and some other exotic schist and amphibolite are within the range of 67-80 Ma [33, 39, 94, 96, 97]. Therefore, the tectono-thermal event of the Tananao Group should be within the range of 166-67 Ma, i.e., from Late Jurassic to Late Cretaceous.

The similarity in lithology, structure and texture of both the exotic blocks and matrix between the Tananao Group and the Tomura Formation or the Yaiyama metamorphic rocks in the southern Ryukyu, as described above, enable us to propose the possibility that the Tananao Group could be the metamorphosed equivalent of the Late Jurassic-Early Cretaceous melange that is widely distributed in the inner belt of Ryukyus and the Chichibu + Kurosegawa + Sambagawa Terranes of the Japanese Islands. Obviously, the only way to test this idea is to find fossils in the melange matrix of the Tananao Group. So far, attempts to determine the age of the matrix have not been successful because metamorphism apparently totally obscures any fossils.

Table 3. Metamorphic and Plutonic Ages of Tananao Group and its Equivalents in Ryukyus

No	Location	Metamorphism	Age (m.y.) & Method	Reference
1	Yuli belt	blue & green sch.	5, 8-14 (Rb-Sr)	Jahn & Liou, 77 Taiwan
2	Yuli belt Taiwan	nephrite, amphibolite	67-80 ($^{39}Ar/^{40}Ar$)	Yen & Rosenblum, 64; Jahn et al., 81; Yui, 84; Yui et al., 88; Juang & Bellon, 86
3	Yuli belt Taiwan	barrositic amphibole	79 (Rb-Sr)	Jahn & Liou, 77; Jahn et al., 81
4	Tananao Taiwan	granite	85-90 (U-Pb)	Jahn et al., 86
5	Tailuko belt Taiwan	pegmatite	72-86 (K-Ar)	Juang & Bellon, 84
6	Tailuko belt Taiwan	pegmatite	87 (K-Ar, Rb-Sr)	Yen & Rosenblum, 64; Shih, 72
7	Tailuko Taiwan	amphibolite	>87 (K-Ar)	Juang & Bellon, 84
8	Tailuko & Yuli, Taiwan	blueschist	87 (Rb-Sr)	Liou et al., 75; Liou, 81
9	Tananao G. Taiwan	marble	166 (Pb-Pb)	Jahn, 88
10	Tananao Taiwan	marble	200-240 ($^{87}Sr/^{86}Sr$)	Jahn et al., 81, 84, 86
11	Tananao Taiwan	granite source	500-650 (Sm-Nd)	Jahn et al., 86
12	Tananao Taiwan	granite	1000-1700 (U-Pb, zircon)	Jiang, 86
13	Yuwan F. N. Ryukyus	camptonite sheet	103.8 (K-Ar)	Osozawa, 84
14	Tomuru F. S. Ryukyus	green & blue sch.	159-175 (K-Ar)	Shibata et al., 86; Nishimura et al., 83
15	Tomuru F. S. Ryukyus	green & blue sch.	195 (Rb-Sr)	Shibata et al., 72

d) The late Mesozoic continental margin in Ryukyus, Taiwan, Philippines and Borneo

Based on the above interpretation, it is thought that an Andean-type continental margin existed during Late Jurassic and Early Cretaceous times when the paleo-Pacific oceanic crust was subducting westward beneath the Precambrian Yangtze-Cathaysian Craton, which includes western Taiwan and the main part of the East China Sea. The Late Jurassic and Early Cretaceous melange was formed along the trench and subjected to different grades of metamorphism during subduction and later tectono-thermal events. The Yuwan, Yonamine and the Iheya Formations in different parts of the northern and central Ryukyus represent the non-metamorphosed part of the melange; the Tomuru Formation of the southern Ryukyus represents the metamorphosed part of the melange; and the Tananao Group of Taiwan might represent the highest grade of metamorphism of the same age of melange along the Yangtze-Cathaysian Continent. Fig. 12 explains the tectonic process and formation of melange along the eastern continental margin of Yangtze-Cathaysian.

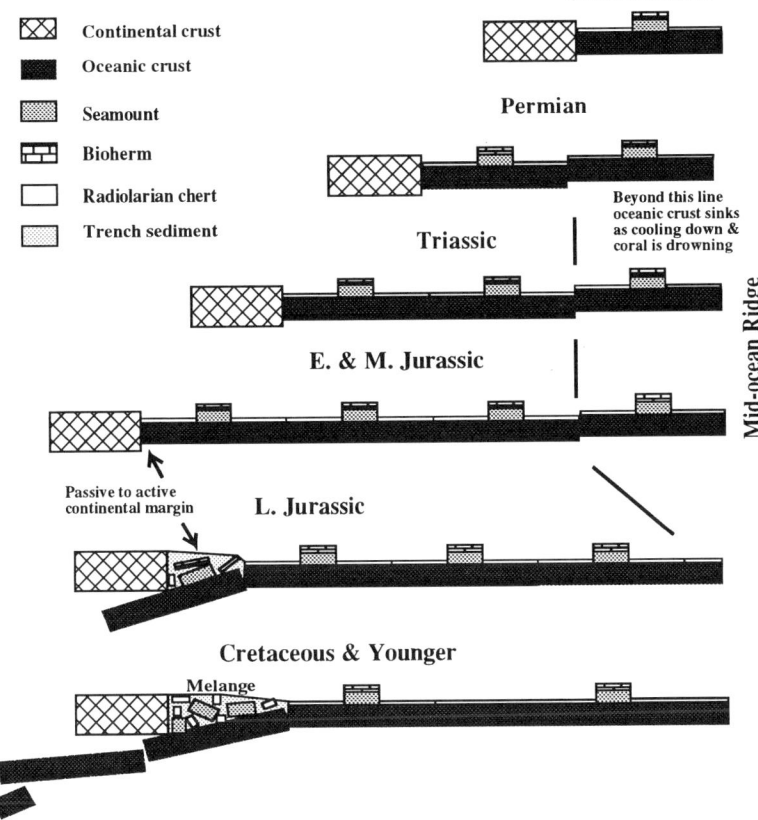

Figure 12 Mechanics of formation of melange along active continental margin. Oceanic crust initiates along mid-ocean ridge and siliceous deposit covers ocean floor. Seamounts grow on ocean floor and are capped by bioherm. Coral is drowning while oceanic crust is cool and sinks. Seamount, capped bioherm and siliceous deposit are transported toward continent and become exotic blocks when continental margin is converted from transform to active. Radiolarian and other fossils found in the matrix of melange mark the age of the sediments in trench and also the time of onset of subduction. The fossil and dating of exotic blocks indicate the ages of down-going oceanic slabs.

The Late Jurassic to Early Cretaceous intermediate and felsic volcanism widely distributed along the coast of Zhejiang, Fujian and the eastern Guangdong is subduction-related magmatism based on geochemical and isotopic studies. (Fig. 13 to 15). Jahn and others [36] pointed out that the Mesozoic granitoids (except for A-type alkali granites) in the coastal region of mainland of China and Taiwan show relatively low initial Sr (0.705-0.710) and high ^{143}Nd / ^{144}Nd ratios, implying that greater amount of mantle components has been added to ancient continental material in the generation of these rocks.

In other words, the inner belt of Ryukyus and the Central Range of Taiwan is a Late Jurassic to Early Cretaceous accretionary belt; the outer belt of Ryukyus connecting the Shimanto belt of the Southwest Japan is a Late Cretaceous-Paleogene accretionary belt. The East Coast Range of Taiwan is a Neogene continent/arc collision assemblage. The younger ages (8-14 Ma and 5 Ma) of the sodic amphibole-bearing schist and greenschist in the Tananao Group represent the overprinting tectono-thermal event relevant to the collision of a volcanic arc in the East Coast Range of Taiwan with the newly accreted continental margin, the Central Range of Taiwan.

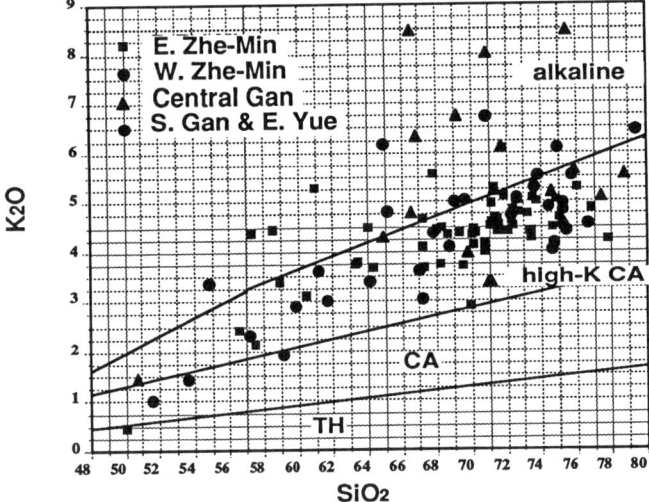

Figure 13 (A) SiO_2 vs. K_2O Harker diagram of Late Jurassic and Early Cretaceous volcanics along the southeast coastal region of China. Zhe = Zhejiang, Min = Fujian, Gan = Jiangxi and Yue = Guangdong. Volcanics show geochemical features of active continental margin.

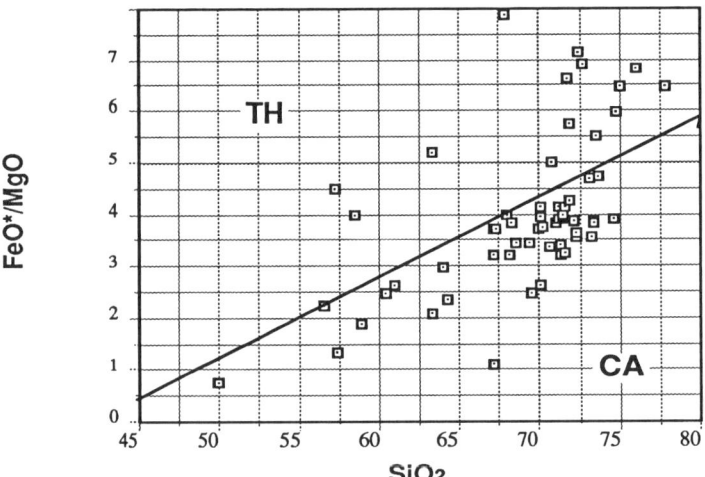

Figure 13 (B) FeO^* / MgO Harker diagram of Late Jurassic and Early Cretaceous volcanics along the southeast coastal region of China, showing that volcanics are mainly calc-alkaline series. Analytic data from Weng and others (1987).

The Jurassic-Early Cretaceous subduction complex can also be traced southwards in the western Philippines and northern Borneo. Faure and Ishida [15] reported that Middle and Late Jurassic, or more precisely Callovian to early Kimmeridgian radiolarians have been found in the turbiditic melange in Malampara Sound of northern Palawan. All those significant radiolarian species have been reported also from the Sanbosan zone of Southwest Japan. Exotic blocks consisting of Permian and Triassic chert and limestone are seen in the melange. The authors pointed out the possibility that the Middle and Late Jurassic melange can be found below the chert-spilite formation of the Sabah area of western Sarawak in northern Borneo (VII in Fig. 11).

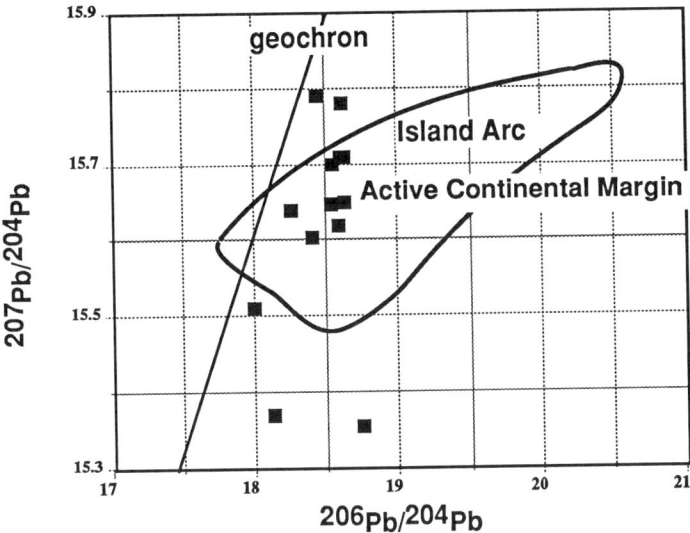

Figure 14 Pb isotope composition of Late Jurassic and Early Cretaceous volcanics in Zhejiang and Fujian Provinces. Data are plotted mainly within the field of island arc and active continental margin volcanics (Wilson, 1989). Analytic data from Weng and others (1987).

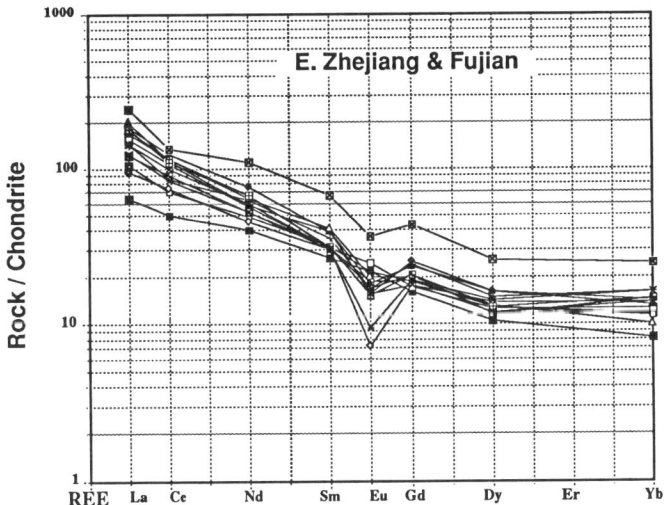

Figure 15(A) Rare earth element abundances in Late Jurassic and Early Cretaceous volcanics in eastern Zhejiang and eastern Fujian.

In summary, a middle Mesozoic (mainly Late Jurassic to Early Cretaceous) subduction zone existed all along the eastern continental margin of Asia from Nadanhada through Japan, Ryukyus, Taiwan and Palawan to northern Borneo. In some sections such as Nadanhada, southern Ryukyu and Taiwan, subduction might have begun earlier in Middle Jurassic. The oceanic crust being subducted along this active continental margin could be as old as Carboniferous, Permian and Triassic as suggested by the oldest chert, seamount basalt and reef limestone capping the seamount now involved in melange as exotic blocks. Therefore, the age range of the subducted ocean crust could be from Carboniferous or Permian to the Middle Cretaceous. Fig. 16 is a tectonic reconstruction of eastern China during this period of time.

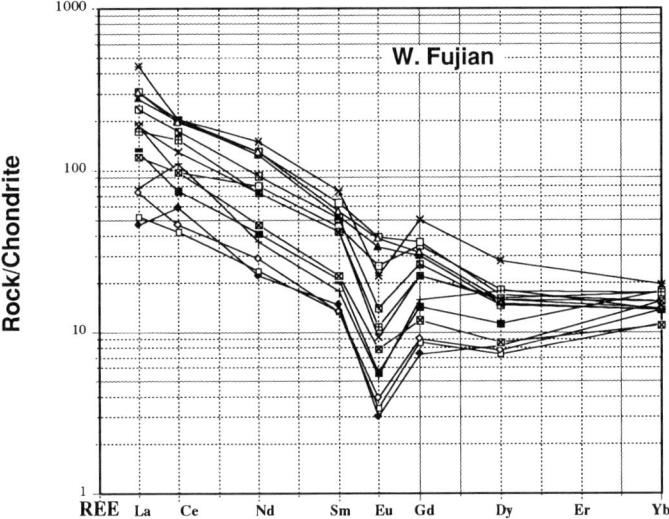

Figure 15(B) Rare earth element abundances in Late Jurassic and Early Cretaceous volcanics in western Fujian Analytic data from Weng and others (1987).

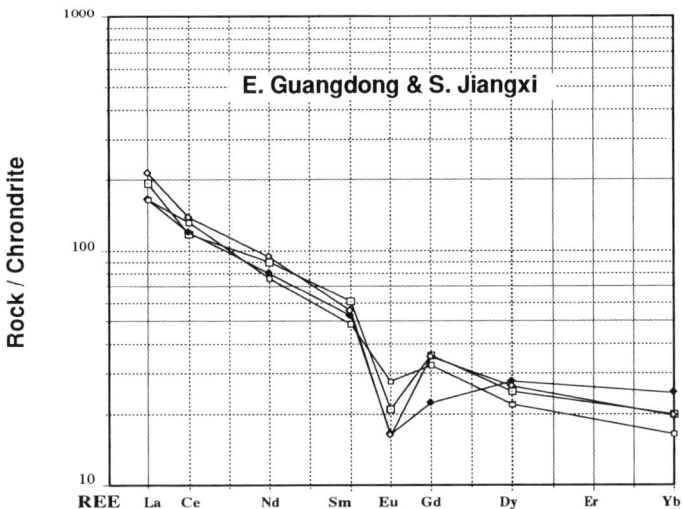

Fig. 15(C) Rare earth element abundances in Late Jurassic and Early Cretaceous volcanics in east Guangdong and south Jiangxi. Note the light REE enrichment and obvious Eu negative anomaly. Analytic data from Weng and others (1987).

e) The Late Cretaceous and Paleogene active continental margin

The Shimanto terrane is the youngest terrane in southwest Japan and consists of Cretaceous, Paleogene and lower Miocene rocks. The Shimanto tectonism in middle Miocene marks the time when the Shimanto terrane accreted to the older part of the Japanese Islands. Some people proposed that the Shimanto belt was deposited along another continent called the Kuroshio Paleoland (Tokuoka et al., 1982) in the Philippine Sea. Therefore, the Shimanto belt is a collision complex between the Kuroshio Paleoland and the main part of Japan. This is a problem concerning Pacifica and will not be discussed here. The Shimanto belt extends into the Ryukyus and forms the outer belt of it. As mentioned before, the outer belt is separated from the inner belt by the Butsuzo Tectonic Line in the northern and central Ryukyus, but this Line is submerged in the southern Ryukyus. The submarine equivalent of the Shimanto terrane exists offshore of the southern Yaeyama Islands [40]. The outer belt of Ryukyus marks the active continental margin of Late Cretaceous and Paleogene age. The current subduction belt is located further oceanward in the Ryukyu trench (Fig. 17).

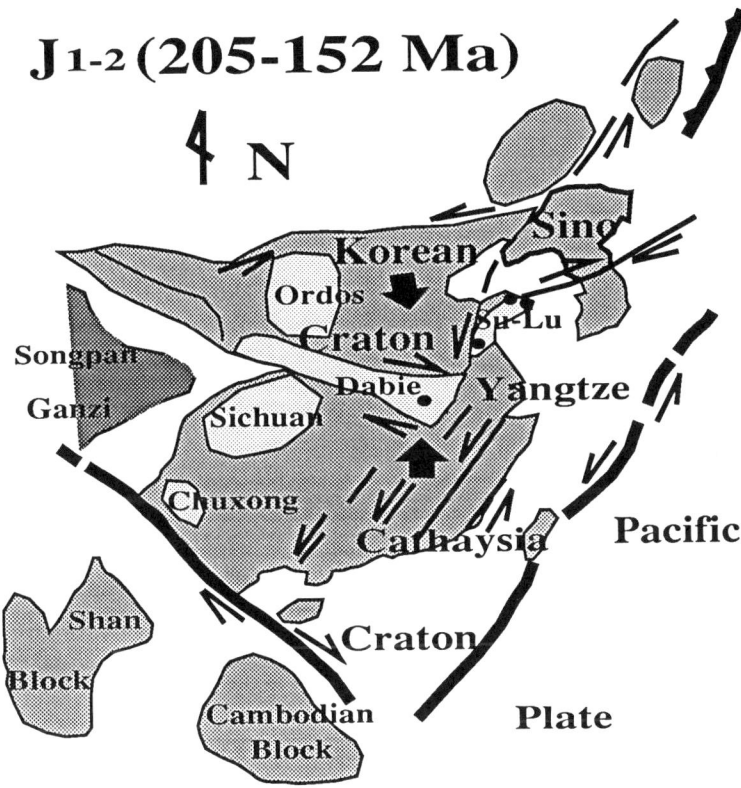

Figure 16 Tectonic reconstruction of eastern China in Late Jurassic and Early Cretaceous (before Albian). Continental margin of Yangtze-Cathaysian was converted from transform to active and subduction-related volcanism was active along the southeast coastal region of the mainland and East China Sea continental shelf as well. Note that Songliao and Erlian became major onshore depocenters in which hydrocarbons source rocks are terrestrial facies.

The East Coast Range of Taiwan is a Neogene island arc which collided with the main part of Taiwan when the oceanic crust between the arc and the continent was completely subducted eastward under the arc [5, 7, 8, 71, 87]. The collision happened during the Pliocene and Quaternary [3], younger than the accretion of the outer belt of Ryukyus. Seismicity study [11] shows that the relative movement between the Philippine Sea and Eurasian plates is 4.9-6.7 cm / yr from the northeast to the southwest along the Ryukyu trench. Back-arc extension in the Okinawa Trough is in E-W direction and includes a left-lateral stress regime in the eastern Asian which might be inherited since the early Mesozoic.

IV. TECTONIC SETTING OF THE EAST CHINA SEA

From the above discussion, it is apparent that the western part of the East China Sea, i.e., the continental shelf basin, overlaps Precambrian basement of the Yangtze-Cathaysian craton. The eastern part of the East China Sea, i.e., the Okinawa Trough Basin is floored by oceanic or transitional crust formed in a back-arc tectonic environment.

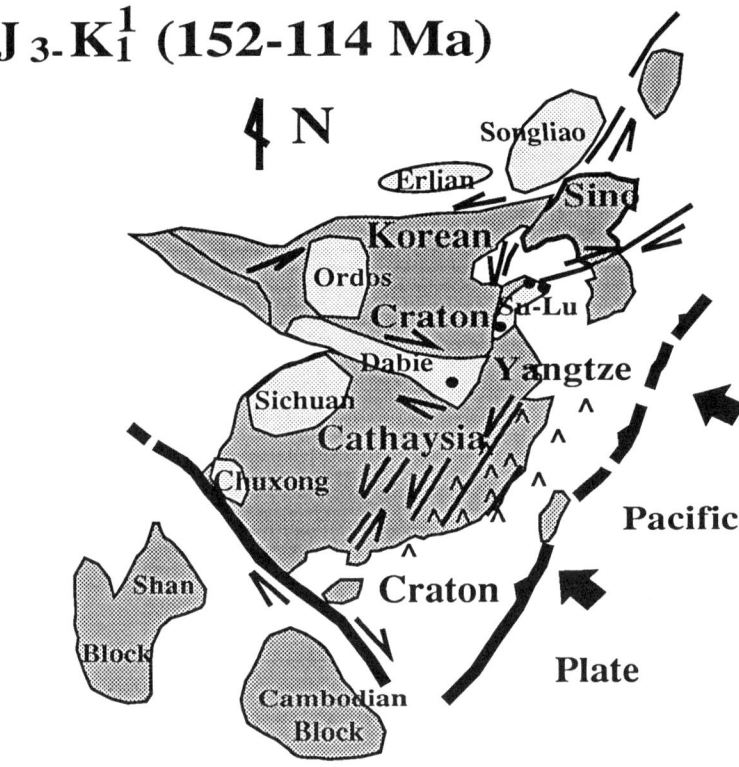

J_3-K_1^1 (152-114 Ma)

Figure 17 Tectonic reconstruction of eastern China in Late Cretaceous (including Albian and Aptian) and Paleogene. Subduction continued along the eastern margin of Yangtze-Cathaysian continent. First phase of back-arc spreading in South China Sea and Okinawa trough started. At the meantime, many large and small terrestrial sedimentary basins took shape in the eastern part of China. Songliao and Erlian faded out. Note the moving directions of South China and North China blocks shifted due to the collision of Indian continent in early Paleogene. The left-lateral strike slip movement between the two blocks caused the rifting of Fen-Wei graben in North China block.

Like many of the sedimentary basins onshore, the continental shelf basin started rifting at the end of the second phase of the Yanshanian orogeny (Fig. 17). The oldest sediments of the basin are Aptian and Albian (Middle Cretaceous for some stratigraphers). In Taixi (West Taiwan) depression, an Aptian ammonite (Philloceratid) together with calcareous nannofossils were first discovered in the PK-2 well of the Beigang area (Huang, 1981). More calcareous nannofossils of late Early Cretaceous (Aptian and Albian) are found in the adjacent wells: PK-3, MLN-1, HP-1, etc. The maximum penetrated thickness is 1539 m (Wh-1 well). Seismic data show that the same reflectors can be traced over almost the entire basin [6]. In Taixinan (Southwest Taiwan) depression, clastic rocks of marine-terrestrial transitional facies exist beneath Oligocene strata in the central part as seen in CFC-1, 2, 3 wells. The maximum penetrated thickness is 600 m. The strata are Early Cretaceous in age, probably Aptian and Albian. From seismic data, the Oligocene is unconformable with the Cretaceous. In the Taibei (North Taiwan) Depression, sandstone, arkose, shale and tuff of marine-terrestrial facies are found beneath the Eocene in the YCC-IX well at the western slope of the Pengjiayu High. The apparent thickness is 1082.6 m. The age of the strata is commonly accepted as Maastrichtian of the latest Cretaceous, but somebody believe that it should be late Early Cretacous in age (e.g., Huo, 1988). In other parts of the continental shelf basin, the oldest fossiliferous strata above the sharp unconformity are upper Paleocene. No well penetrates this unconformity so far. But it is believed that, at least, Upper Cretaceous-lower Paleocene should exist. In general, the continental shelf basin started to subside in late Paleocene and continued to Pliocene and Quaternary. The older strata are the infillings in the initial stage of rifting. The Okinawa Trough basin started to subside in Miocene soon after the accretion of the outer belt of Ryukyus.

V. CONCLUDING REMARKS

The tectonic evolution of the eastern Asia can be summarized as follows:
1. A single Precambrian continent existed in South China. It is called Yangtze-Cathaysian craton or South China block. Wudang-Tongbuo-Dabie-Su-Lu is a microcontinent between North China and South China blocks.
2. Paleo-ocean between North China block and South China block was closed before Devonian in the middle section (North Huaiyang to North Jiaodong) and Devonian turbidite basins in East Qinling and Imjingang disappeared before Carboniferous. Along the boundary between North China and South China blocks, a narrow sedimentary basin call the Central Latitudinal basin developed and received shallow marine and terrestrial deposits in Carboniferous time. This boundary was a right-lateral transform fault since Devonian and Carboniferous until early Paleogene.
3. At the end of Late Paleozoic, the Siberian craton collided with North China block. A combined super-continent moved southeastward, while South China block kept moving westward. Severe contraction between North China and South China blocks enabled the northern margin of the intervening Wudang-Tongbuo-Dabie-Su-Lu microcontinent to underthrust beneath North China crust at a "spur" (or syntaxis) where is the intensive point-source of stress and formed ultrahigh pressure metamorphism in Early and Middle Triassic time. There is no real evidence of oceanic crust subduction leading to continent-continent collision. Triassic oceanic existed only in West Qinling and Songpan-Ganzi area.
5. The collision complex consists of very low-angle thrust sheets. At the climax of the continental collision in Late Triassic, the Tan-Lu fault was initiated. As Tan-Lu is primarily a hinge fault with eastern block uplifting higher than western block, and with minor left lateral strike-slip component, the apparent offset of the coesite-bearing Dabie belt and Su-Lu belt is exaggerated to be 450 km.
6. During Early and Middle Jurassic (the early Yanshanian), southeastern China was under a compressive-shearing stress field. A series of left-lateral strike-slip faults was active. S-type granites of the same ages in the southeastern China were generated from anatexis and intruded along these faults.
7. The continental margin of the Yangtze-Cathaysian craton dramatically changed from a transform-type to subduction-type at the end of the Middle Jurassic. Late Jurassic and Early Cretaceous (the middle Yanshanian) melange in the inner belt of Ryukyus is the first evidence of this Mesozoic west-directed subduction and accretionary folded belt. Tananao Group is the metamorphosed equivalent of the melange of the same age. Subduction-related calc alkaline magmatism became active along the coastal area of the mainland and the continental shelf of East China Sea for the first time.
8. The outer belt of Ryukyus is Late Cretaceous to Neogene accretionary folded belt, which accreted to the earlier formed inner belt in mid-Miocene, contemporary with the Shimanto or Takachiho in Japanese Islands. The East Coast Range of Taiwan collided with the continent during Pliocene-Quaternary.
9. From late Early Cretaceous to Late Cretaceous (the late Yanshanian), extension basins initiated in western Taiwan and in the north, west and southwest offshore Taiwan, together with those sedimentary basins in eastern China (except for Songliao and Erlian).
10. The main period of extension and subsidence of the continental shelf basin of the East China Sea is Paleogene, when the main petroleum source rocks were formed.
11. The Okinawa Trough basin of the East China Sea developed since the Miocene, after the accretion of the outer belt of Ryukyus. The Okinawa Trough was created by back-arc spreading associated with the youngest subduction along the western Pacific.

ACKNOWLEDGMENTS

This manuscript has been critically reviewed and materially improved by R. G. Coleman, S. A. Graham and B. R. Hacker. The authors also benefit from the discussion with W. G. Ernst, J. G. Liou, P. Molnar and An Yin. This research was supported by Stanford-China Geoscience Industrial Affiliates Program and is under the auspices of Department of Geology, Stanford University and IGCP-283.

REFERENCES

1. Anhui Bureau of Geology and Mineral Resources, *Regional Geology of Anhui Province.* Geology Publishing House, Beijing. (1987). In Chinese

2. S.L. Bai, *Devonian Biostratigraphy of Guangxi and Adjacent Areas.* Beijing University Press, Beijing. (1982). In Chinese

3. E. Barrier and J. Angelier. Active collision in eastern Taiwan, the coastal range, *Mem. Geol. China.* (7), 135-159. (1986).

4. C.C. Biq. A fossil subduction zone in Taiwan, *Proc. Geol. Soc. China.* (14), 146-154. (1971).

5. C.C. Biq. Transcurrent buckling, transform faulting and transpression: Their relevance in eastern Taiwan kinematics, *Petrol. Geol. Taiwan.* (10), 1-10. (1972).

6. W. Bosum, G.D. Burton, S.H. Hsien, E.G. Kend, A. Schreiber, and C.H. Tang. Aeromagnetic survey of offshore Taiwan, *CCOP Tech. Bull.* (3), 1-34. (1970).

7. C. Bowin, R.S. Lu, C.S. Lee, and H. Shouten. Plate convergence and accretion in Taiwan-Luzon region, *Am. Assoc. Petrol. Geol. Bull.* **62**, 1645-1672. (1978).

8. B.H.T. Chai. Structural and tectonic evolution of Taiwan, *Am. Journ. Sci.* **272**, 389-422. (1972).

9. J.F. Chen, K.A. Foland, F.M. Xing, X. Xu, and T.X. Zhou. Magmatism along the southeast margin of the Yangtze block: Precambrian collision of the Yangtze and Cathaysia blocks of China, *Geology.* **19**, 813-818. (1991).

10. R.G. Coleman and X.M. Wang, Overview of the geology and tectonics of UHPM, In: *Ultrahigh-pressure Metamorphism,* R.G. Coleman and X.M. Wang (Ed.), (in press), Cambridge University Press, Cambridge. (1993).

11. T. Eguchi and S. Uyeda. Seismotectonics of the Okinawa Trough and Ryukyu Arc, *Mem. Geol. Soc. China.* (5), 189-210. (1983).

12. M. Enami and Q. Zang. Quartz pseudomorph after coesite in eclogites from Shangdong Province, east China, *Am. Mineralogist.* **75**, 381-386. (1990).

13. W.G. Ernst and B.M. Jahn. Crustal accretion and metamorphism in Taiwan, a post-Paleozoic mobile belt, *Phil. Trans. R. Soc.* **A321**, 129-161. (1987).

14. W.G. Ernst, G. Zhou, J.G. Liou, E. Eide, and X.M. Wang. High-pressure and superhigh-pressure metamorphic terranes in the Qinling-Dabie mountain belt, central China: early- to mid-Phanerozoic accretion of the western Paleo-Pacific rim, *Pacific Science Association Information Bulltin.* **43**(1-2), 6-15. (1991).

15. M. Faure and K. Ishida. The Mid-Upper Jurassic olistostrome of the west Philippines: a distinctive key-marker for the North Palawan block, *Jour. Southeast Asia Earth Sciences.* **4**(1), 61-67. (1990).

16. H. Fujita. The "Paleozoic" formations in the Motobu Peninsula of Okinawa jima, Ryukyu Islands, *Mem. Geol. Soc. Japan.* (22), 3-13. (1983).

17. B.R. Hacker and S.M. Peacock. Creation, preservation and exhumation of coesite-bearing, ultrahigh-pressure metamorphic rocks, In: *Ultrahigh-pressure Metamorphism,* R.G. Coleman and X.M. Wang (Ed.), (in press), Cambridge University Press, Cambridge. (1993).

18. C.G.A. Harrison. Rates of continental erosion and mountain building, , (in press). (1993).

19. Henan Bureau of Geology and Mineral Resources, *Regional Geology of Henan Province.* Geology Publishing House, Beijing. (1989). In Chinese

20. C.S. Ho, *An Introduction to the Geology of Taiwan.* The Ministry of Economic Affairs, Taipei, Taiwan. (1975).

21. C.S. Ho, *Tectonic Evolution of Taiwan. Explanatory Text of the Tectonic Map of Taiwan.* The Ministry of Economic Affairs, Taipei, Taiwan. (1982).

22. C.S. Ho, *An Introduction to the Geology of Taiwan.* The Ministry of Economic Affairs, Taipei, Taiwan. (1986).

23. K.L. Hsü. Melange and the melange tectonics of Taiwan, *Proc. Geol. Soc. China.* **31**(2), 87-92. (1988).

24. K.L. Hsü, J.L. Li, H.H. Chen, Q.C. Wang, S. Sun, and A.M.C. Sengör. Tectonics of South China: Key to understanding West Pacific geology, *Tectonophysics.* **183**(9-39), . (1990).

25. J.Q. Huang, J.S. Ren, C.F. Jiang, Z.K. Zhang, and D.Y. Qin, *The Geotectonic Evolution of China.* Science Press, Beijing, China. (1980).

26. W. Huang and Z.W. Wu. Evolution of the Qinling orogenic belt, *Tectonics.* **11**(2), 371-380. (1992).

27. Hubei Bureau of Geology and Mineral Resources, *Regional Geology of Hubei Province.* Geology Publishing House, Beijing. (1990). In Chinese

28. K. Ichikawa. Pre-Cretaceous terranes of Japan, In: *Pre-Cretaceoous Terranes of Japan,* K. Ichikawa (Ed.), Nippon Insatsu Shuppan Co., Osaka. (1990).

29. T. Ishibashi. Stratigraphy of the Triassic formation in Okinawa-jima, Ryukyus, *Mem. Fac. Sci. Kynshu University, Dept. Geology.* (19), 373-385. (1969).

30. T. Ishibashi. Triassic system of Hedo-misaki, Okinawa Prefecture, *Jour. Geol. Soc. Japan.* (80), 329-330. (1974).

31. Y. Isozaki and Y. Nishimura. Fusake Formation, Mesozoic complex of Ishigake-jima, Yaeyama Islands (Preliminary report), *High Pressure Metamorphic Belt of the Inner Belt.* (4), 23-26. (1987).

32. B.M. Jahn and J.G. Liou. Age and geochemical constraints of glaucophane schists of Taiwan, *Mem. Geol. Soc. China.* (2), 129-140. (1977).

33. B.M. Jahn, J.G. Liou, and H. Nagasawa. High-pressure metamorphic rocks of Taiwan: REE geochemistry, Rb-Sr ages and tectonic implications, *Mem. Geol. Soc. China.* (4), 497-520. (1981).

34. B.M. Jahn, F. Martineau, and J. Cornichet. Chronological significance of Sr-isotopic compositions in the crystalline limestones of the Central Ranges, Taiwan, *Mem. Geol. Soc. China.* (6), 295-301. (1984).

35. B.M. Jahn, F. Martineau, J.J. Peucat, and J. Cornichet. Geochronology of the Tananao Schist complex and crustal evolution of Taiwan, *Mem. Geol. Soc. China.* (7), 383-404. (1986).

36. B.M. Jahn, X.H. Zhou, and J.L. Li. Formation and tectonic evolution of southeastern China and Taiwan: Isotopic and geochemical constraints, *Tectonophysics.* **183**, 145-160. (1990).

37. Jiangxi Bureau of Geology and Mineral Resources, *Regional Geology of Jiangxi Province.* Geology Publishing House, Beijing. (1984). In Chinese

38. F.Q. Jin. Carboniferous paleogeography and paleoenvironment between the North and South China blocks in eastern China, *Jour. Southeast Asia Earth Sciences.* **3**(1-4), 219-222. (1989).

39. W.S. Juang and H. Bellon. Potassium-argon ages of the Tananao Schist in Taiwan, *Mem. Geol. Soc. China.* (7), 405-416. (1986).

40. K. Kizaki, Pre-Cretaceous in the Ryukyus, In: *Pre-Cretaceous Terranes of Japan,* K. Ichikawa (Ed.), 217-224, Nippon Insatsu Shuppon Co., Osaka. (1990).

41. K. Konishi. Geologic note on Tanaki-jima and width of Motobu Belt, Ryukyu Islands, *Sci. Pub. Kanazawa University.* (9), 169-188. (1964).

42. R.L. Larson and W.C.I. Pitman, *The Bedrock Geology of the World.* W.H. Freeman and Company, Inc., New York. (1985).

43. C.Y. Li, Q. Wang, X.Y. Liu, and Y.Q. Tang, *Explanation of tectonic map of Asia.* Cartographic Publishing House: Beijing, China. (1982) In Chinese

44. S. Li and D. Liu. Isotopic chronological evidence for Indosinian orogeny in Dabie Mountains, *Geotectonica et Metallogenia.* **14**, 159-163. (1990). In Chinese

45. X.H. Li, Z.H. Zhao, X.T. Gui, and J.S. Yu. Sm-Nd isotopic and zircon U-Pb constraints on the age of formation of the Pre-Cambrian crust in southeastern China, *Geochemica.* (3), 255-264. (1991). In Chinese

46. J.L. Lin. The apparent polar wander path for the South China block and its geological significance, *Scientia Geologica Sinica.* **4**, 306-315. (1987). In Chinese

47. J.L. Lin, M. Fuller, and W.Y. Zhang. Preliminary Phanerozoic polar wander paths for North and South China blocks, *Nature.* **313**, 444-449. (1985).

48. J.G. Liou and W.G. Ernst. Summary of Phanerozoic metamorphism in Taiwan, *Mem. geol. Soc. China.* (6), 133-152. (1984).

49. G.D. Liu. Geology and petroleum exploration in the East China Sea, *Acta Geophysica Sinica.* **31**(2), 184-197. (1990). In Chinese

50. X.H. Liu and H. Hao. Structure and tectonic evolution of the Tongbai-Dabie Range in the east Qinling collisional belt, China, *Tectonics.* **8**, 637-645. (1989).

51. C.H. Lo and C. Wang Lee. Marble enclave in the Chipan gneiss, Hualien, eastern Taiwan, *Proc. Geol. Soc. China.* **24**, 40-55. (1981).

52. C.H. Lo and C. Wang Lee. Mineral chemistry in some gneiss bodies in the Hoping-Chipan area, Hualien, eastern Taiwan, *Proc. Geol. Soc. China.* (22), 137-140. (1981).

53. C.Y. Lu. The origin of the lithic blocks in the Tienhsiang Formation between Loshao and Tzemuchiao, eastern Taiwan, *Proc. Geol. Soc. China.* (29), 87-97. (1986).

54. W.P. Ma. The Carboniferous at the northern foot of the Dabie Mountains and its tectonic implication, *Acta Geologica Sinica.* **65**(1), 17-26. (1991). In Chinese

55. X.Y. Ma, (Ed.). *Lithospheric Dynamics Atlas of China.* China Cartographic Publishing House. Beijing, China. (1989). In Chinese

56. S. Maruyama and J.G. Liou. Tectonic evolution of ultrahigh- and high-P/T metamorphic complex from central China, *Geology.* , (in press). (1993).

57. M. Mattauer, P. Matte, J. Malavieille, P. Tapponnier, H. Maluski, Z. Xu, L.Y. Lun, and Y.Q. Tang. Tectonics of the Qinling Belt: build-up and evolution of the eastern Asia, *Nature.* **317**, 495-500. (1985).

58. M.W. McElhinny, B.J.J. Embleton, X.H. Ma, and Z.K. Zhang. Fragmentation of Asia in the Permian, *Nature.* **293**, 212-216. (1981).

59. Y. Nishimura, Y. Itaya, Y. Isozaki, and A. Kameya. Depositional age and metamorphic history of 220 Ma high P/T type metamorphic rocks: an example of the Nishiki-cho area, Yamaguchi Prefecture, Southwest Japan, In: *High-pressure Metamorphic Belts and Tectonics of the Inner Zone of Southwestern Japan,* Y. Nishimura (Ed.), Vol. pp. 143-166, Geologic Society of Japan, Tokyo, Japan. (1989).

60. Y. Nishimura, Y. Matsubara, and E. Nakamura. Zonation and K-Ar ages of the Yaeyama metamorphic rocks, Ryukyu Islands, *Mem. Geol. Soc. Japan.* (33), 27-37. (1983).

61. A.I. Okey and A.M.C. Sengör. Evidence for intracontinental thrust-related exhumation of the ultra-high-pressure rocks in China, *Geology.* **20**, 411-414. (1992).

62. F. Osozawa. Geology of Amami-Oshima, Central Ryukyu Islands, with special reference to effect of gravity transportation on geologic structure, *Science Report, 2nd Series, Tohoku University.* (52), 165-189. (1984).

63. J.S. Ren, T.Y. Chen, B.G. Niu, Z.G. Liu, and F.R. Liu, *Tectonic Evolution of the Continental Lithosphere and Metallogeny in Eastern China and Adjacent areas.* Science Press, Beijing, China. (1990). In Chinese

64. A.M.C. Sengör, *The Cimmeride orogenic system and tectonics of Eurasia.* Geol. Soc. Am. Spec. Paper, Vol. 195, (1984).

65. Shaanxi Bureau of Geology and Mineral Resources, *Regional Geology of Shaanxi Province.* Geology Publishing House, Beijing. (1989). In Chinese

66. K. Shibata, K. Konoshi, and T. Nozawa. K-Ar age of muscovite from crystalline schist of the northern Ishigake-jima, Ryukyus, *Bull. Geol. Surv. Japan.* **19**(529-533), (1968).

67. K. Shibata, R.K. Wanless, H. Kano, H. Yoshida, T. Nozawa, S. Igi, and K. Konishi. Rb-Sr ages of the basement rocks of the Japanese Islands, *Bull. Geol. Surv. Japan.* **23**, 505-510. (1972).

68. L.S. Shu and G.Q. Zhou, The first discovery of the high-pressure minerals in the collage zone of Proterozoic terranes in north Jiangxi and its tectonic significance, In: *Journal of Nanjing University, Natural Sciece Section*, Vol. 24, 421-429, Nanjing University, Nanjing, China. (1988). In Chinese

69. T. Shui. Tectonic framework of the continental basement of southeast China, *Scientia Sinica, Ser. B.* **XXI**(7), 885-896. (1988).

70. T. Shui, B.T. Xu, R.H. Liang, and Y.S. Qiu. Shaoxing-Jiangshan deep-seated fault zone, Zhejiang Province, *Kexue Tongbao.* **31**(18), 1250-1255. (1986). In Chinese

71. J. Suppe. Mechanics of mountain-building and metamorphism in Taiwan, *Mem. Geol. Soc. China.* (4), 67-90. (1981).

72. A. Taira, Y. Saito, and M. Hashimoto. The role of oblique subduction and strike-slip tectonics in the evolution of Japan, *Am. Geophys. Union Geodyn. Ser. 11.*, 303-316. (1983).

73. H. Ujiie and Y. Hashimoto. Geology and radiolarian fossils in the inner-zone of "Motobu Belt" of Okinawa Islands, *The Earth Monthly.* **5**, 706-712. (1983).

74. H.Z. Wang, (Ed.). *Atlas of the Palaeogeography of China.* Cartographic Publishing House. Beijing, China. (1985). In Chinese

75. C. Wang Lee, C. Chen, Y. Wang, T.F. Yui, C.Y. Lu, and C.H. Lo. Relic of ancient oceanic crust in the Changchun formation of eastern Taiwan, *Proc. Geol. Soc. China.* **28**, 10-22. (1985).

76. W.C. Wang, S. Sun, J.L. Li, D. Zhou, K.J. Hsü, and G.W. Zhang. The tectonic evolution of the Qinling Mountain belt, *Sci. Geol. Sinica.* (2), 129-142. (1989). In Chinese

77. X.M. Wang, Y.J. Jing, J.G. Liou, W. Pan, W. Liang, M. Xia, and S. Maruyama. Field occurrences and petrology of eclogites from the Dabie Mountains, Anhui, central China, *Lithos.* **25**, 119-131. (1990).

78. X.M. Wang and J.G. Liou. Regional ultrahigh-pressure coesite-bearing eclogite terrane in central China: Evidence from country rocks, gneiss, marble and metapelite, *Geology.* **19**, 933-936. (1991).

79. X.M. Wang, J.G. Liou, and H.K. Mao. Coesite-bearing eclogite from the Dabie Mountains in central China, *Geology.* **17**, 1085-1088. (1989).

80. X.M. Wang, J.G. Liou, and S. Maruyama. Coesite-bearing eclogites from the Dabie Mountains, central China: petrogenesis, P-T paths, and implications for regional tectonics, *Jour. Geology.* **100**, 231-250. (1992).

81. Z. Wang, On division of the metamorphosed strata in the Jiaonan Uplift and its tectonic evolution, In: *Proceeding of International Symposium on Precambrian Crustal Evolution (tectonics)*, Geological Society of China, 208-217, Geological Publishing House, Beijing, China. (1986). In Chinese

82. Z.D. Wang and Z.G. Shan. Geological characteristics of the ancient Qinling rift zone and its significance in tectonics, *Reg. Geol. China.* (15), 29-39. (1985). In Chinese

83. Z.D. Wang, R. Tang, K.Y. Yi, and F.X. Wang. The discovery of Early Cambrian fossils in the "Taowan Group" in Shaanxi and their geological significance, *Reg. Geol. China.* (2), 113-122. (1989). In Chinese

84. P.M. Watson, A.B. Hayward, D.N. Parkinson, and Z.M. Zhang. Plate tectonic history, basin development and petroleum source rock deposition onshore China, *Marine and Petroleum Geology.* **4**, 205-225. (1987).

85. S.J. Weng, Q.S. Kong, and H. Huang, *Late Mesozoic volcanism of Zhejiang, Fujian, Jiangxi and Guangdong.* Geol. Publishing House, Beijing, China. (1987). In Chinese

86. M. Wilson, *Igneous petrologenesis - A global tectonic approach.* Unwin Hyman, London. (1989).

87. F.T. Wu. Recent tectonics of Taiwan, *Jour. Phys. Earth.* **26**, 5252-5299. (1978).

88. Xian Institute of Geology and Mineral Resources, *Atlas of Paleontology of Northwest China.* Geol.ogy Publishing House. Beijing, China. (1983). In Chinese

89. S.T. Xu, W. Su, Y.C. Liu, and L.L. Jiang. Diamondiferous high-pressure metamorphic rocks in Dabie Shan, Anhui: mineral paragenesis and metamorphic conditions, *Geology of Anhui.* 1(1), 3-18. (1991). In Chinese

90. Z.Q. Xu, Y.L. Lu, Y.Q. Tang, and Z.T. Zhang, *Formation of the composite eastern Qinling of China.* Environmental Sciences Press, Beijing, China. (1988). In Chinese

91. Z.J. Yang. New progress in the study of the Jiaodong block, *Reg. Geol. China.* (1), 43-50. (1992). In Chinese

92. T.P. Yen. On the occurrence of the late Paleozoic fossils in the metamorphic complex of Taiwan, *Bull. Geol. Surv. Taiwan.* **4**, 23-26. (1953).

93. T.P. Yen. The metamorphic belts within the Tananao schist terrane of Taiwan, *Proc. Geol. Soc. China.* (6), 72-74. (1963).

94. T.P. Yen and S. Rosenblum. Potassium-argon ages of micas from the Tananao Schist terrane of Taiwan, a preliminary report, *Proc. Geol. Soc. China.* (7), 80-81. (1964).

95. A. Yin and S.Y. Nie. An indentation model for the North and South China collision and the development of the Tan-Lu and Honam fault systems, eastern Asia, *Tectonics.* , (in press). (1993).

96. T.F. Yui. Stable isotope studies on serpentinites and nephrite deposits in the Fengtien area, Hualien, Taiwan. National Taiwan University: 1984) 163pp

97. T.F. Yui, C.H. Lo, C.Y. Lan, and C. Wang Lee. Nephrite/Rodingite in metamorphosed ophiolite, Nanao, northwest Taiwan, *Proc. Geol. Soc. China.* **30**, 22-29. (1988).

98. D.Q. Zhang and G.Y. Sun, *Granites in eastern China.* China Geologic College Press, Beijing, China. (1988). In Chinese

99. R.Y. Zhang, B. Cong, T. Hirajima, and S. Banno. Coesite eclogite in Su-Lu region, eastern China, *EOS, Transactions Am. Geophy. Union.* **71**, 1708. (1990).

100. Z.M. Zhang, Tectonostratigraphic terranes of Japan that bear on the tectonics of mainland Asia, In: *Tectonostratigraphic Terranes of the Circum-Pacific Region*, D.G. Howell (Ed.), 409-420, Circum-Pacific Council for Energy and Mineral Resources, Houston, Texas, U.S.A. (1985).

101. Z.M. Zhang, J.G. Liou, and R.G. Coleman. An outline of plate tectonics of China, *Geol. Soc. Am. Bull.* **95**(3), 295-312. (1984).

102. Z.M. Zhang, J.G. Liou, and R.G. Coleman, The Mesozoic and Cenozoic tectonism in eastern China. In: *The Evolution of the Pacific Ocean Margins,* Z. Ben-Avraham (Ed). 124-142, Oxford University Press and Clarendon Press, New York and Oxford. (1989).

Comparison Of Arc-Trench Systems In The Early Paleozoic Gorny Altai And The Mesozoic-Cenozoic Of Japan

TERUO WATANABE[1], MIKAIL M. BUSLOV[2] AND SHIGEKAZU KOITABASHI[3]

[1] Dept. Geol. & Mineral., Hokkaido University, Sapporo, Japan,
[2] United Institute of Geology, Geophysics and Mineralogy, Novosibirsk, Russia
[3] Sapporo Branch of Meiji Consultant Co., Ltd, Chuo-ku, Sapporo, Japan

ABSTRACT -Study of the tectono-stratigraphic units composing the Early Paleozoic Gorny Altai arc-trench system reveals a multi-stage history; primitive volcanic arc with boninite magmatism or tholeiitic magamtism and a more evolved volcanic arc characterized by calc-alkaline magmatism. The composition of the tectono-stratigraphic units is comparable with that of the Izu-Bonin linear arc-trench system. However, the tectono-stratigraphic units in the Gorny Altai presently are distributed into fragmented blocks. An application of a model of subduction zone infancy to the Izu-Bonin and Jurassic of California is also extended to the Early Paleozoic Gorny Altai arc.

In a modern, active continental margin arc like the Japanese islands, both linear and fragmented block patterns for the distribution of the tectono-stratigraphic units comprising the arc-trench system are recognized. Study of the structure of the Japan arc yields insights as to whether the motion of micro-plates is an essential factor for determining if an original linear structure is preserved for a long time. The examples of Japanese islands suggest that the Early Paleozoic Gorny Altai arc-trench system suffered complicated fragmentation processes from Paleozoic-Early Mesozoic, large-scale displacement related to complex collision events among Precambrian blocks.

KEY WORDS - Gorny Altai arc, trench, arc trench system, Izu-Bonin arc, calc-alkaline mgmatism, boninite

INTRODUCTION

The comparison of fold belts of different ages, which have been preserved at various stages of development related to global plate motion, allows the construction of a model of fold belt development from the initial stage of an active continental margin through the linear collision-subduction structure to a post-collisional mosaic block structure, Such an approach allows us to solve the reverse problem concerning the decoding of paleo-ocean basin evolution, for example, Paleozoic Asian-ocean. In this paper we present our preliminary results of the comparison between the arc-trench systems of the Early Paleozoic Gorny Altai(Russia) and the Mesozoic- Cenozoic in Japan.

According to the compilation of [6], the ophiolitic rocks in the Gorny Altai are distributed in the Caledonides orogen near a boundary with the Hercynides. Although they did not refer to the occurrence of boninite, recently [1] and [2] reported the occurrence of boninite in the Gorny Altai. The subduction and collisional processes forming the boninitic and calc-alkaline volcanic rocks in Gorny Altai are reconstructed by the analysis of the tectono-straigraphic units comprising the arc-trench system, such as the fore-arc basin(trough), accretionary wedge, volcanic arc, and other parts [1,2]. In the Early Paleozoic Gorny Altai, the tectono-stratigraphic units indicating subduction-collisional processes are exhibited by fragments and are divided by large-scale Late-Paleozoic - Mesozoic strike-slip faults of hundreds of kilometers in length. The results of subduction-collisional processes which have formed a classic linear structure for both supercrustal and deep seated rocks are observed in the Mesozoic-Cenozoic fold belts in southwest Japan. An advanced stage of formation into a block-structure is also observed in northeast Japan. Magnesian andesite also occurs not only in a primitive stage of arc-volcanism, but also in the advanced stage of arc-volcanism that formed block structure in Japan.

The processes of fragmentation of the tectono-stratigraphic units in Gorny Altai are discussed below with comparison of the Mesozoic-Cenozoic arc-trench system in Japan. This comparison of the structure of arc-trench systems identifies an important role of motion of micro-plates.

REVIEW OF THE TECTONO-STRATIGRAPHIC UNITS OF AN ACTIVE CONTINENTAL MARGIN OF THE JAPANESE TYPE

The modern arc-trench system of the western Pacific region is a narrow linear structure 250-300 km wide and more than one thousand kilometers in length. Structures include trench, accretionary wedge, fore-arc trough, volcanic arc and back arc trough (or basin). The accretionary wedge has an imbricated structure resulting from being thrust towards the trench [11]. Numerous blocks (oceanic islands, guyots, and others.) moving into the subduction zone are incorporated in the accretionary wedge and produce swells in overriding forearc wedge [41]. An outer non-volcanic arc such as Mentawai Islands, Indonesia is a typical example [21].

If the subducted block (terrain) is large enough, its collision with the arc results in the formation of reverse flow in the accretionary wedge with the uplift of high-pressure rocks (blueschists, eclogites, and other rock types.) to the surface. One such example may be the collision of seamounts in the Kamuikotan belt [39] in Hokkaido,Japan. The collision of the seamounts with the Yezo volcanic arc may have caused the uplift of the Kamuikotan metamorphic rocks.

Since study by [3] reverse flow in an accretionary wedge using two dimensional model has been discussed by researchers [4]. Otsuki [25] .examined uplift and shear deformation of a high pressure metamorphic belt using a quasi-3-dimensional lubricant model and pointed out that the degree of opening of the prism outlet to the wedge mantle and the obliquity of plate subduction against the trench axis are the two key parameters for the uplift dynamics of the high pressure metamorphic belt.

Due to the terrain collision with a volcanic arc the subduction zone may move outwards towards the oceanic basin. Terrains are incorporated into accretionary wedges as trapped fragments of oceanic crust such as the Cretaceous Sorachi greenstones in Hokkaido[13]. The fore-arc troughs are usually filled with turbidites which are composed of the detritus from the adjacent island arcs.

The time and spatial regularity in magmatism displayed in volcanic arcs is characteristic of this tectonic setting. The early stages of arc development are usually characterized by boninite magmatism (Bonin island, Izu-Mariana arc, Tonga arc, and others [26]), the later stages by calc-alkaline and alkaline magmatism. In volcanic arcs, zonation of volcanic types perpendicular to the length of the arc from tholeiites in the frontal regions through calc-alkaline volcanics in the central parts to alkalic or shoshonites in the back regions is well know [17]. Volcanic arcs are distinguished by granitoid bodies which are co-magmatic with the volcanics and are exposed only after significant erosion. They can be easily studied in the eroded ancient arc structures of Japan and Gorny Altai. The granitoids and volcanogenic rocks are typically distributed from gabbro-plagiogranite tholeiites on the fore arc side, through the calc-alkaline granite-granodiorite batholiths of the central part, to diorite-monzonites of the latite group in the back arc.

EARLY PALEOZOIC ARC-TRENCH SYSTEM OF GORNY ALTAI

Distribution of the main pre-Devonian structural units of the arc-trench system of Gorny Altai, which formed on the boundary of Paleo-Asian ocean and Siberian continent in Early Paleozoic, is shown in Fig. 1, .and the 520 - 640 ma high-pressure rocks occur in the surrounding area [5].

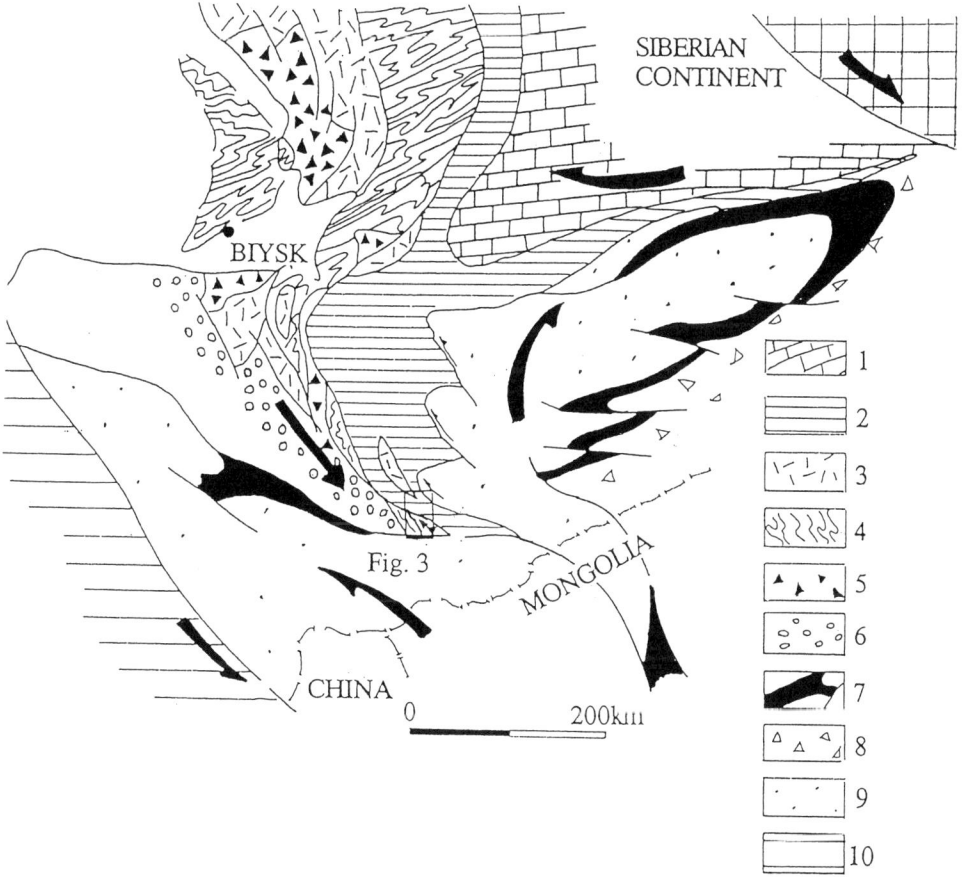

Figure.1 Outline of tectonic setting for the Early Paleozoic complexes of arc-trench system of Gorny Altai.1- Tomsk micro-continent;active continental margin of Siberian continent. 2-Primitive island arc (Uymen-Lebed). 3- Developed island arc (Ust-Syoma). 4- Guyots (Barnaul, Biya-Katun and Baratal). 5- Chaotic subduction compelxes. 6- Fore-arc and back-arc basins and complexes of active margin of Chinese continent. 7- Ophiolite sutures(Uymon-Dzhebash-Kurtushuba). 8- Accretionary prism (Khemchik Sistigkhem). 9- Fore-arc trough (south-Altai-Sayan). 10- Hercynian complexes of Ob-Zaisan ocean

Archaeocyathus indicating Upper Middle Cambrian is found in carbonate sequences(Berzin, pers. comm.), which are considered to have deposited on rocks forming a primitive volcanic arc. A recent K-Ar study of the Kuray metamorphic rocks in this area indicates an uplift episode during Devonian (Buslov et al., in prep.). Future detailed geochronological study should reveal a complex deformation sequences in the study area.

The following geodynamic complexes are distinguished from northeast to southwest, although uncertainity of the ages still presently exists:
1. Tomsk micro-continent. Late Riphean carbonate sequences.
2. Uimen-Lebedsk primitive island arc. Vendian-Early
   Cambrian ophiolites of tholeiite-boninite composition.
3. Accretionary wedge. Early Cambrian olistostromes/melange
   with rocks comparable in composition of oceanic crust and
   guyots. The large blocks of Vendian-Early Cambrian guyots
   (Barnaul, Biya-Katun, Kadrin, Baratal) are incorporated into
   olistostrome/melange. The garnet amphibolites,
   blueschists and eclogites, metaperidotites and serpentinite
   melanges with blocks of high pressure rocks (Chagan-Uzun complex)
   are observed.
4. Ustr-Seminsk mature island arc. Early-Middle Cambrian
   tholeiite and calc-alkaline volcanics, tonalite-gabbro-diorite
   plutons. Volcanos and plutons formed both on the Uimen-
   Lebed primitive island arc and on the accretionary wedge.
5. Fore-arc and inter-arc troughs. Middle-Late Cambrian flysch
   and molasse of the Anui-Chuya zone.

A possible model for the formation of the Early Paleozoic arc-trench systems of Gorny Altai based on Buslov [2] is illustrated in Fig 2.
Five stages are distinguished.

Vendian : Paleo-asian Ocean has reached its maximum size, a subduction zone has been formed near the Siberian continent. Ophiolites of the Uimen-Lebed-Kuznetzk primitive island arc formed on crust of oceanic type. At the same time volcanic islands of Hawaiian type are formed in the central parts of Paleo-asian Ocean.

Vendian-Early Cambrian : in the subduction zone, high-pressure rocks are formed, central-type volcanos develop in the primitive island arc, non-volcanic island arc grows incorporating debris of the upper layer of oceanic crust. Volcanic islands (Barnaul, Biya-Katun, Baratak, and others) approach the subduction zone and are overlain by a thick cover of siliceous-carbonate sediments.

Middle - Early Cambrian : in Gorny Altai, the collision of the Biya-Katun, Kadrinsk and Baratal guyots with the island arc occurs and results in the closing of the subduction zone and theappearance of reverse flows in the accretionary wedge. After this collision, the uplift of serpentinite melange and high-pressure rocks towards the arc-slope surface occurs. Guyots are incorporated into the accretionary wedge, a new subduction zone is formed more outboard.

End of Early Cambrian - beginning of Middle Cambrian : subduction processes cause the formation of volcanos and batholiths of thedeveloped island arc, the basement is transitional-type crust, consisting of the accretionary wedge and the Uimen-Lebed primitive island arc.

Middle-Late Cambrian : subduction processes in Gorny Altai are nearing the end, the trench is filled with debris from eroding Early Cambrian olistostromes of the accretionary wedge, guyots, and the primitive and developed island arcs.

Fig. 3 shows the present-day structure of the units of the Early Paleozoic arc-trench system in the Kuray region of the Gorny Altai. They are intensely imbricated, folded and divided by strike-slip faults into several blocks. The Kuray ophiolites of the island arc type and the Chagan-Uzun ophiolites of the oceanic type [4] are distinguished.

The Kuray ophiolites of the Uimen-Lebed primitive arc consist of layered pyroxene-gabbro complex, parallel dykes and sub-intrusive dyke-sill complex (sheeted dyke complex), and a

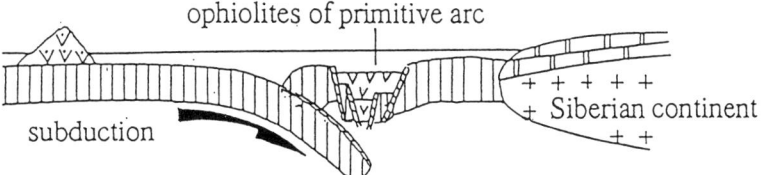

I stage (Vendian) - subduction. Formation of ophiolites with boninite - tholeiite magmatism.

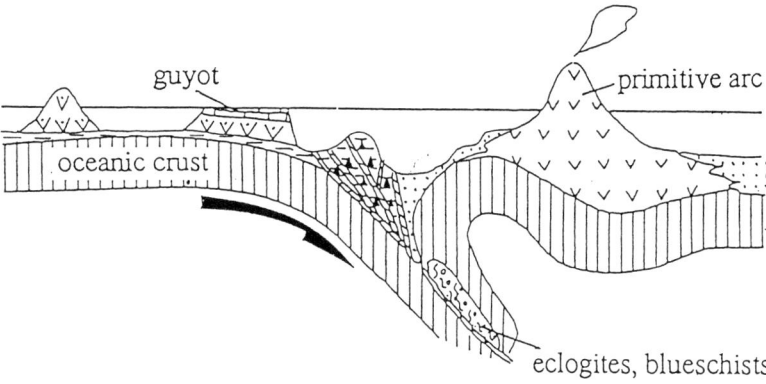

II stage (Vendian-Early Cambrian) - subduction. Formation of accretionary prism and subduction high pressure rocks.

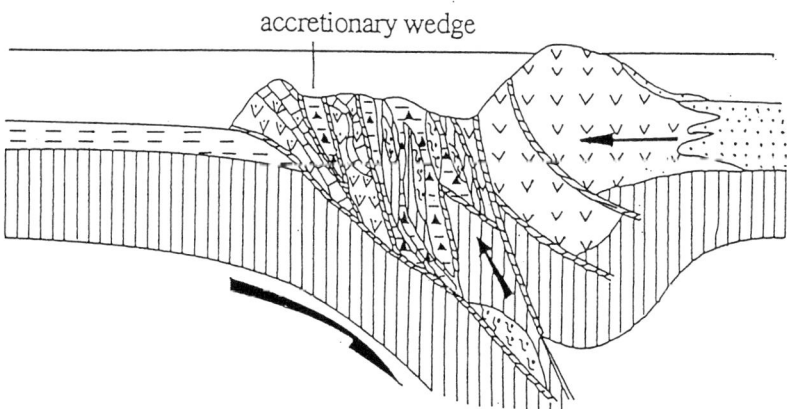

III stage (Early Cambrian) - collision of guyots with island arc. Closing of subduction zone and reverse flows in accretionary wedge.

Figure.2a Model for Early Paleozoic arc-trench systems of Gorny Altai (based on [2]

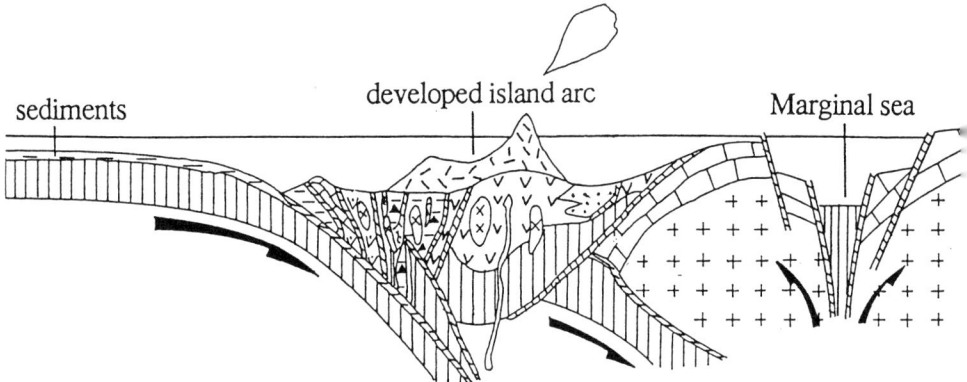

IV stage (Early-Middle Cambrian) - formation of a new subduction zone, accretionary prism and developed island arc with calc-alkaline magmatism.

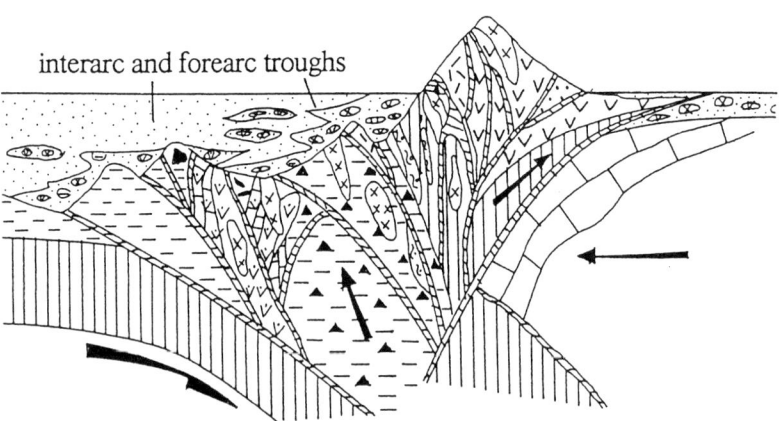

V stage (Middle-Upper Cambrian) - collision of island arc with continent. Formation of molasses and flysch, forearc and interarc troughs.

Figure 2 b Model for Early Paleozoic arc-trench systems of Gorny Altai (based on [2]

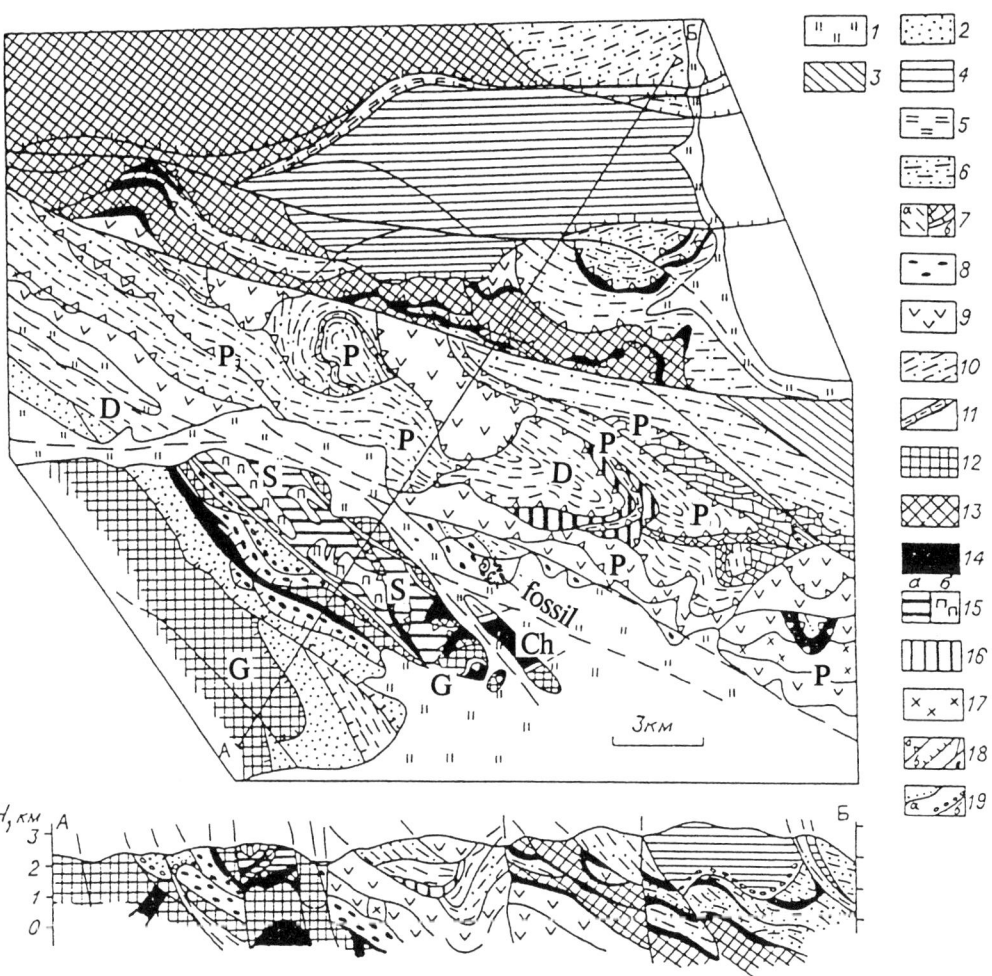

Figure.3. Geological map and cross section showing nappe-imbricated structure in the central part of the Kuray region (details: see Fig.18 in Buslov, 1992). 1- Tertiary-Quaternary deposits. 2 to 4 - Devonian rocks, 5- Lower Silurian rocks. 6- Lower-Middle Cambrian flysch-like psammitic schists. 7-8- Lower-Middle Cambrian rocks (7- Turbidites, 8- carbonates). 9 to 12- Vendian-Lower Cambrian rocks (9-Greenstones, 10- Turbidites, 11- Carbonates). 12- Volcanic-sedimentary rocks. 13- Kuray complex (gneisses,amphibolites, & granites). 14 to 16- Serpentinite melanges 17-Vendian-Lower Cambrian gabbro-diorites, 18- Fault contacts 19- Stratigraphic contacts (a. conformity, b.unconformity). P- primitive arc, D- developed arc, G- seamounts (guyots) S- subduction complexes, Ch- Chagan-Uzun massif.

Table 1
Representative chemical compositions of bonitic and volcanic rocks in the Early Paleozoic arcs of Gorny Altai.

	1 C-106a	2 C-106b	3 C-107b	4 12814	5 12815	6 12815	7 575
SiO_2	54.70	54.55	54.66	49.25	49.70	50.85	69.35
TiO_2	0.23	0.24	0.34	0.58	0.61	0.33	0.34
Al_2O_3	11.28	11.19	9.41	17.12	18.36	13.62	10.39
Fe_2O_3	10.57	11.02	10.85	5.62	3.93	0.31	4.95
FeO	-	-	-	6.87	7.52	9.28	2.53
MnO	0.21	0.22	0.25	0.19	0.19	0.21	0.05
MgO	10.22	10.21	10.23	5.71	5.22	9.34	3.98
CaO	7.89	7.37	9.78	7.83	8.29	11.60	1.01
Na_2O	1.13	1.90	2.15	2.64	2.46	2.40	4.94
K_2O	0.54	0.24	0.32	0.36	0.34	0.39	0.12
P_2O_5	0.03	0.03	0.03	0.06	0.06	0.05	0.07
H_2O/CO_2	3.04	3.30	1.92	4.24	3.05	1.55	1.79
Total	99.84	100.27	99.94	100.46	99.73	99.93	99.52

1-3 Boninite, 4-5 Tholeiite series from the Kuray Ophiolites, 6-7 Calc-alkaline series from the Kuray Ophiolites.

Table 2
Chemical compositions of magnesian andesites in the Kabato and Etaidake-Hamamasu area. Data from [22,23]

	1 Kabato-9	2 Hamamasu-12	3 Etaidake-18	4 Etaidake-19	5 Etaidake-20
SiO_2	52.52	52.75	53.65	53.66	54.46
TiO_2	0.61	1.19	0.94	0.68	0.90
Al_2O_3	12.46	15.79	15.76	16.78	16.13
Fe_2O_3	9.90	3.99	3.36	2.79	2.41
FeO	-	4.23	5.30	4.86	4.92
MnO	0.24	0.21	0.10	0.16	0.15
MgO	9.12	6.04	7.30	6.96	7.05
CaO	7.85	8.64	8.06	8.18	8.37
Na_2O	4.20	3.23	2.99	2.95	3.09
K_2O	0.71	1.42	1.49	1.48	1.47
P_2O_5	0.19	0.36	0.19	0.19	0.21
$H_2O(+/-)$	3.32	1.74	0.85	1.16	0.75
Total	101.12	99.59	99.93	99.85	99.91

1 Hornblende dolerite dyke[22]. 2-5 Basaltic andesite[23]

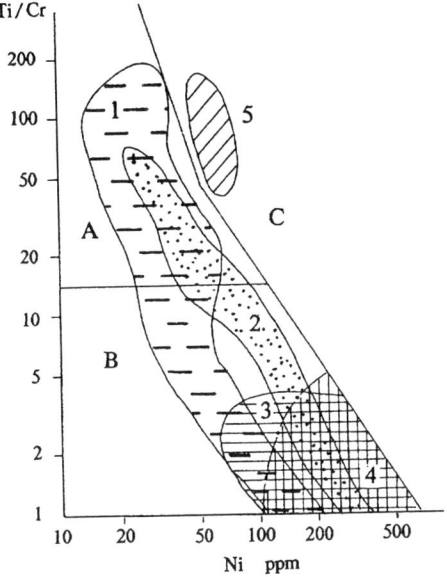

Figure.4 Diagram of Ti/Cr vs Ni, based on Dobretsov et al.(unpublished paper). A-Moderate Ti tholeiite of island arcs. B-Low-Ti boninites of island arcs. C- High-Ti tholeiite of ocean and marginal sea. 1- Early Paleozoic island-arc ophiolites of Gorny Altai. 2- Early Paleozoic island-arc ophiolites of east Sayan. 3- Early Paleozoic boninite of Mongolia. 4-boninite of Bonin islands and Mariana arc. 5-garnet amphibolite and eclogite in the Chagan-Uzun massif.

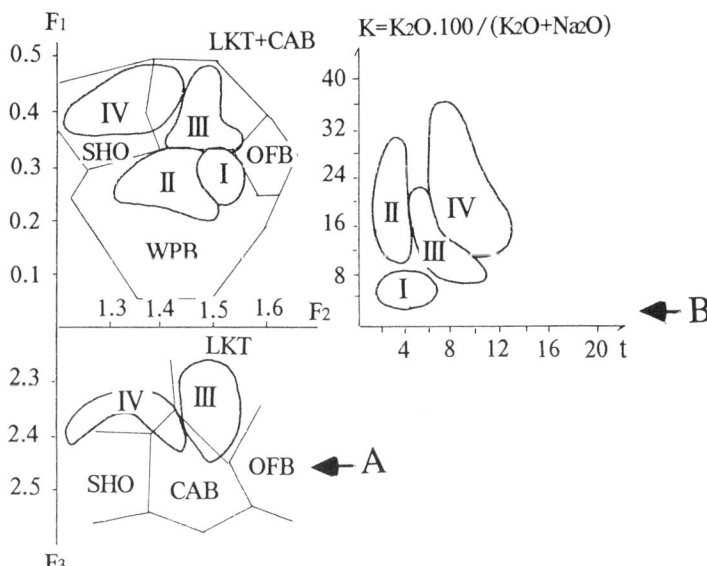

Figure.5 Diargam of chemical composition of volcanic rocks from Early Cambrian complexes of Gorny Altai [16,7]. I- Garnet-amphibolites and blueschists from Cgagan-Uzun massif and Uimon suite. II- Guyot basalts (Biya-Katun,Baranaul). III- Basalts of the primitive Uimon-Lebed arc. IV- volcanic rocks of the Ust-Syoma developed arc. A- Diagram by Pearce, Jour. Pet. 17:14-19 (1976). B- Diagram by Piskunov in Geol. and petrol. pecularities of island arc volc. Nauk Moscow (1987)

Vendian-Lower Cambrian effusive - volcaniclastic sequence. Early Cambrian black shale turbidites overly the Kuray ophiolites. These dykes and effusives are characterized by tholeiite-boninite magmatism (Table 1).

The Chagan-Uzun ophiolites consist of peridotites, garnet amphibolites, eclogites and sheets of serpentinized melanges composing the structure of accretionary wedge. The wedge consists of olistostromes derived from Baratal guyot.

Island-arc and oceanic ophiolites are distinguished on a diagram of Ti/Cr vs Ni (Fig. 4). The composition of the dykes and the effusives from the island-arc ophiolites plot in the field of low-Ti boninites and moderate-Ti tholeiites for island arcs of the Bonin type. They also plot close to Vendian-Cambrian ophiolites of the East Sayan and Mongolia [29]. While the Chagan-Uzun eclogites and the garnet amphibolites lie within the field of basalts from the high-Ti series of mid-oceanic ridges.

Fig.5 shows the composition of the effusives of the Baratal and Biya-Katun guyots. Over 380 XRF analyses indicate their similarity with the effusives of modern oceanic islands. Over 300 XRF analyses from the Lower-Middle Cambrian effusives of the developed island arc of Gorny Altai belong to the tholeiite and calc-alkaline series of modern island arcs in the eastern part of Asia.

The position of the Early Paleozoic arc-trench system of Gorny Altai, Salair and Kuznetzky Alatau within the structure of central Asia is shown in Fig.1. It is bounded by strike-slip faults into a 100 km long and 300 km wide area within the outward units of the active continental margin of the Chinese continent (south) and the Hercynian complexes of the Ob-Zaisan ocean (west). Preliminary paleomagnetic data (Buslov, in prep) confirm that ophiolites of the primitive island arc of Gorny Altai were formed close to the Siberian continent. Analogues of the Vendian-Cambrian ophiolites of the primitive island arc of Gorny Altai are known in the East Sayan and Mongolia [1]. They should be the fragments of a previously intact arc-trench, linear structure. The occurrence of fragments in Early Paleozoic arc-trench system relates to the rotation of the Siberian continent during the Mesozoic 180° counterclockwise, resulted in block displacements around it [43].

MESOZOIC - CENOZOIC FOLD BELTS OF JAPAN

Japan consists of five arc-trench systems of the Kurile, northeast Japan, southwest Japan, Izu-Bonin, and Ryukyu arc. The structural framework for the tectono-stratigraphic units (or terranes) in Japan is shown in Fig.6. The present-day tectono-morphological setting of Izu-Bonin and northeast Japan form a continuous active volcanic arc system; southwest Japan and Ryukyu arcs form another system [38]. We have chosen the following three arc systems because they have different structures in their tectono-stratigraphic units; Izu-Bonin arc, southwest Japan and northeast Japan.

The structural framework of the Izu-Bonin(Ogasawara) arc, the type locality of boninite, is shown in Fig.7. Recent ODP scientific results [37] reveal that middle and late Eocene boninites, island-arc tholeiite, their differentiates and their plutonic roots formed an enormous province of supra-subduction zone magmatism, similar to the widespread occurrence of many ophiolites.

Boninite occurs in the Ogasawara ridge, the older and frontal arc region [9], in the Izu-Ogasawara arc. Intense tholeiitic and calc-alkaline volcanism post dating the boninite magmatism occurred along an Oligocene arc (the present frontal arc) up to 27ma [36]. In the Quaternary, a more mature volcanic arc formed in the back-arc area.

Along-arc(from north ot south), variations in rock chemistry, from low alkali tholeiite to alkaline and variation in the Sr isotopic ratios are recognized [42]. Serpentinite seamounts (diapirs) are also reported [36]. The overall tectonic framework in the Izu-Bonin (Ogasawara) arc shows a multistage volcanic history evolving from primitive boninitic

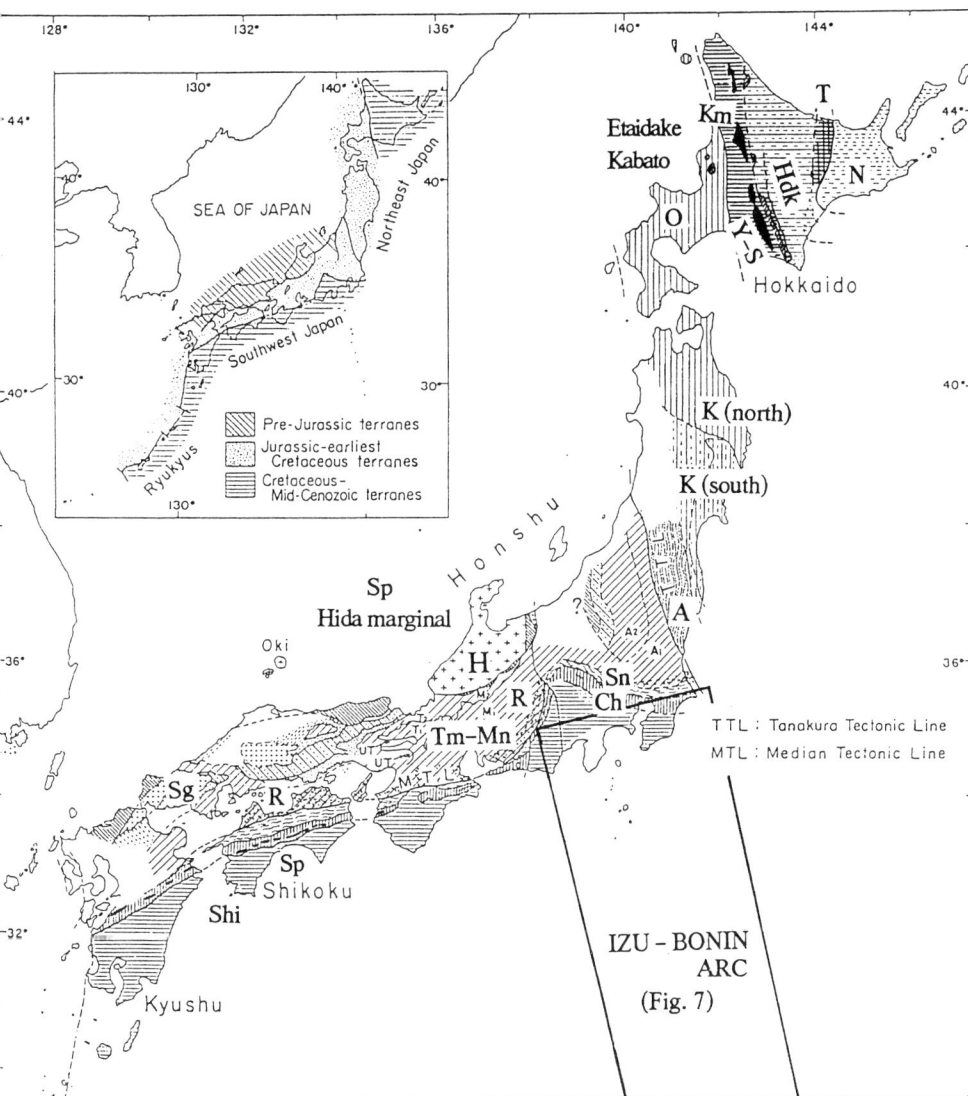

Figure.6 Pre-Tertiary tectonic division of Japanese islands (modified from Ichikawa and others Pre-Cretaceous terranes of Japan in *Pre-Jurassic Evolution of Eastern Asia*, IGCP Project 285, Osaka, 1990). O-Oshima belt, Y-S- Yezo-Sorachi belt, Km- Kamuikotan rocks, Hdk- Hidaka belt, T- Tokoro belt, N- Nemuro belt, K-Kitakami massif, A- Abukuma massif, Tm-Mn- Tamba-Mino(-Ashio) belt (Tamba-Mino (-Ashio) belt is further divided by nappe units). R.- Ryoke belt, H- Hida belt, Sg- Sangun rocks, Sn- Sanbagawa belt, Ch- Chichibu belt, Shi:Shimanto belt, Sp- Serpentinite melange (Hida marginal & Kurosegawa).

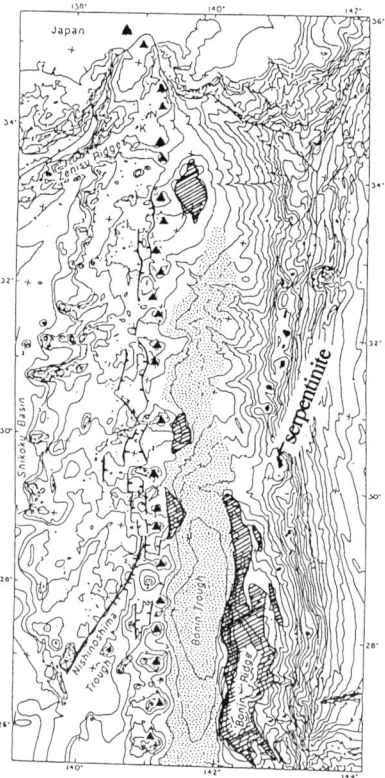

Figure. 7 Tectonic framework in the northern part of the Izu-Bonin arc-trench (modified from [36]. Boninite and related arc-tholeiitic rocks occur in the older frontal arc(shaded area). Ticked heavy lines- normal faults, Filled triangle- frontal arc volcanoes, Stippled area- thickly sedimented forearc basin, Filled bathymetric contours- serpentinite seamounts.

volcanic arc to a calc-alkaline volcanic arc. This is very similar to that of supercrustal rocks of the early Paleozoic Gorny Altai.

Located in southwest Japan are parallel oriented Cretaceous paired metamorphic belts of the Ryoke and the Sanbagawa [20]. The Sanbagawa rocks and the Chichibu rocks on the south are the metamorphosed products of Mesozoic accretionary complexes. The Izumi group, composed of Upper Cretaceous turbidites, was deposited in the fore-arc trough related to buckling and uplift of the Sanbagawa rocks. Continental margin type igneous activity, mainly granitic and dacitic-rhyolitic magmatism, is recognized in the Cretaceous to Paleogene rocks of the inner side of southwest Japan. However, a small amount of Mesozoic magnesian andesite has been described from some areas in the inner side of southwest Japan [8,12]. This occurrence may suggest a genetic relation between granitic magma and magnesian andesite, but this is beyonds the scope of our manuscript. Thus, southwest Japan is one example illustrating that deep seated rocks are distrubuted linearly, separate from older Paleozoic units of the inner region, which have been fragmented into blocks [40].

Since the Neogene, arc-type magmatism has occurred in southwest Japan and is the result of the initiation of subduction of the Philippine sea plate and the opening of Japan Sea. The Paleogene is a comparatively stable time (for 35 my). Chemically diverse magmatism is recognized in this stage. One of the characteristic magma types is middle Miocene magnesian andesite (Sanukite) activity of the Seto Inland Sea area. This magmatism is considered to have a genetic relation with the newly formed and hot Philippine sea plate subduction [35]. Sanukite and similar magnesian andesite associated with calc-alkaline rocks was called Sanukite type magnesian andesite [26]. Thus, since the Cretaceous, southwest Japan records a complicated history of arc magmatism. The reason why southwest Japan retains its original linear structure(see Fig. 6) is that subduction and rifting has occurred normal or parallel to the older arc framework. Therefore, due to compression normal to the southwest Japan arc, it has produced a nappe structure [10] that suggests shorterning normal to the arcs- linear framework.

In northeast Japan, the pre-Neogene units are divided into block structures (Fig.6). Primitive magnesian basalt and Sanukite type mangesian andesite ocur in northeast Japan [28,24]; magmatism probably related to the opening of the Japan Sea. Northeast Japan has suffered strong lateral displacements produced by the opening of the Japan Sea. This situation may explain why only typical oceanic crust develops in the back-arc side of northeast Japan [32] Extrusion tectonics may be significant for northeast Japan [14]. Magnesian andesite is reported [31] from the Izu peninsula, the northen end of the Philippine Sea plate which collided with the Honshuu arc from the south. Exceptional linear structures of arc-trench elements, Oshima and Sorachi-Yezo belts of the Mesozoic arc are preserved in Hokkaido (Fig.6). Magmatism of this Mesozoic arc is mainly tholeiitic and calc-alkalic in the Oshima belt, but we recognize a Cretaceous dyke-rock chemically equivalent to magnesian andesite (Table 2) in the Kabato area. The magnesian andesitic rock occurred together with arc-tholeiite in the frontal volcanic arc side of Sorachi-Yezo-Oshima arc-trench system during the Cretaceous [22]. The Sorachi-Yezo belt consists of Yezo fore-arc sediments in the mainly Upper Cretaceous and Sorchi accretionary complex also containing some trapped oceanic crust and high-pressure metamorphic rocks.

Because the collision of the Kurile arc occurred normal to the Mesozoic arc in Hokkaido, the linear structure was not destroyed. This collision created a nappe structure in the Yubari Creatceous - Tertiary zone and the southern Kamuikotan zone at the Niikappu River area [31,59].

Around the Kabato area of the Cretaceous magnesian andesite dyke, magnesian andesite activity occurred in late Neogene (Table 2). This allows us to speculate that there has been a magma-source zone of peculiar composition beneath the Kabato and adjacent areas since the Cretaceous.

DISCUSSION AND CONCLUSIONS

Since boninite was re-evaluated by Kuroda and Shiraki [18], boninite and related magnesian andesite have been noted as a primary andesitic magma in incipient subduction (the case for the Izu-Bonin arc) and the opening of the back-arc or collision of the arc(the case for NE Japan)[26,27,34]. We describe mainly the mode of distribution of the tectono-stratigraphic units of the arc-trench system in the Gorny Altai, Izu-Bonin, southwest Japan and northeast Japan and in all arc system magnesian andesite is now recognized (boninite: only in the Izu-Bonin arc in Japan).

Boninite occurs typically in the western Pacific oceanic arc with thin arc crust. Nascent subduction zone magmatism development of boninite has recently been suggested for the infant-arc model applied to the Late Jurassic ophiolite of California [30]. They adopt an infant-arc origin for the ophiolites, i.e., the Klamaths, Smartville and the Jurassic Coast Range Ophiolite. Incidentally, the tectonic setting for the infant arc ophiolites [30 (in Fig11)] is astonishingly coincident with the first stage of the Gony Altai arc (Fig.2). The tectonic setting for the formation of the Coast Range Ophiolite is still controversial and earlier its origin was dicussed as a trapped oceanic crust (see [30]). This discussion is reminiscent of the that used for the trapped oceanic crust to explain the Mesozoic Sorachi greenstone in Hokkaido, northern Japan. Thus it is important to re-examine the tectonic setting of ophiolites situated in the fore arc areas such as the Sorachi greenstones to establish whether they are trapped oceanic crust or a nascent arc. We will discuss this problem separately in the future, as we wish to concentrate our present discussion on the comparison between the overall tectonic setting of the Early Paleozoic Gorny Altai with the Mesozoic-Cenozoic of Japan.

The diverse occurrrences of Sanukite type magnesian andesite in Japan tells us that the magnesian andesite occurs in an evolved volcanic arc with or without the presence of a back arc. Looking backward to the Gorny Altai, we assume that the primitive arc magmatism is boninitic, allowing us to speculate on the presence of an oceanic arc as shown in Fig.2. If we assume the presence of an evolved volcanic arc and a back arc as shown in Fig.2, the occurrence of sanukite-type magnesian andesite, as menioned above for southwestern Japan, may be expected in the Gorny Altai. In Fig.8, we show a diagram of temporal and spatial distribution of igneous rocks in the Chugoku region (Inner side of southwest Japan). This region suffered intense continental margin type magmatism since Cretaceous. In Miocene, island arc type magmatism including sanukite magmatism occurred on the both continental side and on the oceanic side in relation to the marginal sea opening.

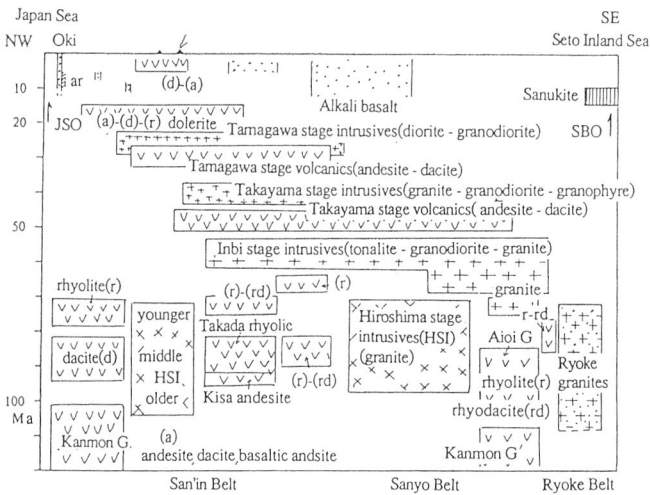

Figure 8. Simplified diagram showing temporal and spatial distribution of rocks produced by continental margin type magmatism in the Chugoku region (based on Igi and others (eds), *Regional Geology of Japan, Part 7, Chugoku, Kyoritsu, Tokyo (1987)*. In Pleogene,the igneous activity shifted toward the Asian continental side and then the Japan sea opening occcured. Present volcanic front(dacite volcano) is located in the San'n district (Japan Sea side). JSO- Japan Sea Opening stage. SBO- Shikoku Basin Opening stage

Thus further detail study of the so-called calc-alkaline rocks is important for full understanding of the Early Paleozoic Gorny Altai arc-trench system.

The mode of occurrence of the tectono-stratigraphic units forming the arc-trench systems in the areas discussed here shows that the preservation of the original linear arc-trench structure in Japan was related to the direction of motion of micro-plates; specifically, normal to and parallel to the older arc structure. Fragmentation of the tectono-stratigraphic units of the arc-trench system is not simply a time factor. For example, the Mesozoic Abukuma metamorphic terrane occurs as a fragmented block in northeast Japan; however, the middle Paleozoic arc-trench system in the New England Fold Belt in eastern Australia exhibits a well defined N-S trending linear structure [19]. Fragmentation of the units in

Gorny Altai is considered to be fundamentally related to its tectonic position at the center of amalgamation of micro-continents. Gorny Altai has also is situated at the northen margin of the zone influenced by collision of India sub-continent with Asia [33]. This same collision may have also influenced the opening of the Japan Sea [33]. If so, study of the process of the fragmentation of linear arc-trench systems holds a key to global plate motion.
Further detailed study on the fragmentation process in the Gorny Altai is significant in undestanding the formation of the Asian continent.

CONCLUSIONS

1. The tectono-stratigraphic unit of the arc-trench system of an active continental margin of the Japanese type are displayed in the mosaic-block area of Gorny Altai. These help to decode the structure and evolution of ancient folded areas.

2. Early Paleozoic Gorny Altai arc-trench system is an ancient analogue of the Izu-Bonin (Ogasawara) arc-trench system, the type location of boninite. This comparison may suggest importance of a further comparative study between the Gorny Altai and the Jurassic California arc-trench system disscussed by [30]

3. Fragmentation of the tectono-stratigraphic units of the arc-trench system in the fold belt is not directly a function of age, but motion of micro-continents in Japanese islands. Older arc-trench systems, however, have suffered many collision episodes of a global tectonic scale indicated by the case of Gorny Altai. Further detailed study on the fragmentation process within the Gorny Altai will be significant to our undestanding of the Asian continent formation.

Acknowledgements
The senior author,Teruo Watanabe, would like to express his thanks to Prof. N. Dobretsov and Dr N. Berzin (United Institute of Geology, Geophysics and Mineralogy, Novosibirsk, Russia) and Dr E. Sklyarov (Institute of Earth Crust, Irkutsk) for their kind guidance with an excursion to the Gorny Altai and their valuable discussions. We especially express our sincere thanks to Prof. Dobretsov for his kind arrangemant of our joint study. Without this support it would have been impossible for us to make this study The Scientific Grants-in-aid Oversea Research by Sawa Soft Science and Meiji Consultant Co. Ltds were used for this study. Especially, Mr T. Sawada, Sawa Soft Science, was very supportive of our exchange project. M.Buslov extends his thanks to the Department of Geology and Mineralogy, Hokkaido University, for the invitation to Japan and to Prof. T. Itaya, Okayama University of Science for his preliminary joint dating study. Teruo Watanabe thanks to Prof. K. Ichikawa, Osaka University of Technology, Assoc. Prof. A. Borovikov, Novosibirsk University, and Ms B. Oksana,Irkusk, for their support in the initiation of our joint study.
Dr C. Herzig, Research Institute for Earth's Interior of Okayama University, critically read and gave kind comments for our manuscript in its early stages. We also would like to acknowleged. Mr T. Hirama and Ms H. Kutsu of Kyoyo-bu, Hokkaido University who assisted us in drafting and typing the manuscript,respectively.
Prof. R.G. Coleman, Stanford University and Prof. X. Xuchang, Institute of Geology, Beijing, kindly edited our joint paper.

REFERENCES

1. N.A. Berzin, M.M. Buslov, and N.L. Dobretsov, Geology and tectonics of Gorny Altai (Guide Book). Acad. Sci. Novosibrsk: (1991)

2. M.M. Buslov,*Tectonic nappes of Gorny Altai*. Nauk, Novosibirsk. (1992). (*in Russian*)

3. M. Cloos. Flow melanges: numerical modelling and geologic constraints on the origin in the Franciscan subduction complex, California, *Geol. Soc Amer. Mem.* **93**, 330-345 (1982).

4. N.L. Dobretsov and A.G. Kirdyashkin. Dynamics of subduction zones: models of accretion wedge formation and transportation of blueschists and eclogites, *Geolog. & Geoph.* **N3**, 5-9 (1991). (*in Russian*).

5. N.L. Dobretsov, N.V. Sobolev, and V.S. Shatsky, *Eclogites and blueschists in folded belts*. Nauka, Novosibirsk (1989). (*in Russian*)

6. N.L. Dobretsov and L.P. Zonenshain. The evolution of pre-Mesozoic ophiolites in N. Eurasia: A comparative review, *Chemical Geology.* **77**, 323-330 (1989).

7. N.I. Gusev, Reconstruction of the geodynamic environments of Precambrian and Cambrian volcanism in southeastern part of Gorny Altai, In: *Paleodynamics and formation of the productive zones of south siberia*, pp. 4-19, Nauk, Novosibirsk (1991). (*in Russian*)

8. T. Imaoka, T. Seki, and K. Nakajima, Chromite and Cr-endiopside in basaltic andesites from the Cretaceous Kanmon Group, In: *Magnesion andesites in Japan*, K. Shiraki (Ed.), pp. 119-123, Tokyo (1989). (*in Japanese*)

9. E. Inoue and E. Honza, Marine geological map around Japanese Island, 1:3,000,000. Geol. Survey of Japan: Tokyo (1983)

10. Y. Isozaki and T. Itaya. Pre-Jurassic klippe in northern Chichibu Belt in west-central Shikoku, Southwest Japan - Kurosegawa terrane as a tectonic outlier of the pre-Jurassic rocks of the inner Zone, *Jour. Geol. Soc. Japan.* **97**, 431-450 (1991). (*in Japanese with English abstract*)

11. D.E. Karig and G.F. Sharman. Subduction and accretion in trenches, *Geol. Soc. Am. Bull.* **86**, 377-389 (1975).

12. M. Kiji, High-magnesian andesites and related dike rocks from the Tamba Belt, southwest Japan, In: *Magnesian andesites in Japan*, K. Shiraki (Ed.), pp. 125-132, Tokyo (1989). (*in Japanese*)

13. K. Kiminami. Cretaceous tectonics of Hokkaido and environs of the Okhotsk Sea, *Monograph Assoc. Geol. Collab. Japan.* **31**, 403-418 (1986). (*in Japanese with English abstract*)

14. G. Kimura and K. Tamaki. Collision, roatation, and back-arc spreading in the region of the Okhotsk and Japan Seas, *Tectonics.* **5**, 389-401 (1986).

15. S. Koitabashi, Greenstones in the Niikappu River area (52 p.). Hokaiddo University (Msc) (1977)

16. L.V. Kungustsev, Paleodynamic complexes and ore deposits of the Kuznetsk-Gorny Altai paleo-island arc, In: *Paleodynamics and formation of the productive zones of south Siberia*(Ed.), Vol. pp. 82-106, Nauka, Novosibirsk (1991). (*in Russian*)

17. H. Kuno. High-alumina basalt, *Jour. Petrol.* **1**, 121-145 (1960).

18. N. Kuroda and K. Shiraki. Boninite and related rocks of Chichi-jima, Bonin Islands, Japan, *Rep. Fac. Sci. Shizuoka Univ.* **10**, 145-155 (1975).

19. E.C. Leitch. The geological development of the southern part of the New England fold belt, *Jour. Geol. Soc. Australia.* **21**, 133-156 (1974).

20. A. Miyashiro,*Metamorphism and metamorphic belts.* George Allen & Unwin, London (1973).

21. G.E. Moore, H.G. Billman, P.E. Hehanussa, and D.E. Karig. Sedimentology and paleobathymetry of Neogene trench slope deposits, Nias Island, Indonesia, *Jour. Geol.* **88**, 101-180 (1974).

22. M. Nagata, N. Kito, and K. Niida. The Kumaneshiri Group in the Kabato Mountains: the age and anture as an Early Cretaceous volcanic arc., *Monograph Assoc. Geol. Collab. Japan.* **31**, 63-79 (1986). (*in Japanese with English abstract*)

23. Y. Oba. Petrological study of the Late Pliocene basalts from the western part of Hokkaido. (2) Petrology of the Late Pliocene basalts of the western part of Hokkaido, *Jour. Fac. Sci. Hokkaido Univ.* **Ser. IV, 15**, 11-25 (1972).

24. S. Okamura and T. Yishida, High-magnesian, plagioclase phyric andesite from Okushiri Island, southwest Hokkaido, In: *Magnesian Andesites in Japan*, K. Shiraki (Ed.), pp. 167-176, Tokyo (1989).

25. K. Otsuki. Uplifting and shear deformation of high-P metamorphic belt by oblique plate subduction : An investigation by quasi-3-dimensional lubricant model, *Jour. Geol. Soc. Japan.* **98**, 435-444 (1992). (*in Japanese with English abstract*)

26. K. Shiraki, Some aspects of the magnesian andesites, In: *Magnesian Andesites in Japan*, K. Shiraki (Ed.), pp. 5-25, Tokyo (1989). (*in Japanese*)

27. K. Shiraki and N. Kuroda. The boninite revisited, *Jour. Geogr.* **86**, 174-190 (1977). (*in Japanese*)

28. K. Shuto and T. Takimoto, Chemical heterogeneity of the upper mantle beneath the volcanic fron side region of the Northeast Japan arc, In: *Magnesian Andesites in Japan*, K. Shiraki (Ed.), pp. 177-187, Tokyo (1989). (*in Japanese*)

29. V.A. Simonov, E.V. Skylarov, and M.M. Buslov. Types of ophiolites and their tectonic settings in the fold belts of the south Siberia, *29th Inter. Geol. Congress Abstract.* **2**, 477(1992).

30. R.J. Stern and S.H. Bloomer. Subduction zone infancy: Examples from the Eocene Izu-Bonin-Mariana and Jurassic California arcs, *Bull. Geol. Soc. Amer.* **104**(12), 1621-1636 (1992).

31. M. Takahashi, High magnesium andesites and basalts in the early Miocene Nishina group, Izu Peninsula and tectonic setting of alkali-rich high magnesium andesite generation, In: *Magnesian andesites in Japan*, K. Shiraki (Ed.), pp. 177-187, Tokyo (1989). (*in Japanese*)

32. K. Tamaki. Nihon-kai no keishi-kikou (Mechanism of formation of the Japan Sea, *Kagaku.* **62**, 729.(1992) (*in Japanese*)

33. P. Tapponnier, G. Peltzer, and R. Armijo, On the mechanics of collision between India and Asia, In: *Collision Tectonics*, M.P. Coward and A. Ries (Ed.), Vol. 19, pp. 115-157, Geol. Soc. of London, London (1985).

34. Y. Tatsumi. Melting experiments on a high-magnesium andesite, *Earth Plan. Sci. Lett.* **54**, 357-365 (1981).

35. Y. Tatsumi and s. Maruyama, Boninites and high magnesium andesites: Tectonics and petrogensis, In: *Boninites*, A.J. Crawford (Ed.), pp. 50-71, Unwyn-Hyman, London (1989).

36. B. Taylor, Rifting and the volcanic-tectonic evolution of the Izu-Bonin-Mariana arc, In: *Proc. Ocean Drilling Program Sci. Results*, B. Taylor, K. Fujioka, ande. al (Ed.), Vol. 126, pp. 709, U.S. Printing Office, College Station, Texas (1992).

37. B. Taylor, K. Fujioka, and others,*Proc. ODP Sci. Results.* Vol. 126., U.S. Printing Office, College Station, Texas.(1992).

38. S. Uyeda, *The new view of the Earth.* Freeman, San Francisco (1978).

39. T. Watanabe and H. Maekawa, Early Cretaceous dual subduction system in and around the Kamuikotan Tectonic Belt, Hokkaido, Japan, In: *Formation of Active Ocean Margins*, N. Nasu and others (Ed.), pp. 677-699, Terrapub, Tokyo (1985).

40. T. Watanabe, T. Tokuoka, and T. Naka, Complex fragmentation of Permo-Triassic and Jurassic accreted terranes in the Chugoku Region, southwest Japan and formation of the Sangun Metamorphic rocks, In: *Terrane accretion and orogenic belts*, E.C. Leitch and E. Scheibner (Ed.), Vol. 19 Geodynamic series, pp. 275-289, American Geophysical Union, Washington, D.C. (1987).

41. T. Yamazaki and Y. Okamura. Subducting seamounts and deformation of overridding forearc wedges around Japan, *Tectonophysics.* **160**, 207-229 (1989).

42. M. Yuasa, Along-arc variations in morphology and rock chemistry of submarine volcanoes in the Izu-Odasawara arc, northwest Pacfic. Hokkaido University, Dsc thesis (109 p.) (1991)

43. L.P. Zonenshain, M.T. Kuzmin, and L.M. Natapovi,*Tectonics of the lithospheric plates of the USSR territory.* Vol. 2., Nedra, Moscow (1990). (*in Russian*)

QUATERNARY ENVIRONMENTAL CHANGES

Editor
E.H. Juvigné

CONTENTS

Preface	191
Peculiarities of formation of flat glaciolacustrine hills *A. Bitinas*	193
What controls the calcium carbonate content in Quaternary sediment cores raised from below the carbonate lysocline depth in the Equatorial Indian Ocean? *P. Divakar Naidu, B.A. Malmgren, and L. Bornmalm*	201
Rates of erosion by the Colorado River in the Grand Canyon, Arizona *W.K. Hamblin*	211
Method of estimating mean annual temperatures from plant fossils in the Pliocene and Pleistocene periods *Y. Hase and A. Iwauchi*	219
Spatial variation of CO_2^- and SO_3^- radicals in massive coral from Ishigaki Island, Japan and its implications *S. Ikeda, M. Furusawa, and M. Ikeya*	225
Late Quaternary paleoceanography of the Japan Sea: a tephrochronological and sedimentological study *K. Ikehara, K. Kikkawa, H. Katayama, and K. Seto*	229
A Pleistocene stratotype at Alzenau (Vorspessart, Germany) *E. Juvigné, R. Geeraerts, F. Geissert, H.-J. Gregor, M. Hottenrott, J.J. Hus, G. Seidenschwann, and R.C. Walter*	237
Analysis of paleo-environments based on electric conductivity of the STICS-water - On five drill cores in Kanto-plain, Central Japan *M. Koarai and Research Group for Alluvium Deposit*	251
Holocene dinoflagellate cyst assemblage in lagoonal lakes along the east coast of Japan *N. Kojima*	263
The late Quaternary environment around Lake Nojiri in Central Japan *Nojiri-ko Excavation Research Group*	269
Vegetation and climate since Last Glacial Maximum in Darjeeling (Mirik Lake), Eastern Himalaya *C. Sharma and M.S. Chauhan*	279
Quaternary environmental changes in the Northeastern Japan *S. Takeuti and K. Manabe*	289
Beginning of a new period - the Technogene *G. Ter-Stepanian*	299
The essential characteristics and facies model of windblown sand deposits during the Last Glacial period in Eastern China *M. Zhang and J. Liu*	309

PREFACE

SYMPOSIUM II -5-6 QUATERNARY ENVIRONMENTAL CHANGE

Kazuyuki KOIKE*

This symposium (Poster Session) was held in the afternoon of 28th, August, 1992 organized by Etienne JUVIGNE (Univ. Liège, Belgium) and Kazuyuki KOIKE (Komazawa Univ., Japan) as one of six symposia in the section of Quaternary Studies (chief organizer, Hiroshi MACHIDA; Tokyo Metro. Univ.).

Among 42 papers preliminarily offered to this symposium, 21 papers (including 1 additional offer) were presented at this session. This symposium dealt with wide range of topics concerned with Quaternary environmental change so that 21 papers would be divided into following 5 subdivisions;

I Deep sea core analysis (No. 1~3, Equatorial Indian Ocean, Japan Sea and Baltic's Gotland deep)
II Coral reef and nearshore environment (No. 4~7, Riukyu Islands, Cook Islands, Great Barrier Reef and Lagoonal bay, east coast of Japan)
III Paleoenvironmental analysis (No. 8~13, 20, Kyushu, Northeast Japan, Lake Nojiri, Kanto Plain, Northern Great Poland Lowland, Finnish Lapland and Northern Mali)
IV Tephras, loess, and pedology (No. 14~17, Middle Europe, French Central Massif, China and Coast of China)
V River morphology and the others (No. 18, 19 & 21, Grand Canyon, Proposal of "Technogene" and Last Interglacial shoreline map of Japan).

Many persons gathered and discussed with the poster presenters. Most of the posters exhibited at this symposium were rich in contents and would contribute to the development of the Quaternary environmental studies. Among them, the alphaspectrometric $^{230}Th/^{234}U$ dating of Pleistocene Riukyu Limestone by Omura & Ota (No. 4) will contribute to establishing reliable chronology in sea level changes during the last 300,000 years.

---

* Department of Geography, Komazawa University
   1-23-1, Komazawa, Setagaya-ku, Tokyo, 154, JAPAN

Peculiarities of formation of flat glaciolacustrine hills

A. BITINAS
Institute of Geology, Ševčenkos 13, 2600 Vilnius, Republic of Lithuania

Abstract--The flat glaciolacustrine hills (or so called zvontsy) are widespread in the regions, that had been covered by the glacier of the last Scandinavian continental glaciation. Their genesis is problematic up to the present days. Results of the large scale geological mapping in Lithuania showed, that the groups of flat glaciolacustrine hills are situated only in the hilly massifs, formed between two glacial lobes during the stade glacial advance. It enabled to think, that flat glaciolacustrine hills were formed under the influence of two glacial streams of different (stadial) age. The flat glaciolacustrine hills of the same massif are of the same age and related to each other genetically. These forms of relief aren't the kind of the glaciolacustrine kames and they should be distinguished into a separate (polygenetical) group.

INTRODUCTION

For a long time Lithuanian Qaternary geologists and geomorphologists have been interested in original relief forms - flat glaciolacustrine hills with till foundation - which are widespread in marginal zones of the last glaciation [1]. Flat glaciolacustrine hills are also widespread in neighbouring Baltic countries [2], Poland, Danmark [3], North-West Russia [4,5], they are known on the other side of the Atlantic ocean as well, e.g. the U.S.A. [6]. In Russia these relief forms have been given special name - zvontsy [7]. Scientists have been trying to explain their formation by different hypoteses. O.M.Tatarnikov, who has been investigating zvontsy in Russia most intensively, even presents classifications of the hypotheses concerning the formation of these relief forms [4]. But all these hypotheses can't explain a number of problems. Why are the flat glaciolacustrine hills found only in groups? Why do their surfaces predominate in the relief? Why are these hills characteristic only to the relief of the last (Nemunas, Valdaj, Weichselian) glaciation?

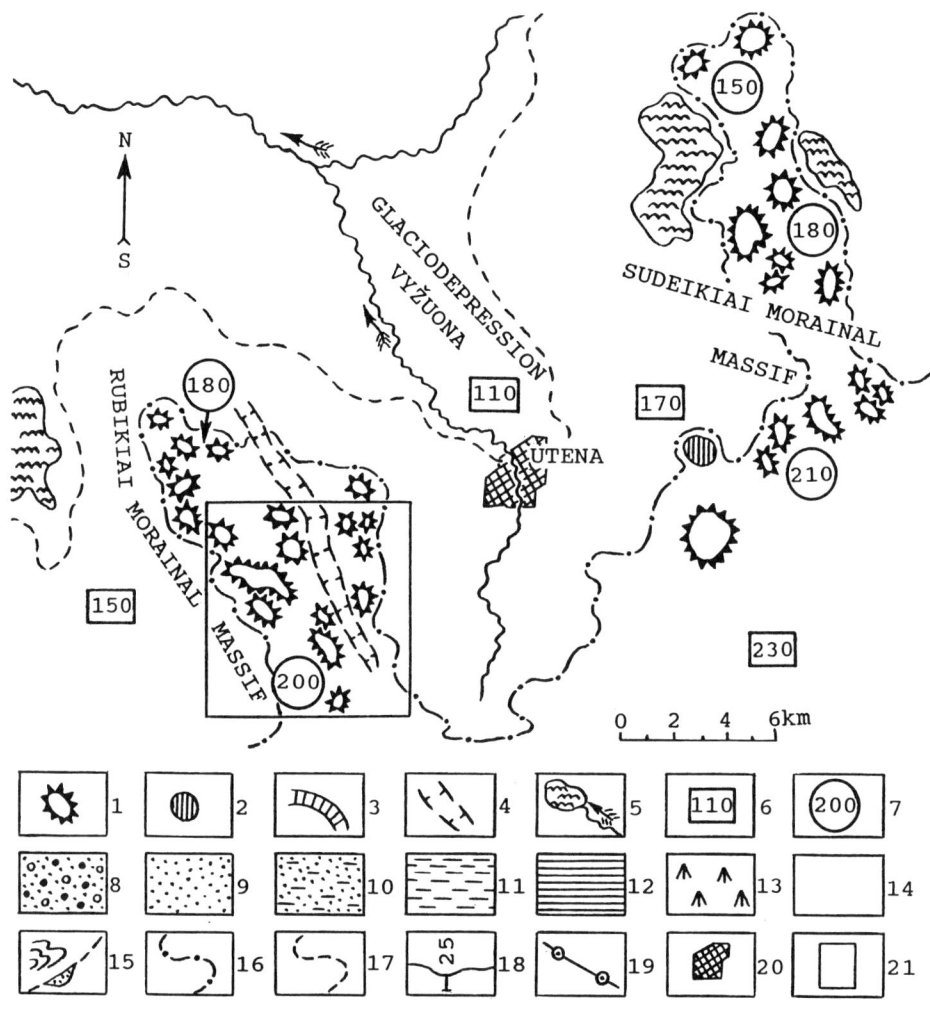

Figure 1. Geological and geomorphological situation of the Baltic marginal highland in North-East Lithuania. 1 - flat glaciolacustrine hills. 2 - kames. 3 - steep slopes. 4 - valey-like forms. 5 - lakes and rivers. 6 - prevailing absolute altitude of relief. 7 - prevailing absolute altitude of surfaces flat glaciolacustrine hills. 8 - mixture of sand and gravel. 9 - sand. 10 - silty sand. 11 - silt. 12 - clay. 13 - peat. 14 - till. 15 - fold-scaly structure of till (glaciodislocations). 16 - boundary of the East Lithuanian stade of last glaciation. 17 - boundary of glacial oscillation. 18 - boreholes. 19 - line of geological profile. 20 - town. 21 - detaled area (Fig. 3). Age of deposits: W - Weichselian glaciation (Wg - Gruda stade, We - East Lithuanian stade, Ws - South Lithuanian stade), S - Saalian glaciation.

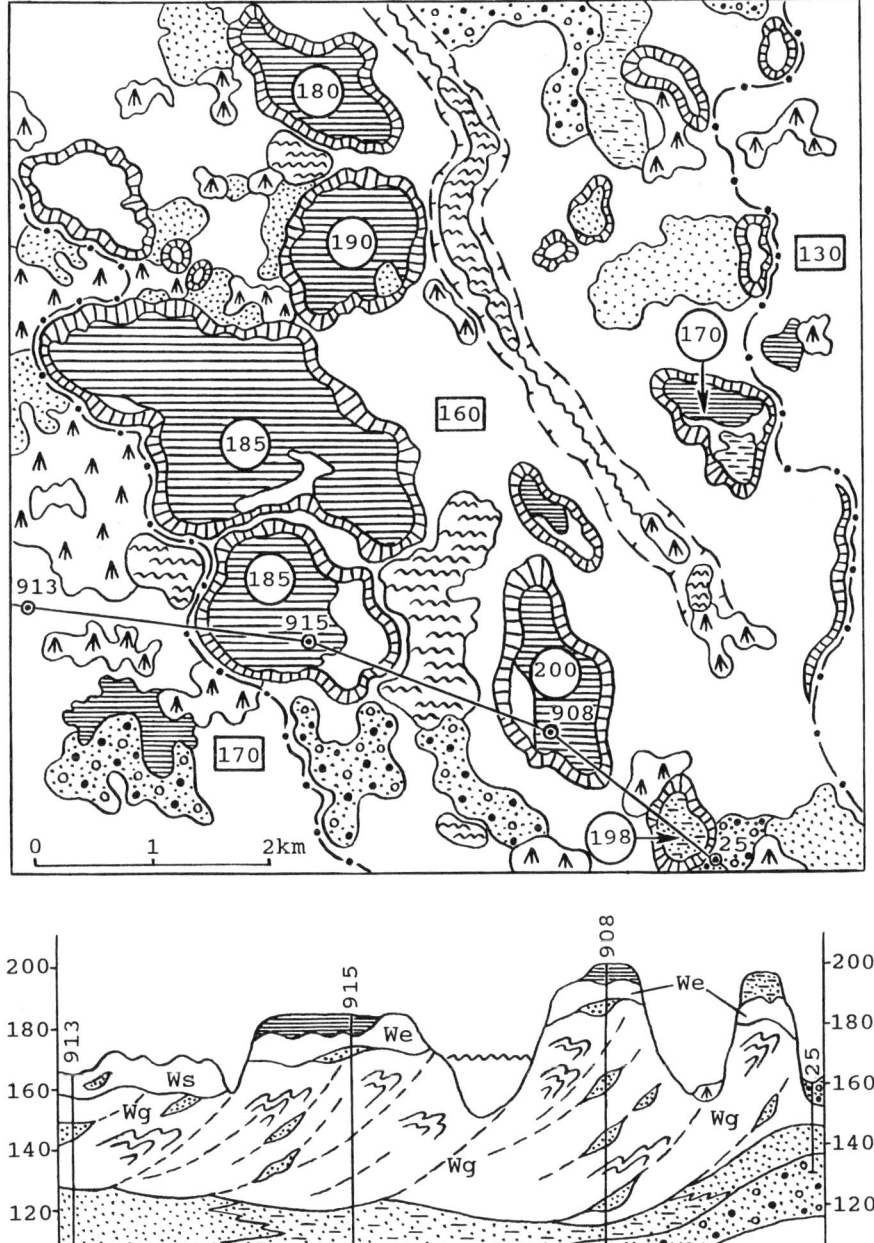

Figure 2. Detailed area of Rubikiai morainal massif. Legend in Figure 1.

INVESTIGATIONS OF FLAT GLACIOLACUSTRINE HILLS

The specific morphological, lithological and structural features of flat glaciolacustrine hills enable to distinguish them from other forms of glacial relief [1-7]. The flat glaciolacustrine hills are found in groups - from several to some tens of hills in one group. Their flat surfaces are in one level or in some different levels, which, as a rule, predominate in the surrounding relief. The surface is plain or slightly waved, sometimes - terraced. The hillsides are steep ($10°$ - $40°$), often engraved by ravines. In the plan the form of flat glaciolacustrine hills can be different, but most often - round or oval. The area of their surfaces varies from less than 1 sq. km to 30 sq. km and more. The relative hight varies from 10-15 m to 40-50 m. The flat glaciolacustrine hills always have elevated till foundations covered by glaciolacustrine deposits. But sometimes flat glaciolacustrine hills without glaciolacustrine deposits cover are found - these are so called "undeveloped" forms. The till foundations have a dislocated structure.

In recent years a lot of new information, obtained by making a large scale geological mapping in Lithuania, suggested some ideas about the peculiarities of flat glaciolacustrine hills formation. The detailed geological and geomorphological analysis of flat glaciolacustrine hill massifs showed them to be located in those heights that caused disintegration of glacier margin into lobes during pulsations (stades) of the last glaciation (Figs. 1,2). It suggested the idea that the mechanism of flat glaciolacustrine hills formation is much more complicated than it was considered till now.

HYPOTHESIS OF FORMATION

One can distinguish 6 stages in the flat glaciolacustrine hills formation (Fig. 3).

I. The height, which caused the change of the plastic ice flow into the moving of ice blocks and repletion of ice with clastic material, was necessary in the underglacial relief. Thus, the thrustfold ice structure was formed [2,8]. The underglacial height had to be reflected on the ice surface as well.

II. At the beginning of deglaciation the abliation process was the most intensive in the most prominent and cracked places of the glacier.

III. It was the ice containing clastic material that first appeared in the clefts and it speeded up the process of abliation. Large ice cracks were formed there first of all.

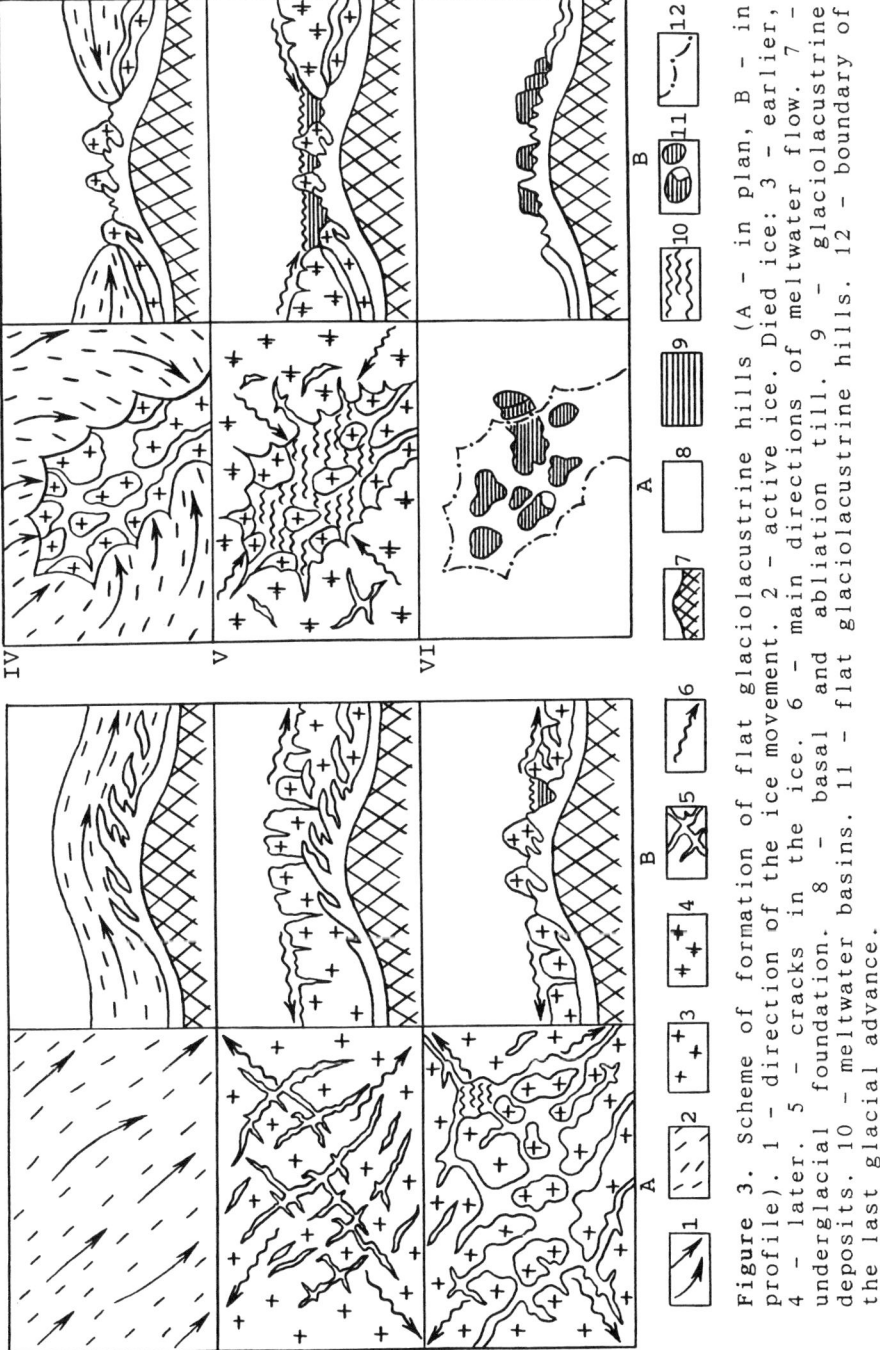

Figure 3. Scheme of formation of flat glaciolacustrine hills (A – in plan, B – in profile). 1 – direction of the ice movement. 2 – active ice. Died ice: 3 – earlier, 4 – later. 5 – cracks in the ice. 6 – main directions of meltwater flow. 7 – underglacial foundation. 8 – basal and abliation till. 9 – glaciolacustrine deposits. 10 – meltwater basins. 11 – flat glaciolacustrine hills. 12 – boundary of the last glacial advance.

The abliation till was uncovered on the bottom of ice cracks. As water was flowing down the cracks to the lower areas of glacier surface, there was no possibility for formation of basins.

IV. After the climate had grown colder, the process of deglaciation stopped. During a new stade glacial advance (already less powerful), the heights of older died ice became as obstacles which split glacier into lobes. These older died ice massifs turned into closed or semi-closed spaces. surrounded by active glacier lobes.

V. After the climate had grown warmer again, deglaciation included ice masses of the last glacial advance as well. But now the heights of older died ice with cracks found themselves in the lower absolute level. Meltwater was flowing there and glaciolacustrine basins were formed.

VI. After the whole ice had melted completely, surface inversion occured. Deposits of glaciolacustrine basins became prevailing in the relief. The glaciolacustrine cover remained only in those places where it had formed on the till foundation. Only the same level of the surface of flat glaciolacustrine hills shows that one or several glaciolacustrine basins had existed in this place.

CONCLUSIONS

Flat glaciolacustrine hills could form only glacier margin being thin and pulsation glacier deglaciation taking place - what was characteristic to the last glaciation. Flat glaciolacustrine hills existing in the same ice dividing massif formed at the same time in one or several glaciolacustrine basins of different level. Thus, flat glaciolacustrine hills are not the variety of glaciolacustrine kames - they are separate, poligenetic forms of the relief.

ACKNOWLEDGEMENT

The author wishes to thank colleagues geologists with whom he worked in North-East Lithuania and Ms. N.Bitinienė who corrected the English text.

REFERENCES

1. A.A. Basalykas. Variety of relief of the glacial accumulation. In: Continental glaciations and glacial morphogenesis. pp. 65-154. Vilnius (1969). (in Russian).

2. O.P. Aboltinš. Formation of island-form higlands. In: Glacial morphogenesis. pp. 51-61. Zinatne, Riga (1972). (in Russian).

3. W. Niewiarowski. Types of kames occuring within the area of the last glaciation in Poland as compared with kames from other regions. In: Reports of the VI International Congress of Quaternary. vol.3, pp. 475-485. Lodz (1963).

4. O.M. Tatarnikov. The morphogenesis of zvontsy and their disposition in the glacial relief of the Valdaj glaciation. Doctoral dissertation. Moscow (1985). (in Russian).

5. I.M. Ekman, V.A. Ilyin and A.D. Lukashov. Degradation of the late glacial sheet on the territory of the Karelian ASSR. In: Glacial deposits and glacial history in Eastern Fennoskandia. pp. 103-117. Academy of Sciences of USSR, Apatity (1981).

6. L. Clayton and J.A. Cherry. Pleistocene superglacial and icewalled lakes of West-Central North America. In: Glacial geology of the Missouri coteau and adjacent areas. pp. 44-56. Grand Fork, North Dacota (1967).

7. D.B. Malachovskij and M.E. Vigdorchik. Some forms of glacial accumulative relief in the north-west of the Russian plain. In: Marginal forms of relief of continental glaciation in the Russian plain. pp. 47-53. AN SSSR, Moscow (1963). (in Russian).

8. J.A. Lavrushin. Structure and formation of basal tills of continental glaciations. Moscow (1976). (in Russian).

WHAT CONTROLS THE CALCIUM CARBONATE CONTENT IN QUATERNARY SEDIMENT CORES RAISED FROM BELOW THE CARBONATE LYSOCLINE DEPTH IN THE EQUATORIAL INDIAN OCEAN?

P. DIVAKAR NAIDU, BJÖRN A. MALMGREN and LENNART BORNMALM
Department of Marine Geology, University of Göteborg Box 7064, S-402 32, Göteborg (Sweden).

ABSTRACT

Swedish Deep-Sea Expedition cores 144, 147 and 154 are raised from below the regional presentday lysocline depth of the equatorial Indian Ocean. In all these cores higher $CaCO_3$ content coincides with less dissolution and lower $CaCO_3$ concentration with greater dissolution. This inverse relationship between $CaCO_3$ and dissolution strongly suggests that dissolution of $CaCO_3$ dictates the Quaternary carbonate cycles in the equatorial Indian Ocean. The very low values of $CaCO_3$ (2-10%) at the troughs in the deeper depth core 154 (4,860 m) and relatively higher $CaCO_3$ content at the troughs in the shallower depth cores 144 (4,417 m) and 147 (4,380 m), indicates that dissolution increases with water depth in this region. These dissolution cycles are discussed in relation with the saturation state of carbonate ions in the water column and the productivity variations between siliceous and carbonaceous organisms.

INTRODUCTION

Calcium carbonate is a large and reactive reservoir of oceanic carbon (Broecker and Peng, 1987). Therefore, the study of Pleistocene carbonate cycles in the world oceans is important for understanding the global geochemical cycle of CO_2. Arrhenius (1952) was the first to point out the relationship between calcium carbonate fluctuations and climate. Subsequently, considerable effort has been devoted to studying the Pleistocene carbonate cycles from the Atlantic Ocean (Olausson, 1965, 1967; Gardner, 1975; Damuth, 1975 and Curry and Lohmann, 1986), Pacific Ocean (Arrhenius, 1952, 1988; Olausson, 1960a, 1965, 1971; Chuey et al., 1987; Farrell and Prell, 1989, 1991; Archer, 1991; Berger, 1992) and the Indian Ocean (Olausson, 1960b, 1971; Oba, 1969; Peterson and Prell, 1985; Naidu, 1991 and Naidu et al., manuscript). However, it is not yet clear whether the surface-water productivity or dissolution of calcium carbonate is responsible for deriving the Pleistocene $CaCO_3$ cycles in the deep-sea sediment cores.

Several investigations have been carried out dealing with preservation patterns of $CaCO_3$ in the Pacific and Atlantic Oceans (Berger, 1973, 1977, 1992; Damuth, 1975; Farrell and Prell, 1989, 1991 and Le and Shackleton, 1992 among others). Limited Pleistocene $CaCO_3$ preservation data are available from the Indian Ocean (Peterson and Prell, 1985; Naidu et al., manuscript). In this paper we are reporting the late Pleistocene $CaCO_3$ variations along with dissolution indices records to demonstrate the possible effect of dissolution on the $CaCO_3$ fluctuations in the equatorial Indian Ocean. We are not dealing with the $CaCO_3$ variations with reference to glacial and interglacial periods; therefore, we are not providing any stratigraphic or time scale for these cores. However, the earlier studies of Olausson (1971) infers that these cores are of a Late Pleistocene age. The main goal here is to examine whether the dissolution of $CaCO_3$ alone is responsible for the $CaCO_3$ fluctuations in these cores, which are retrieved below the presentday regional lysocline depth in the area.

MATERIAL AND METODS

Cores 144, 147 and 154 were collected at water depths of 4417 m, 4380 m and 4860 m, respectively, from the equatorial Indian Ocean (figure 1). The cores were sampled approximately at 10 cm intervals. The details of $CaCO_3$ estimation and quantitative measures of planktonic foraminifera are described elsewhere (Naidu, et al., manuscript).

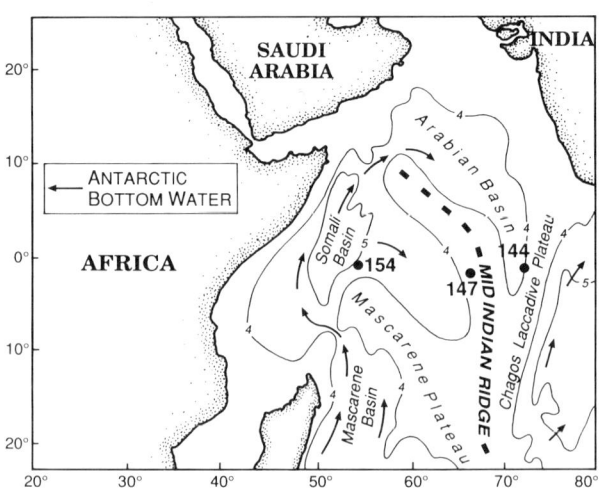

Figure 1. Physiography and bathymetry of the Indian Ocean showing the location of the cores analyzed here. Bathymetry contours are in kilometers.

It was pointed out by Malmgren (1978) that the employment of standard correlations analysis techniques to timeseries curves could lead to false correlations. In order to avoid this possibility, partial rank correlation analysis was used to determine the relationship between $CaCO_3$ and dissolution indices curves over time (Gordon, 1982). In this technique depth or time is held constant by partialling out depth dependent trends in either or both curves that could cause false correlation.

Berger Dissolution Index (BDI)

The BDI was calculated using the formula derived by Le and Shackleton (1992). The ranking of different planktonic foraminifer species was considered based on their susceptibility to dissolution. The ranking of species used in this study are reported earlier by Naidu et al., (manuscript). An increase in BDI values represent increase in $CaCO_3$ dissolution and vice versa.

RESULTS AND DISCUSSION

It is known that major controlling factors of $CaCO_3$ in any ocean are productivity, dissolution and dilution. Among these three factors dilution plays an important role in the sediments off continental margins and in other higher siliceous productivity areas. A detailed study of the calcium carbonate distribution in surficial sediments indicates that the dilution effects are very insignificant in the equatorial Indian Ocean (Kolla et al., (1976). Therefore, this area is suitable to test the hypothesis whether productivity or dissolution or both are playing a role in creating the observed $CaCO_3$ fluctuations.

The primary productivity maps of the world oceans prepared by Berger et al., (1987) show that the productivity varies from 60 - 90g C m^{-2} y^{-1} in the area, and that no productivity difference exists among the three core sites. Therefore, we assume that during the Pleistocene there was also no significant productivity difference between these core sites.

The calcium carbonate distribution studies of Kolla et al., (1976) and composite dissolution index (CDI) data of Peterson and Prell (1985) from the Indian Ocean suggested a lysocline depth at 3800 m in the equatorial Indian Ocean. It is thus clear that all the three cores of the present study are raised from below the regional lysocline depth in the area.

Core 144: The $CaCO_3$ content varies from 51 to 88% and the BDI value ranges from 0.38 to 1.61 (figure 2).

Core 147: The $CaCO_3$ and BDI vary from 44 to 87% and 0.57 to 1.21, respectively (figure 3).

Core 154: A maximum of 88% and a minimum of 2% $CaCO_3$ are recorded. The BDI

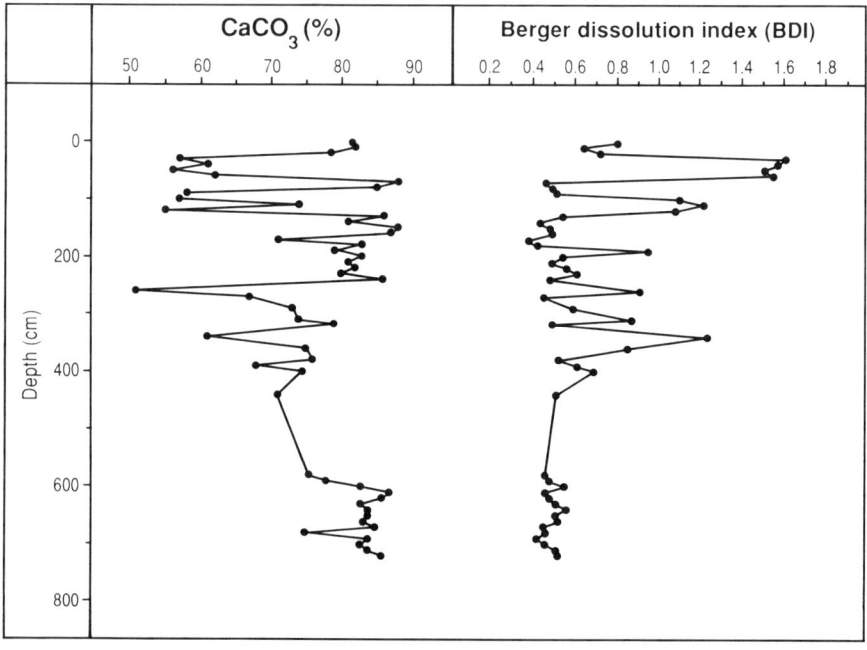

Figure 2. Profiles of variations in $CaCO_3$ and the Berger dissolution index against depth in the core 144.

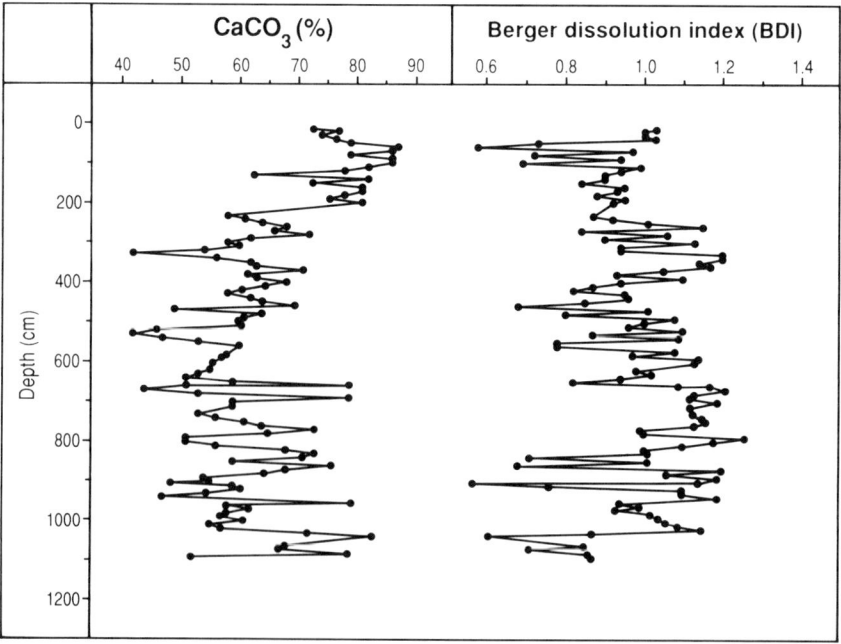

Figure 3. Profiles of variations in $CaCO_3$ and the Berger dissolution index against depth in the core 147.

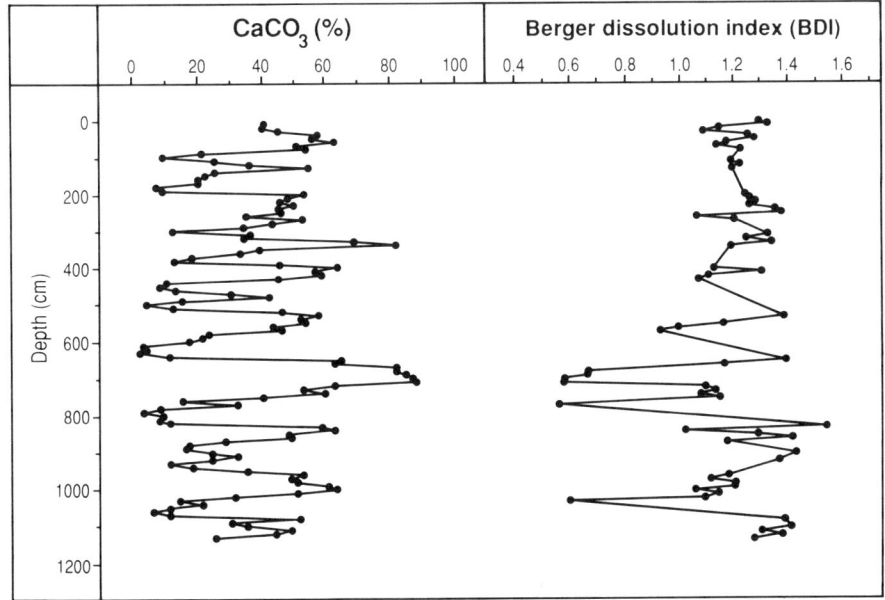

Figure 4. Profiles of variations in CaCO₃ and the Berger dissolution index against depth in the core 154.

value ranges from 0.55 to 1.53 in this core (figure 4).

Examination of $CaCO_3$ and BDI data in cores 144, 147 and 154, reveals that high $CaCO_3$ content coincides with low values of BDI, indicating low dissolution effects and vice versa (figures 2, 3 and 4). We have computed the partial rank correlations between $CaCO_3$ and BDI, and we found that there are significant negative correlations between them in all three cores (Table 1). The good agreement between carbonate maxima and less dissolution and carbonate minima and more dissolution provides conclusive evidence that dissolution of calcium carbonate is controlling the Quaternary $CaCO_3$ cycles in the equatorial Indian Ocean. Therefore, the carbonate cycles are thought to be dissolution cycles.

Table 1

Partial rank correlation coefficients between $CaCO_3$ and Berger dissolution index (BDI) in cores 144, 147 and 154. All partial rank correlations are significant at least at the 0.01 level.

Core no.	Water depth (m)	Partial rank correlation coefficient
144	4417	-0.4846
147	4380	-0.3745
154	4860	-0.3501

In these three cores, the deeper core 154 shows less concentration of $CaCO_3$ at the troughs compared with the troughs of the shallower cores 144 and 147 (figures 2, 3 and 4). This is probably a reflection of the fact that core 154 is from deeper water depth compared with 144 and 147, and hence stronger dissolution of $CaCO_3$ could have caused the lower $CaCO_3$ content at these depths. This suggests that dissolution rate changes with water depth in the equatorial Indian Ocean. Kolla et al. (1976) suggested that the Holocene calcite compensation depth (CCD) is located at 5,100 m in the equatorial Indian Ocean. Our dissolution indices data from all three cores demonstrate that the CCD depth has shown cyclic variations in the Late Pleistocene (figures. 2, 3 and 4), which means that CCD at several periods was shallower than 5,100 m in the western equatorial Indian Ocean.

The studies by Olausson (1967, 1969, 1971), Oba (1969), Naidu (1991), Naidu et al., (manuscript) and Berger (1992) suggests that the Pacific type of carbonate cycles are recorded in the Indian Ocean, i.e. greater $CaCO_3$ content during glacials and lesser in interglacials. Since limited carbonate preservation data exist from the Indian Ocean, some of the following arguments are based on the Pacific Ocean data also.

As indicated above the $CaCO_3$ cycles are mainly caused by dissolution in the equatorial Indian Ocean. The next question is what processes are driving these dissolution cycles. Two alternative hypothesis have been offered (i) dissolution hypothesis: the carbonate dissolution cycles are driven by the carbonate ion concentration and bottom-water chemistry (Farrell and Prell, 1989; Berger, 1992), (ii) productivity hypothesis: the dissolution cycles reflect the changes in rain rates of calcite and organic carbon to the sediments, which depend on the surface-water productivity (Olausson, 1971, 1985; Arrhenius, 1952, 1988; Archer, 1991).

The significant negative correlation between the $CaCO_3$ and BDI, and the depth dependent increase of $CaCO_3$, suggest that dissolution plays an important role in regulating the $CaCO_3$ fluctuations in the Late Pleistocene sediments of the equatorial Indian Ocean. Therefore, these carbonate cycles are thought to be dissolution cycles. The dissolution of carbonate is controlled by the degree of saturation of the sea water with respect to the biogenic carbonate phase. Broecker and Takahashi (1978) showed that the state of preservation does correspond to saturation on a global scale, at least in the vicinity of the lysocline and away from continental margins. Therefore, it is logical to believe that the carbonate ion concentration of the bottom water is responsible for the first order variation in the depth of the lysocline between and within oceans.

Emerson and Bender (1981) argued that changing productivity can also change carbonate preservation by changing the flux of organic matter to the sea floor. However, recently, Grötsch et al. (1991) concluded that the carbonate ion saturation fluctuations are the main factors responsible for carbonate cycles.

Strong evidence exists that saturation level of the deep ocean did change during the Pleistocene (Berger, 1977; Peterson and Prell, 1985; Droxler, 1988). A comparison of calcium carbonate preservation records of the Indian Ocean (Peterson and Prell, 1985) with the calculated carbonate ion saturation in the water (Gröstch et al., 1991), indicates that the preservation state is tied to saturation levels.

The other array of workers (Arrhenius 1952, 1988; Olausson, 1965, 1985; Archer, 1991) argued that productivity is playing a significant role in regulating the $CaCO_3$ cycles. Arrhenius (1988) postulated a link between productivity and preservation. In contrast to the productivity and supply hypothesis of Arrhenius (1988), decreasing productivity can parallel increased dissolution in the Pacific Ocean (Keir and Berger, 1985). Therefore, the surface-water productivity and the carbonate dissolution are not tied together as it was believed earlier.

The $\delta^{13}C$ data (a proxy productivity measure) of the Swedish Deep-Sea Expedition core 154 show a negative relationship with $CaCO_3$ indicating that higher productivity in fact causes dissolution, instead of more carbonate accumulation (Naidu et al., manuscript), which supports the hypothesis of Emerson and Bender (1981). The productivity hypothesis of Arrhenius (1988) has its merit, as long as one views the carbonate productivity alone. The fundamental difference in the controls of silica and carbonate sedimentation should make one cautious in expecting parallel behavior (Berger, 1970). There is a growing body of evidence infering that siliceous and carbonate productivity variations are decoupled in the Quaternay period (Pisias and Prell, 1985; Rea et al., 1986; Lyle et al., 1988; Le and Shackleton, 1992), as well as in the present day (Sautter and Sancetta, 1992).

As discussed, it is very important to deal with the productivity of carbonate and silica separately. If silica productivity increases, the ratio of organic carbon to carbonate ion increases, which may cause dissolution and dilution, and in turn lower $CaCO_3$ concentration. On the other hand, if carbonate productivity increases, the organic carbon to carbonate ion ratio decreases, which causes better preservation and higher $CaCO_3$ content. Therefore, the productivity hypothesis of Arrhenius (1988) holds well, if one considers only carbonate productivity.

The increased carbonate productivity during glacials (Berger, 1992) could have increased the lysocline depth in the Indian Ocean, and the increased siliceous productivity during interglacials might have shallowed the lysocline in the Indo-Pacific oceans.

SUMMARY

A study of the variation in calcium carbonate and dissolution indices in Quaternary sediment cores from the equatorial Indian Ocean reveals that $CaCO_3$ fluctuations in the western equatorial Indian Ocean are mainly derived from calcium carbonate dissolution. Lower $CaCO_3$ content at the troughs of the deeper depth core 154 (4860 m) than at the shallower cores 144 (4417 m) and 147 (4380 m) suggests that deeper depth cores experienced higher dissolution. The dissolution indices data further infer that dissolution of $CaCO_3$ is depth dependent in the Indian Ocean.

ACKNOWLEDGEMENTS

We thank Eric Olausson for supplying us with his unpublished planktonic foraminifer data and for his comments on the manuscript. This research is supported by the Swedish Natural Science Research Council (NFR) grant no. G-GU 4076-313.

REFERENCES

D.E. Archer. Equatorial Pacific calcite preservation cycles: Production or dissolution? *Paleoceanography*, **6**, 561-571 (1991).

G. Arrhenius. Sediment cores from the East Pacific. *Reports of the Swedish Deep Sea Expedition, 1947-1948*, **15**, 1-227 (1952).

G. Arrhenius. Rate of production, dissolution and accumulation of biogenic solids in the ocean. *Palaeogeography, Palaeoclimatology, Palaeoecology*, **67**, 119-146 (1988).

W.H. Berger. Biogenous deep-sea sediments: fractionation by deep-sea circulation. *Geological Society of America Bulletin* **81**, 1385-1402 (1970).

W.H. Berger. Deep-sea carbonates: Pleistocene dissolution cycles. *Journal of Foraminiferal Research*, **3**, 187-195 (1973).

W.H. Berger. Deep-sea carbonate and deglaciation preservation spike in pteropods and foraminifera. *Nature*, **269**, 661-663 (1977).

W.H. Berger, K. Fischer, C. Lai, and G. Wü. Ocean productivity and organic carbon flux, Part I overview and maps of primary production and export production, Uni. of California, sandiego, SIO references 87-30:1 (1987).

W.H. Berger. Pacific carbonate cycles Revisited: Arguments far and against productivity control. In: *Centenary of Japanese Micropaleontology*. K. Ishizaki and T. Saito (Eds). pp 15-25. Terra Scientific Publishing Company, Tokyo (1992)

W.S. Broecker, and S. Takahashi. The reationship between lysocline depth and insitu carbonate ion concentration. *Deep-Sea Research*, **25**, 65-95 (1974).

W.S. Broecker, and T.H. Peng. The role of $CaCO_3$ compensation in the glacial to interglacial atmospheric CO_2 change, *Global Biogeochemical Cycles*, **1**, 15-29 (1987).

J.M. Chey, D.K. Rea, and N.G. Pisias. Late Pleistocene paleoclimatology of the central equatorial Pacific: A quantitative record of eolian and carbonate deposition. *Quaternary Research*, **28**, 323-339 (1987).

W.B. Curry, and G.P. Lohmann. Late Quaternary sedimentation at the Sierra Leone Rise (eastern equatorial Atlantic Ocean). *Marine Geology*, **70**, 223-250 (1986).

J.E. Damuth. Quaternary climatic change as related by calcium carbonate fluctuations in western eqautorial Atlantic sediments. *Deep-Sea Research*, **22**, 725-743 (1975).

A.W. Droxler. World wide late Pleistocene climatically induced fluctuations of aragonite in periplatform ooze: a possible carbonate preservation record at intermediate water depth. 1988 Ocean Science Meeting, New Orlean. *EOS Trans*. AGU, Vol.**68**, no. 50, p. 1777 (1988).

S. Emerson, and M. Bender. Carbon fluxes at the sediment-water interface of the deep-sea calcium carbonate preservation. *Journal of Marine Research* **30**, 139-162 (1981).

J.W. Farrell and W.L. Prell. Climatic change and CaCO$_3$ preservation: An 800.000 year bathymetric reconstruction from central equatorial Pacific Ocean. *Paleoceanography*, **4**, 447-466 (1989).

J.W. Farrell and W.L. Prell. Pacific CaCO$_3$ preservation and δ^{18}O since 4 Ma: Paleoceanic and paleoclimatic implications. *Paleoceanography*, **6**, 485-498 (1991).

J.V. Gardner. Late Pleistocene carbonate dissolution cycles in the eastern Equatorial Atlantic. In: *Dissolution of Deep-Sea Carbonates*, W.V. Sliter, A.W.H. Bé and W.H. Berger (Eds), Cushman Foundation for Foraminiferal Research, Special Publication, **14**, 129-141 (1975).

A.D. Gordon. On measuring and modelling the relationship between two stratigraphically-recorded variables In: *Quantitative Stratigraphic Correlation*, J.M. Cubitt and R.A. Reyment (Eds), pp. 241-248 (1982).

J. Grötsch, G. Wü, and W.H. Berger. Carbonate cycles in the Pacific: Reconstruction of saturation fluctuations. In: *Cycles and Events in Stratigraphy*, G. Einsele, W. Ricken, A. Seilacher (Eds), Springer-Verlag Berlin, Heidelberg, pp. 110- 125 (1991).

R.S. Keir, and W.H. Berger. Late Holocene carbonate dissolution in the equatorial Pacific: reef groth or Neoglaciation ?. In: *The carbon cycle and atmospheric CO$_2$: Natural Variations Archean to Present*, E. Sundquist and W. Broecker (Eds), Geophysical Monographic Series, **32**, p 208-219 (1985).

V. Kolla, A.W.H. Bé, and P.E. Biscaye. Calcium carbonate distribution in the surface sediments of the Indian Ocean. *Journal of Geophysical Research*, **81**, 2605-2616 (1976).

J. Le, and N.J. Shackleton. Carbonate dissolution fluctuations in the western Equatorial Pacific during the late Quaternary. *Paleoceanography*, **7**, 21-42 (1992).

M. Lyle, D.W. Murray, B.P. Finney, J. Dymond, J.M. Robbins, and K. Brooksforce. The record of late Pleistocene biogenic sedimentation in the eastern tropical Pacific Ocean. *Paleoceanography*, **3**, 39-59 (1988).

B.A. Malmgren. Comparison of Visual and Statistical Correlation in Time Series Curves, *Mathematical Geology*, **10**, 103-106 (1978).

P.D. Naidu. Glacial to interglacial contrasts in the calcium carbonate content and influence of Indus discharge in two eastern Arabian Sea cores. *Palaeogeography, Palaeoclimatology, Palaeoecology*, **86**, 255-263 (1991).

P.D. Naidu, B.A. Malmgren, and L. Bornmalm. Quaternary history of calcium carbonate fluctuations in the western equatorial Indian Ocean (Somali Basin), Submitted to *Deep-Sea Research*.

T. Oba. Biostratigraphy and isotopic paleotemperature of some deep-sea cores from the Indian Ocean. Science Reports, Tohoku University, *2nd Serie (Geology)*, **41** (2), pp. 129-195 (1969).

E. Olausson. Description of sediment cores from central and western Pacific with the adjacent Indonesian region. *Reports of the Swedish Deep Sea Expedition 1947-1948*, **6**, 161-214 (1960a).

E. Olausson. Description of sediment cores from the Indian Ocean. *Reports of the Swedish Deep Sea Expedition (1947-1948)*, **9**, 53-88 (1960b).

E. Olausson. Evidence of climatic changes in North Atlantic deep-sea cores with remarks on isotopic paleotemperature analysis. *Progress in Oceanography*, **3**, 221-252 (1965).

E. Olausson. Climatological, geoeconomical and paleoceanographical aspects on carbonate eposition. In: *Progress in Oceanography*, **4**, M. Sears (Ed) Pergamon, Norwich, pp. 221-252 (1967).

E. Olausson. On the Würm-Flandrian boundary in deep-sea cores. *Geologie en Mijnbouw*, **48**, 349-361 (1969).

E. Olausson. Quaternary correlations and the geochemistry of oozes. In: *The Micropaleontology of Oceans*. B.M. Funnel (Ed). Cambridge University Press, Cambridge, pp. 375-398 (1971).

E. Olausson. The glacial oceans. *Palaeogeography, Palaeoclimatology, Palaeoecology*, **50**, 291-301 (1985).

L.C. Peterson and W.L. Prell. Carbonate preservation and rates of climatic change In: *The carbon cycle and atmospheric CO$_2$: Natural Variations Archean to Present*, E. Sundquist and W. Broecker (Eds), Geophysical Monographic Series, **32**, pp. 251-269, AGU, Washington D.C. (1985).

N.G. Pisias, and W.L. Prell. Changes in the calcium carbonate accumulation in the equatorial Pacific during the late Cenozoic evidence from HPC Site 572. In: *Geophysical Monographic Series*, E. Sundquist and W. Broecker (Eds), **32**, pp. 443-454, AGU, Washington D.C. (1985).

D.K. Rea, L.W. Chambers, J.M. Cuey, T.R. Janecek, M. Leinen, and N.G. Pisias. A 420,000 year record of cyclicity in oceanic and atmospheric processes from the eastern equatorial Pacific. *Paleoceanography*, **1**, 577-586 (1986).

L.R. Sautter and C. Sancetta. Seasonal associations of phytoplankton and planktonic foraminifera in an upwelling region and their contribution to the sea floor, *Marine Micropaleontology*, **18**, 263-278 (1992).

Rates of Erosion by the Colorado River in the Grand Canyon, Arizona.

W. Kenneth HAMBLIN
Department of Geology, Brigham Young University, Provo, Utah, U.S.A.

Abstract-- More than 150 basaltic lava flows poured into the western Grand Canyon during Quaternary time and formed a sequence of 13 major dams ranging from 60 to 700 meters high. Radiometric dates on the formation and destruction of these dams provide important new insights concerning rates of erosion, canyon cutting, and equilibrium in river systems.
    Prior to the formation of the first dam, slightly more than 1 million years ago, the Colorado river had cut down to its present gradient, and the size, shape, and form of the Grand Canyon was essentially the same as that which we see today. After each lava dam was formed the Colorado River cut through the dam, down to its original profile, but no further, and slope retreat removed all but small thin remnants next to the canyon wall. The cumulative thickness of the 13 lava dams was nearly 6 km and the cumulative amount of slope retreat on each side of the river was 1.3 km. The erosion and removal of the 13 dams took place in less than 250,000 years. Thus, the Colorado River has actually cut through a sequence of rock 6 km thick in 250,000 years. Yet the Grand Canyon upstream from the dams was neither deepened nor enlarged during this time.
    This study clearly indicates that the Colorado River has the capacity to erode with remarkable speed through any barrier placed in its way. It then establishes a profile of equilibrium. When this equilibrium is disturbed by any means, such as the formation of a lava dam, regional uplift, or movement along faults, readjustment back to equilibrium takes place almost instantaneously in a geological time frame. The question of the age of the Grand Canyon, (how long it took the Colorado River to form the Canyon), is not how fast did the river erode, but rather what was the rate of tectonic uplift which forced canyon cutting.

INTRODUCTION

    The late Cenozoic basalts in the western Grand Canyon constitute some of the most unique and spectacular displays of volcanic activity in North America (Figure 1). Numerous flows cascaded over the rim of the inner gorge and formed spectacular frozen lava falls 1,000 meters high. These basalts formed a sequence of lava dams impounding the water of the Colorado River to form a series of temporary lakes upstream. The lakes soon filled with water and sediment and, as they overflowed, new gorges were eroded through the lava barriers leaving only small remnants of basalt clinging to the steep canyon walls. In all, during late Cenozoic time, more than 150 lava flows have poured into the western Grand Canyon and formed a sequence of 13 major dams ranging from 70 to more than 700 meters high. Radiometric dates on the formation and destruction of the lava dams provide important new insights concerning rates of erosion, mechanics of canyon cutting, and equilibrium in river systems.
    It is apparent from the sequence of basalts preserved in the inner gorge of the Grand Canyon that four different types of dams were constructed during the period of Late Cenozoic volcanic activity in the area (Figure 2). The simplest type of barrier in the Grand Canyon was a dam constructed of a single lava flow, 50 to 200 meters thick. This type of dam was exceptionally long, extending downstream for several tens of km (in one instance, more than 137 km).
    A distinctly different type of dam was formed in the canyon by several abnormally thick, massive basalt flows. These flows were over 250 meters thick. Because these are the

Figure 1. Volcanic features in the Western Grand Canyon. Numerous lava cascades originate from fissures marked by cinder cones high on the Uinkaret Plateau on the skyline. They spill over the outer rim of the Canyon and fill the major tributary valley (on the left) and spread out over the broad Esplanade Platform, the major terrace in this part of the canyon. Some flows cascade over the rim of the Esplanade Platform and plunge down into the inner gorge of the Canyon. Remnants of lava dams form cliffs and terraces above the river in the center foreground.

oldest dams to form in the canyon, only a few remnants are visible, but the elevation of the upper surfaces of the remnants indicates that these barriers probably were not more than 30 km long.

Other dams were constructed by a sequence of numerous thin flows (units 1 to 10 meters thick), all of which were deposited in rapid succession. These dams ranged from 100 to 270 meters in height. The overall geometry of the dams built from numerous thin flows was much like the single-flow dams, but they were more steeply inclined (downstream slope).

A forth type of dam was built from several flow units 10 to 50 meters thick. The general characteristics of each individual flow are similar to those of the single-flow dams, but locally, deep channels filled with younger flows, ash, and sediment are cut in the flows. Lenses of river gravel, sand, and in some cases, ash are found separating the major units of basalt. The erosion of the flows, together with interstratified river gravels, clearly indicates that the lake which formed behind the dam overflowed during the period of extrusion. The complex dams were quite high, ranging from 180 to 420 meters above the present gradient of the Colorado River. The steep gradient on the preserved remnant suggests that the complex dams were no more than 20 km. In many cases, it may have been much less because the lower end of each flow was undoubtedly eroded by upstream migration of waterfalls and rapids before the subsequent flows were extruded.

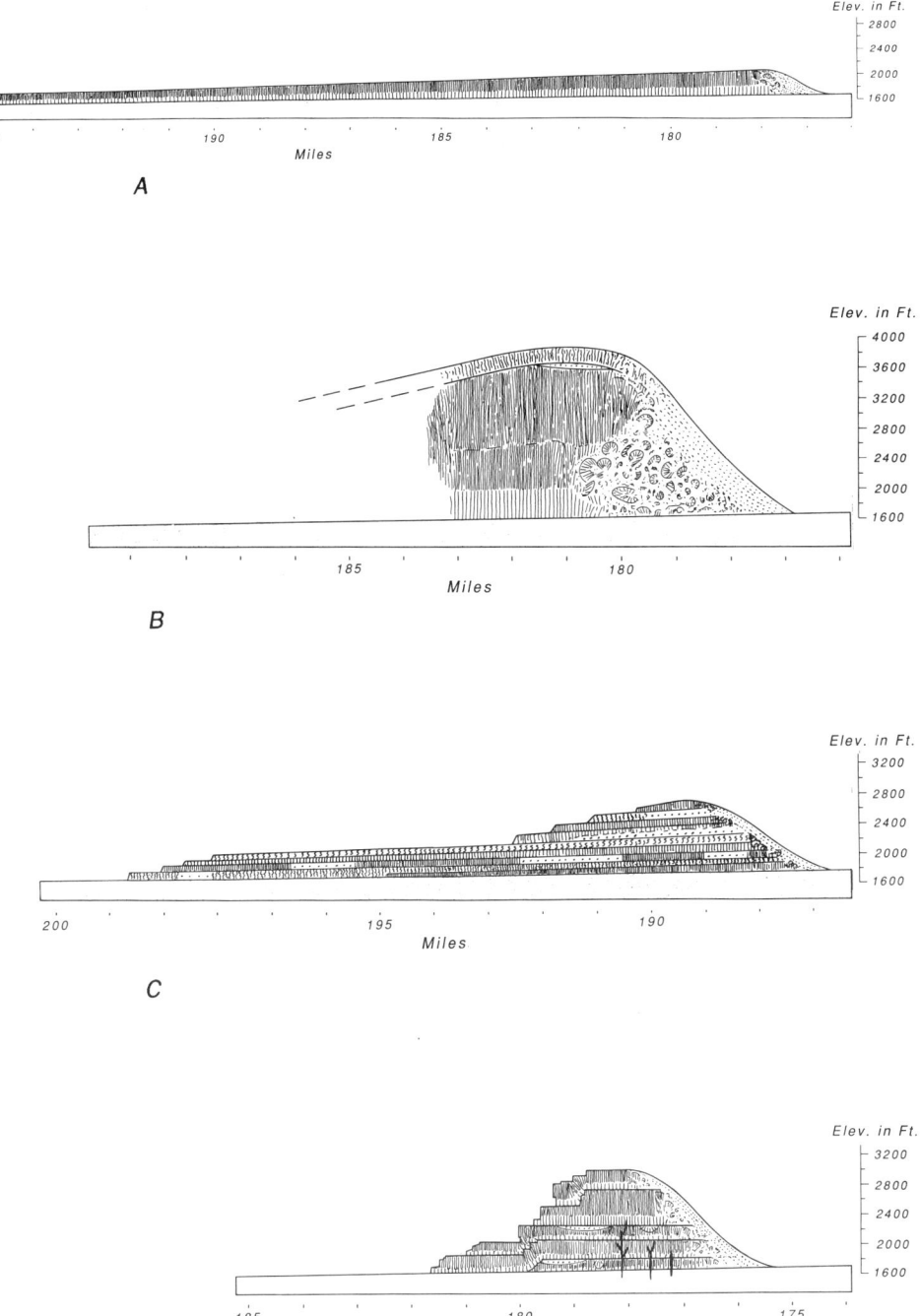

Figure 2. Types of lava dams. A. Single flows 50 to 200 meters thick. B. Massive flows over 250 meters thick. C. Numerous thin flows 1 to 10 meters thick. D. Several flows 10 to 50 meters thick.

RATES OF FORMATION AND DESTRUCTION OF LAVA DAMS

Although the formation of a lava dam in the Grand Canyon was a significant event which briefly changed the canyon morphology, the time necessary for the construction of a lava dam was remarkably short by any standard and would certainly be considered instantaneous in a geologic time frame. Observations of basaltic eruptions in historic times in Iceland, Hawaii, and elsewhere, indicate that many basaltic extrusions occur in a matter of days or weeks. The major flows in the Grand Canyon, which were 30 to 50 meters thick, probably moved tens of km down the Colorado River in only a few days. The construction of a single flow lava dam 700 meters high and tens of km long could be completed within a few months.

Dams built from multiple flow units would be different, in that they involved cycles of extrusion punctuated by periods of erosion, but the short time necessary for the erosion of a dam puts definite time constraints on the lifetime of any barrier to the Colorado River.

The time required for lakes or natural reservoirs to form behind lava dams was also extremely short. Based on hydrologic data from modern discharge measurements of the Colorado River in the Grand Canyon, we have calculated the time necessary for lakes to fill behind dams of a given height. Given modern flow regimes of the river, the lakes formed behind the smaller barriers (30 to 200 meters high) would fill and overflow in 2 to 17 days. Lakes formed behind the higher dams (200 to 300 meters high) overflowed within a year. The highest dam, 770 meters high, overflowed in 22 years.

These data also indicate that reservoirs behind the lava dams became "silted-up" in only a few hundred years at most; many of the smaller reservoirs became filled in a few months. Thus, a normal sediment load was soon transported over the dam, causing channel downcutting by abrasion soon after the dam was formed. Based on modern flow regimes of the Colorado River, a reservoir formed behind the dam 30 meters high would be filled with sediment in 10.3 months. A dam 400 meters high would be full of sediment in only 345 years. The highest dams would be filled with sediment in less than 3,000 years.

Two important erosional processes were undoubtedly initiated soon after the water impounded behind the dam overflowed the barrier. One is the process of downcutting of the channel floor by the normal traction of the sediment load transported over the dam after the lake silted up. A second erosional process is the upstream migration of the rapids or a waterfall which would develop at the toe, or downstream end of the lava flow. This type of erosion would begin immediately after water overflowed the top of the dam. If the migration of Niagara Falls is typical for the order of magnitude of migration of a waterfall on a large river, the lava dams in the Grand Canyon, which were generally less than 40 km long, would take a maximum of 40,000 years to be completely destroyed by headward migration of waterfalls alone.

It seems safe to conclude that the time interval for the various phases of the build-up and destruction of the dams would be in the following order of magnitude: Single flow dams would be formed in a matter of several days, whereas the higher complex lava dams would possibly take several years to be constructed. Water would fill the reservoir of these dams in a matter of months, and at the most, only several years would be required for the water to completely fill the reservoir behind the highest dams. Sediment would completely fill the small reservoirs within one to seven years, whereas in the deeper reservoirs it would take between 100 to 3,000 years to completely fill the lake with sediment. Most of the dams would be completely destroyed 1,000 to 40,000 years after they were formed.

RATES OF EROSION

The data concerning the formation and destruction of the lava dams in the Grand Canyon provide an unusual opportunity to document rates of erosion, slope retreat, and the adjustment of streams to equilibrium. This is possible because the relative ages of the thirteen lava dams have been determined by juxtaposition, and radiometric dates provide a series of time benchmarks from the earliest extrusions to the latest. In addition, we have measured the thickness, length, and volume of the lava dams so that we can calculate the rates at which erosion has cut through the dams. What we found is that the Colorado River was able to erode through the lava dams at an astounding rate (Figure 3). Single flow dams existed as a barrier to the Colorado River an average of only 10,000 years. Thus as shown in Figure 3, the total time that lava dams existed in the Grand Canyon was approximately 250,000 years.

When we consider the total thickness of all the rock through which the river has cut its channel in the last million years, it is clear that the river has the capacity to erode through any rock type almost instantaneously and that the profile of the Colorado River essentially in equilibrium. In addition, slopes of the Grand Canyon appear to be in quasi-equilibrium and receed at a significant rate only when equilibrium is upset. The rationale is as follows:

Rates of Downcutting

All available evidence indicates that prior to the extrusions of lava into the Grand Canyon, some one to two million years ago, the Colorado River had cut down to its present gradient and stratigraphic position. The size and shape of the canyon walls at the time the basalts were extruded were essentially the same as that which we see today. One of the most significant conclusions of our studies, with respect to processes of erosion and canyon cutting, is that after each lava dam was formed, the Colorado River eroded through the dam, down to its original profile *but no farther*. This process of reexcavating the canyon took place at least thirteen times during the last million years. The cumulative thickness of basalts in the thirteen lava dams was 3,550 m. Thus, the Colorado River has actually cut through a cumulative vertical thickness of more than 3 km of rocks (Figure 4). In all probability, the actual downcutting through the basalts took place in less than a million years because there were large time intervals between periods of dam destruction when the Colorado River was not influenced by the presence of lava. The best estimates (based on the time necessary for the destruction of the lava dams) would be that actual downcutting through all of the dams

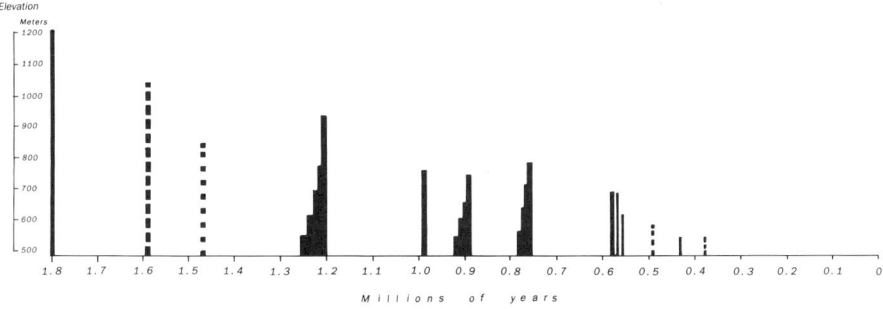

Figure 3. History of the formation and destruction of lava dams in the Grand Canyon. The height of each dam is indicated by vertical lines and the time interval during which the dam occupied the canyon is indicated by the width of the line. The age of the dams is indicated by its position on the graph (dashed where approximate).

(vertical distance of 3 km) took place in approximately 250,000 years. During the intervening time, the Colorado River was essentially at equilibrium flowing at its normal gradient with neither large-scale erosion nor deposition taking place.

This study clearly indicates that the amounts and rates of erosion (downcutting) by rivers are determined by the amount and rate of uplift. The fundamental question of rates of erosion and the age of the Grand Canyon (how long has it taken the Colorado River to cut through a sequence of Paleozoic strata 2 km deep) is not a question of how fast could the river erode, but how fast was the region uplifted. Based on the rates at which the river has been able to cut through the sequence of lava dams (or the Paleozoic strata when its channel was displaced beyond the lava flows), it is clear that the Colorado River has the capacity to cut down faster than tectonic processes can produce uplift.

This conclusion is supported by the fact that recurrent movement on major faults has displaced the strata more than 50 meters in slightly more than one million years. There is

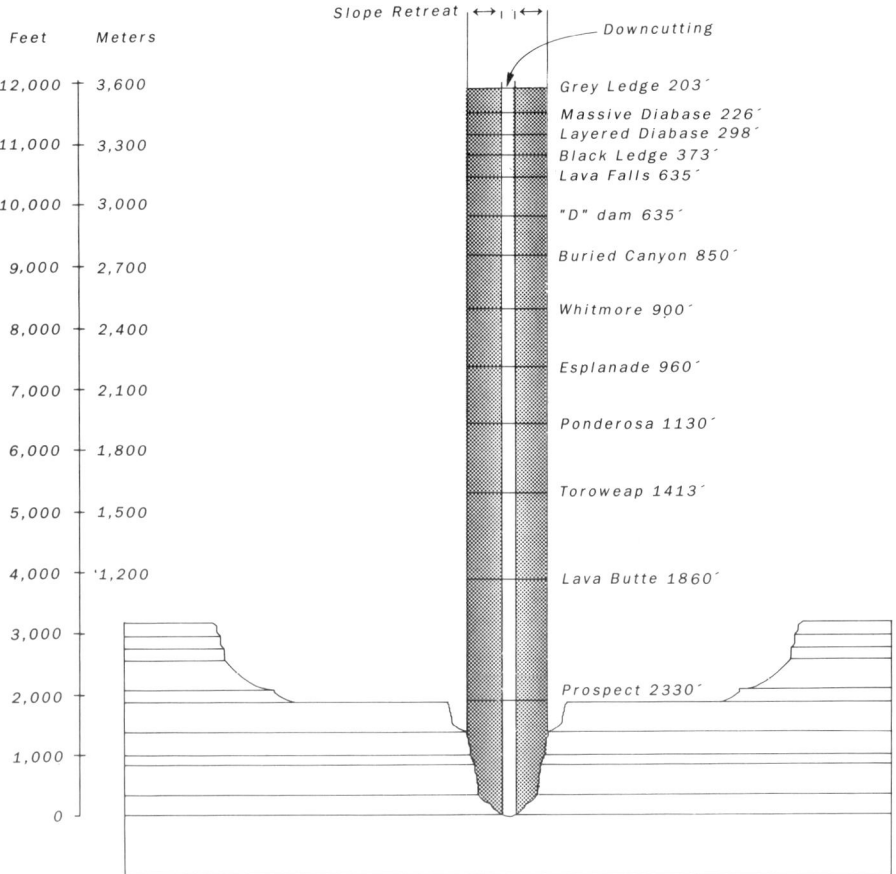

Figure 4. **Diagram showing the rate of downcutting of the Colorado River.** The cumulative thickness of basalt eroded by the Colorado River was a minimum of 3,550 meters. After each dam was formed the Colorado River cut down to its original profile, but no further. When we consider cumulative thickness of all dams and the time they existed as barriers in the Grand Canyon, it is obvious that the Colorado River has actually cut through a vertical thickness of 3,550 meters of rock in approximately 250,000 years.

no indication of rapids or waterfalls associated with the fault scarp, indicating that the escarpment produced by recurrent movement along the fault is erased by erosion immediately after it forms.

Rates of Slope Retreat. In addition to providing an insight into the capacity of a river to downcut and to maintain a profile of equilibrium, the remnants of the lava dams provide important documentation concerning rates of slope retreat and the relationship between slope profiles, stream gradient, and tectonic uplift.

Throughout the sections of the western Grand Canyon where remnants of lava dams still remain, one fact is completely clear - the remnants of lava dams are preserved only in the more protected parts of the canyon; i.e., on the insides of meander bends, in protected alcoves, and as hanging valleys in the mouths of minor tributaries. These remnants clinging to the canyon walls are preserved only as thin slivers, commonly only a few meters wide.

In considering the problem of slope retreat it should be emphasized that actual canyon cutting by abrasion along the river channel would produce a vertical gorge only 50 to 75 meters wide (the average width of the Colorado River channel) (Figure 3). Slope retreat has been responsible for widening the rest of the inner gorge. At the present time, the inner gorge is roughly 600 meters wide. This means that approximately 270 meters of slope retreat occurred on each side of the canyon after each lava dam was formed (Figure 5).

With the destruction of each lava dam and the rapid re-establishment of the river gradient to its original profile, there also was a rapid and contemporaneous retreat of the canyon walls back to their original profile. This process occurred so fast that the original a slope profile of the canyon was re-established before the formation of the next lava dam.

In every case, after a lava dam was eroded, the basalt retreated to within a few meters of the original canyon wall. Then, the processes of slope retreat essentially stopped. In many places, the basalt which formed the dams has been completely removed, but the process of slope retreat did not enlarge the canyon and go back beyond the original canyon walls.

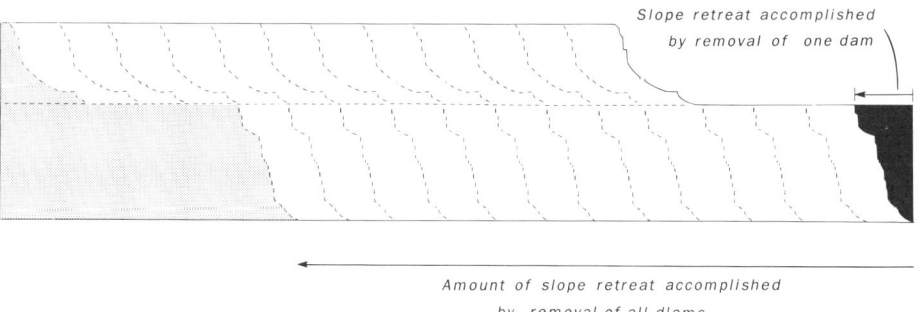

Figure 5. Diagram showing the amount of slope retreat associated with erosion of the lava dams. Actual downcutting of the river channel by abrasion and waterfall migration would produce a gorge 50 to 75 meters wide (the width of the river channel). Slope retreat is responsible for widening the rest of the inner gorge (dark area shown in Figure 3). The inner gorge is roughly 600 meters wide. This means that approximately 270 meters of slope retreat has occurred on each side of the canyon during the erosion of each dam. The cumulative amount of slope retreat that has actually occurred on each side of the canyon in approximately 250,000 years was 3.5 km. Yet the canyon is the same size and has the same profile as it had before the onset of volcanic activity. Obviously after each dam was breached the slopes receded back to their original profile and then remained in dynamic equilibrium.

Several important conclusions are obvious from these facts: (1) the energy system that causes slope retreat in the Grand Canyon has the capacity to erode canyon walls at a rate of 2 to 5 km per million years or more. (2) After uplift (or other processes which alter the rivers gradient) slopes recede rapidly to establish a profile of quasi-equilibrium. They then recede at an extremely slow rate. (3) Slope retreat is delicately balanced with the stream gradient. Renewed downcutting of the stream channel is accompanied by renewed slope retreat. (4) The profile of the Grand Canyon apparently is in a state of quasi-equilibrium. (5) Periods of major slope retreat as well as downcutting are controlled by tectonic uplift.

Method of Estimating Mean Annual Temperatures from Plant Fossils in the Pliocene and Pleistocene Periods

Y. HASE and A. IWAUCHI
Department of Geology, Faculty of General Education, Kumamoto University, Kurokami 2-4-1, Kumamoto 860, JAPAN
JSPS Fellowship for Japanese Junior Scientists, Noguchi-machi 508-4, Kumamoto 860, JAPAN

Abstract--We propose a method for the estimation of paleo-temperature in the Pliocene and Pleistocene periods in Kyushu, Japan. We combined Wolfe's method by using the type of leaf margin and Hase's method based on the distribution of the living plant species.

INTRODUCTION

On the estimation of paleo-temperature by macro-plant fossils, Wolfe (1978) proposed the method, based on the physiognomic character of foliage, that the type of leaf margin was either entire or non-entire (lobed or toothed). The plots for mean annual temperature of many localities of Eastern Asia correlate best with the percentages of entire-margined species in whole species of broad-leafed trees in the forests (Figure 1).

Wolfe's method is very valuable for the estimation of paleo-temperature. His method leads to reliable results based on using fossil flora composed of at least 30 species. We often could not get enough species from each locality in the fossil flora to apply Wolfe's method. In cases, where there were several species, but less than 30 species, we had to estimate by another method to get mean annual paleo-temperature.

Other estimation of the paleo-temperature proposed by Hase (1988) was based on the distribution of living species closely related with the fossil species.

Concept of "climate elements" and "climatic zone's elements"

We think that the climate elements mean the climatic components of the living plants distributed under the environmental conditions: That is the climatic factors concerning climate such as temperature, precipitation, humidity, insolation, snowfall, wind, fog, etc.

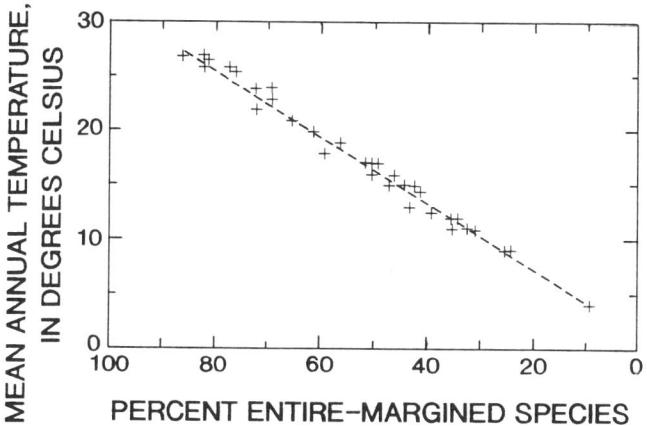

Figure 1 Correlation of percentages of entire-margined species with mean annual temperature. Data based on local floras in the humid to mesic forests of eastern Asia. (Wolfe,1979)

Hase (1988) showed the concept that "climatic zone's elements" were recognized in a plant species distributed in some climatic zones. For example, if a species is distributed in the cool-temperate zone, we can say the species has the cool-temperate zone's element. And if another species is distributed in the warm-temperate and subtropic, we also regard it has the warm-temperate zone's and subtropic zone's elements.

Counting of "climatic zone's elements" and "warm element"

Hase (1988) showed the calculation of the climatic zone's elements in Table 1. Namely, on some living flora, if a plant species (a') is distributed over three climatic zones(subtropic, warm-temperate and cool-temperate), we count one third in each climatic zone's column. Their sum becomes 1 (one). In the same way, if another species (b') is distributed over two climatic zones such as cool-temperate and subarctic, we count one half in each climatic zone's column. Their sum becomes 1 (one). And we count it about c' in the same way. After that, add the figures in the same climatic zone's elements respectively. In this case, one third in the subtropic zone, two third in the warm-temperate zone, seven sixths in the cool-temperate zone and five sixths in the subarctic zone. And we calculate the percentage for each climatic zone's elements, such as subtropic 11.1% , warm-temperate 22.2%, cool-temperate 38.9% and subarctic 27.8%.

Table 1 Counting and calculating methods of climatic zone's elements and percentage of warm elements.

Fossil flora		Climatic zone's elements						
Fossil species	Living species	Tropic	Subtropic	Warm-temperate	Cool-temperate	Subarctic	Arctic	Total
a	a'		1/3	1/3	1/3			1
b	b'				1/2	1/2		1
c	c'		1/3	1/3	1/3			1
Total of climatic zone's elements			1/3	2/3	7/6	5/6		3
Percentage of climatic zone's elements			11.1%	22.2%	38.9%	27.8%		100%
			33.3%		66.7%			100%
		Percentage of warm elements						

The combined percentage of tropic, subtropic and warm-temperate zone's elements is called the percentage of "warm elements" by Hase (1988). The percentage of the warm elements of some flora correlates with the mean annual temperature of the area covered by the forests composed of the species (Figure 2).

Relationship of "warm elements" and mean annual temperature

Figure 2 shows the relationships between the percentage of the warm elements and mean annual temperature on the living flora from some localities in Kyushu (Table 2). We recognize that if the percentage increases, the mean annual temperature becomes higher.

Method of estimating paleo-temperature based on fossil plants

Practically, we combined Wolfe's method and Hase's method for the estimation of paleo-temperature based on fossil flora. The relationships between Wolfe's and Hase's methods on the fossil floras from Pliocene and Pleistocene sediments in central Kyushu is shown in Figure 3. In Figure 3, the left side vertical line shows the percentage of entire-margined species and the right hand vertical line shows the mean annual temperature in degrees Celsius. We graduated on two vertical axes based on the correlation of percentages of entire-margined species with annual temperature shown by Wolfe (1978). The horizontal axis shows the percentage of warm elements. The large squares in Figure 3 show the fossil floras in which we got more than 30 broad-leafed species. The small circles in Figure 3 show the fossil floras in which we got fewer than 30 broad-leafed species. Thirty (30) or more species are enough to calculate the mean annual temperature using only Wolfe's method. We drew a line through these large squares. (On this line, only Wolfe's method is applied.) When the point of small circles became out of this line, we drew the perpendicular from the point on this line and read the measure on the right hand vertical axis for the point of intersection of the line and the perpendicular; thus we could get the mean annual paleo-temperature for the fossil flora.

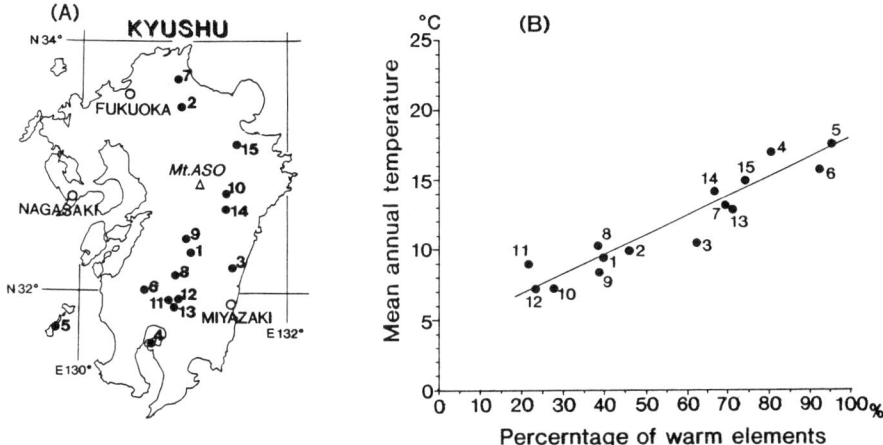

Figure 2 Relationship between percentage of warm elements and mean annual temperature of the living flora in Kyushu. (A) Locality map. (B) Regression line. The numbers in A and B show the localities in Table 2.

Table 2 Percentages of climatic zone's elements at some localities in Kyushu (Miyawaki. 1981). Definition of the climatic zones of living woody plants in Japan is based on Kitamura and Murata (1971 and 1979).

No.	Locality	Altitude (m)	MAT (°C)	Climatic zone's elements					
				Arc.	Suba.	Cool.	Warm.	Subt.	Trop.
1	Ichifusayama	1050	9.4			60.9	37.8	1.3	
2	Hikosan	870	9.7			54.2	44.2	1.7	
3	Kijo-machi	865	10.3			37.9	56.1	6.1	
4	Sakurajima	50	16.7			20.0	70.0	10.0	
5	Shimokoshikijima	60	17.7			4.6	68.5	26.9	
6	Okuchi-shi	160	15.8			7.4	70.6	22.1	
7	Kawaradake	420	12.8			29.1	56.7	14.2	
8	Shiragadake	900	10.3			61.9	38.1	2.4	
9	Kunimiyama	1220	8.4		1.4	60.1	38.4		
10	Kosobosan	1400	7.3		4.0	68.3	27.8		
11	Kirishimayama(1)	1270	9.0		3.6	75.0	21.4		
12	Kirishimayama(2)	1550	7.3		2.9	73.5	23.5		
13	Kirishimayama(3)	580	13.1		1.1	29.6	58.6	10.8	
14	Takachiho-cho	290	14.5		0.6	33.1	57.5	8.8	
15	Shonai-machi	120	14.9			26.3	65.9	7.9	

MAT: Mean annual temperature

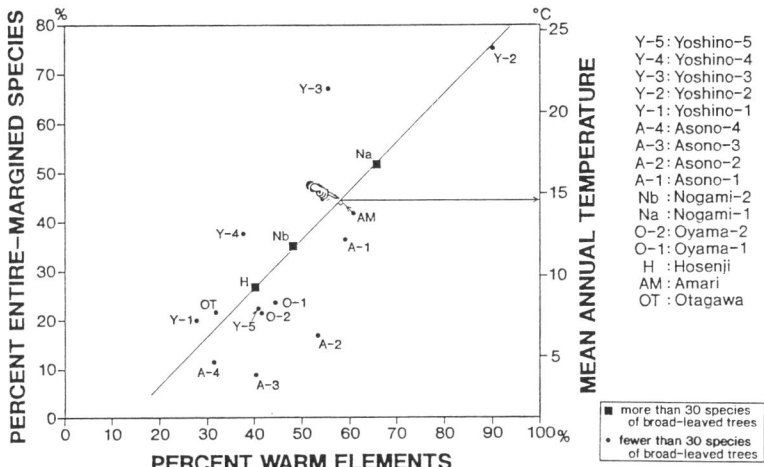

Figure 3 Estimation of mean annual temperature based on the percentages of entire-margined species and warm elements of the fossil flora in Pliocene and Pleistocene in central Kyushu. (Hase & Iwauchi, 1985; Iwauchi & Hase, 1986, 1987, 1989, 1992)

We would like to show a typical example for the estimation of paleo-temperature and change of paleoenvironment in the middle Pleistocene in central Kyushu (Figure 4). Figure 4 shows the mean annual paleo-temperature curve in the middle Pleistocene in Kyushu, based on five florules (Yoshino -1, -2, -3, -4, -5) from the Yoshino Formation, two florules (Na, Nb) from the Nogami Formation and four florules (Asono -1, -2, -3, -4) from the Asono Formation. We plot the points of mean annual paleo-temperature for each fossil florule. Then we calculate the mean annual paleo-temperature at sea level a using a lapse rate of 0.6 C/100m and today's altitude of each lake deposit. After that, we draw a curve through those points of the mean annual paleo-temperature at the sea level according to the paleo-vegetational changes estimated by fossil pollen flora. This curve is shown by the bold line in Figure 4. It is thought that central Kyushu is an uplifting area at a rate of about 500m/2 million years. So we make a zone for 3 C along this mean annual paleo-temperature curve like the dotted area in Figure 4. It means that if we consider the amount of uplifting after the lake deposits were formed, the mean annual paleo-temperature curve moves only in this zone.

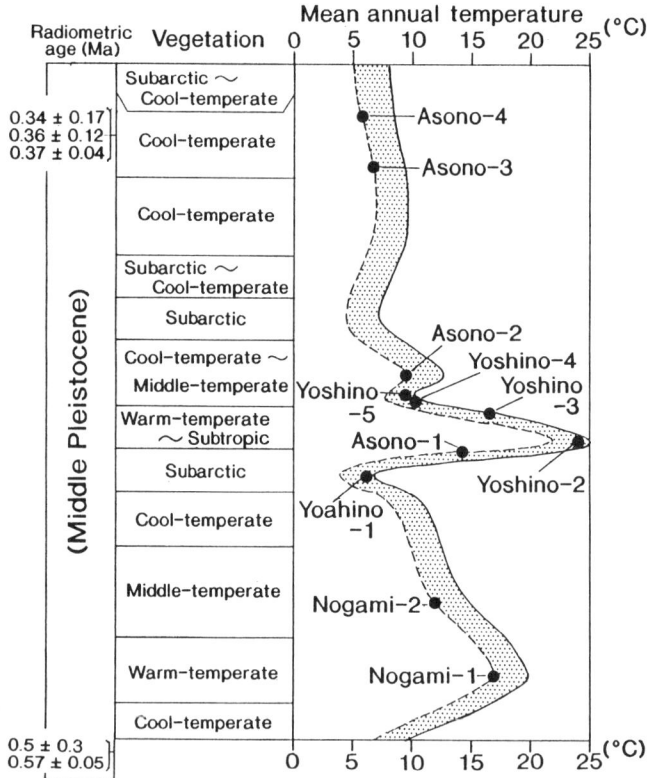

Figure 4 Mean annual temperature curve in the middle Pleistocene in central Kyushu.

In general, the plant species in Pliocene are related with the living species; therefore, we can estimate the mean annual temperature by using the living species related to the fossil species. Iwauchi and Hase (1992) showed the climatic changes based on estimation from plant fossils during the last 3 million years in Kyushu, Japan.

REFERENCES

Y. Hase. Late Cenozoic history and paleoenvironment of southern Kyushu, Japan, Mem. Fac. Gen. Educ. Kumamoto Univ. Nat. Sci. 23, 37-82(1988).

Y. Hase and A. Iwauchi. Late Cenozoic vegetation and paleoenvironment of northern and central Kyushu, Japan -Part 1 Asono area-, Jour. Geol. Soc. Japan, 91, 753-770(1985).

A. Iwauchi and Y. Hase. Late Cenozoic vegetation and paleoenvironment of northern and central Kyushu, Japan -Part 2 Ajimu-Innai area(Upper Pliocene)-, Jour. Geol. Soc. Japan, 92, 591-598(1986).

A. Iwauchi and Y. Hase. Late Cenozoic vegetation and paleoenvironment of northern and central Kyushu, Japan -Part 3 Southern part of Kusu basin (Lower and Middle Pleistocene)-, Jour. Geol. Soc. Japan, 93, 496-489 (1987).

A. Iwauchi and Y. Hase. Late Cenozoic vegetation and paleoenvironment of northern and central Kyushu, Japan -Part 4 Oyama and Tsuetate areas(Lower Pleistocene)-, Jour. Geol. Soc. Japan, 95, 63-75 (1989).

A. Iwauchi and Y. Hase. Late Cenozoic vegetation and paleoenvironment of northern and central Kyushu, Japan -Part 5 Yoshino areas (Middle Pleistocene)-, Jour. Geol. Soc. Japan, 98, 205-221 (1992).

A. Iwauchi and Y. Hase. Late Cenozoic climatic changes in Kyushu. 29th International Geological Congress, Abstracts, vol.2, 248 (1992).

S. Kitamura and G. Murata. Coloured Illustrations of Woody Plants of Japan, I, II, Hoikusha Publishing CO., LTD, Tokyo (1971, 1978).

A. Miyawaki. Vegetation of Japan, 2, Kyushu, Shibundo CO., LTD. Publishers, Tokyo (1981)

J. A. Wolfe. A paleobotanical interpretation of Tertiary climates in the Northern Hemisphere. American Scientist, 66, 694-703(1978).

J. A. Wolfe. Temperature parameters of humid to mesic forests of Eastern Asia and relation to forests of other regions of the Northern Hemisphere and Australasia., Geol. Surv. Professional Paper 1106, 1-37 (1979)

Spatial variation of CO_2^- and SO_3^- radicals in massive coral from Ishigaki Island, Japan and its implications

S. IKEDA, M. FURUSAWA, and M. IKEYA

Department of Earth and Space Science, Faculty of Science, Osaka University, Toyonaka, Osaka 560, Japan

Abstract-- Electron spin resonance (ESR) signal intensities of a massive coral colony from Ishigaki Island, in the Ryukyu Islands, Japan are measured along the growth direction with a microwave scanning ESR imaging system. The annual high intensity regions of CO_2^- radicals are the same as the skeletal high density regions. Periodic high intensity of the CO_2^- radicals with an interval of 11 year is also observed. The intensity of SO_3^- radicals has not distinct seasonal variation. This seems to correlate the lack of fluorescent bands in the sample, which means low influence of terrestrial water causing the variation of the SO_3^- intensity.

INTRODUCTION

Electron spin resonance (ESR) spectroscopy allows the detection of unpaired electrons at radiation-induced radicals. The ESR intensities of the radicals in various geological and archaeological materials have been used for ESR dating. Coral is one of the materials in which ESR dating has been successful [1-4], and the signals of SO_2^-, SO_3^-, and CO_2^- radicals are generally observed in naturally irradiated coral [5-7]. The ESR intensities of the CO_2^- and SO_3^- radicals increase with artificial gamma-ray irradiation.

Recently, ESR of coral has been shown to be an useful indicator of paleoenvironment [8]. Ikeda et al. (1992) measured the ESR intensities of the CO_2^- and SO_3^- radicals in the last 15 years' growth of a coral from Hudson Island, in the Great Barrier Reef using a microwave scanning ESR imaging system [9] after gamma-ray irradiation. The sample coral possesses yellow-green fluorescent bands [10] due to terrestrial humic acids incorporated into the skeleton. The intensities of both CO_2^- and SO_3^- radicals varied seasonally but the periods of high intensity were different; the high intensity regions of the SO_3^- radicals deposited earlier than those of CO_2^-. The annual intense regions of the CO_2^- radicals correlated with the dense bands in the skeleton. The fluctuation of the CO_2^- radicals is attributable to the change in effective sample weight due to the skeletal density variation. The annual high intensity regions of the SO_3^- radicals are the same as the yellow-green fluorescent bands which correlate with summer, monsoonal rainfall seasons [10]. The SO_3^- radicals are derived from SO_3^{2-} ions incorporated into aragonite. However, it is still unknown what caused the change of the SO_3^- intensity. A longer cycle of ~12 year was also detected by maximum entropy method (MEM) for the ESR intensities.

In the present study, we measured the ESR intensities of the CO_2^- and SO_3^- radicals in a longer coral core of a massive *Porites* spp. from Ishigaki Island, Japan to know the pattern of the variation of the ESR intensities and to compare with that of the coral of Hudson Island [8]. The reason of the variation is explained by the above comparison and the correlation between the ESR intensities and data of annual sunshine duration, rainfall and mean temperature.

SAMPLE

The core of a massive coral, *Porites* spp., was taken from Ishigaki Island in the Ryukyu Islands in August, 1990. Fringing reefs are well developed around the island, particularly in the east coast (windward side). The sampling site is in the moat adjacent to the inner reef flat in Shiraho area, in the southeast coast. Many living massive *Porites* spp. including microatolls are observed in this area [11]. The mouth of the Todoroki River, the discharge of which is 8×10^5 m^3 / yr, is located 1 km north of the site, however, the terrestrial water flows mainly to the north from the mouth (contrary to the Shiraho area)[12]. No fluorescent band due to humic acids is observed in the sample. The last growth of 0.45 m (1947-1990) of a massive *Porites* colony, which is about 2-3 m in diameter, was used for the study.

EXPERIMENTAL

The sample coral was sliced parallel to the growth direction and was irradiated at ~50 kGy with a ^{60}Co gamma source. The sample was placed on the aperture of a TE$_{111}$ mode microwave cavity and was scanned by every 1 mm step. ESR intensities of the CO$_2^-$ radicals (g = 2.0007) and the SO$_3^-$ radicals (g = 2.0031) at each step were obtained with a microwave scanning ESR imaging system along the growth direction of the coral.

RESULTS AND DISCUSSION

Figure 1 shows X-radiograph (positive) and the measured ESR intensities of the CO$_2^-$ and SO$_3^-$ radicals for (a) the growth of 1947-1965 and (b) the growth of 1966-1990 of the coral. Data of annual rainfall, temperature and sunshine duration are also shown. The white part of the X-radiograph corresponds to skeletal low density bands and the black part corresponds dense bands. The relative intensity of the SO$_3^-$ radicals denotes SO$_3^-$ intensity / CO$_2^-$ intensity.

The positions of annual high and low ESR intensity of the CO$_2^-$ radicals correlate with the skeletal high and low density bands, which are distinguished by the X-radiograph of the sample. The result confirms that the variation of CO$_2^-$ intensity is caused by changes of skeletal density. The sample coral deposited dense skeletons in summer, which seems to be the optimum season of growth. This is contrary to the many cases that light bands deposit during optimum conditions [13].

The annual fluctuations are not apparent in the intensity of the SO$_3^-$ radicals. This is different from the result of the coral from Hudson Island [8], the seasonal variation of which is very obvious. The lack of fluorescent bands in this sample, i. e., relatively low influence of terrestrial runoff may be related with indistinct seasonal variation of the SO$_3^-$ intensity. This indicates that SO$_3^{2-}$ ions, from which the SO$_3^-$ radicals are formed, are supplied both from terrestrial and marine waters, and the change of the SO$_3^-$ intensity is due to the variation of terrestrial water.

There is not clear relation between the ESR intensities and annual sunshine duration, rainfall, and mean temperature. But high values of sunshine duration in 1967 and 1977 seems to correlate with the high CO$_2^-$ intensity. Periodic high CO$_2^-$ intensity in ~1955, ~1966, ~1977, and ~1988 at 11 years' interval may be related to the solar activity which has 11.1 years' cycle [14]. Similar periodicity was also detected in the Hudson Island's coral [8].

CONCLUSIONS

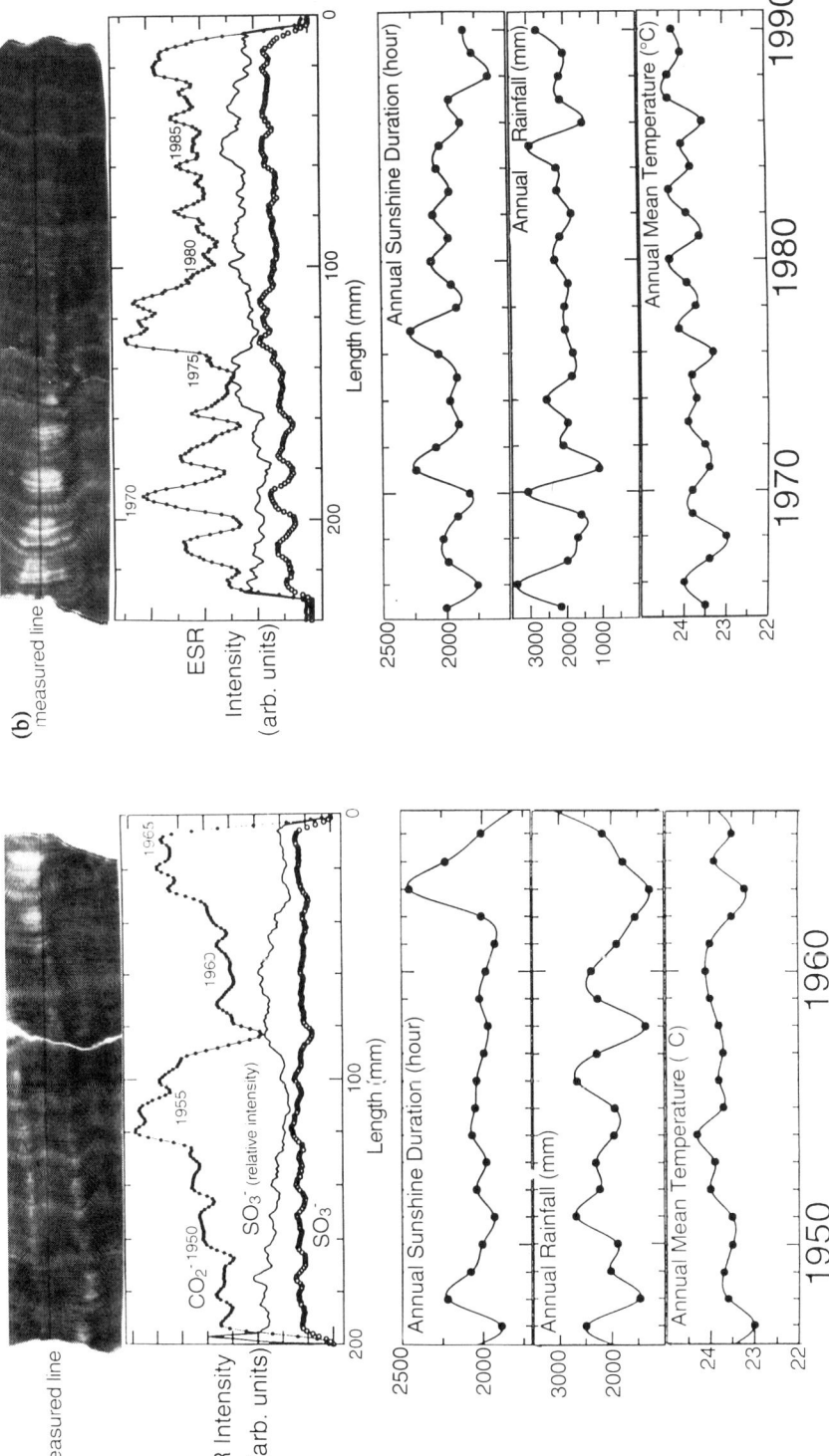

Figure 1. X-radiograph (positive) and the ESR intensities of the CO_2^- and SO_3^- radicals for (a) the growth of 1947-1965 and (b) the growth of 1966-1990 of the coral with the data of annual rainfall, temperature and sunshine duration.

CONCLUSION

The annual high intensity regions of CO_2^- radicals in the massive *Porites* spp. from Ishigaki Island are the same as the dense bands of the skeletons. Periodic high intensity of the CO_2^- radicals with an interval of 11 year is also observed. The intensity of SO_3^- radicals does not change seasonally. The result corresponds to the lack of fluorescent bands in the sample, which means low influence of terrestrial water.

Acknowledgements
We are grateful to Prof. S. Mezaki, Dr. T. Nakamori, Mr. T. Yamada and Mr. I. Saito for providing the sample and for discussion, and to Dr. M. Akaboshi for the gamma-ray irradiation. Dr. C. Yamanaka and Mr. M. Yamamoto are also acknowledged for preparing the figure. This work is partially supported by the grant of the JSPS fellowships for Japanese Junior Scientists.

REFERENCES

1. M. Ikeya and K. Ohmura. Comparison of ESR ages of corals from marine terraces with ^{14}C and $^{230}Th/^{234}U$ ages, *Earth Planet. Sci. Lett.* **65**, 34-38 (1983).
2. M. Koba, M. Ikeya, T. Miki and T. Nakata. ESR ages of the Pleistocene coral reef limestones in the Ryukyu Islands, Japan. In: *ESR Dating and Dosimetry* M. Ikeya and T. Miki (Eds). pp. 93-104. Ionics, Tokyo (1985).
3. U. Radtke, R. Grün and H. P. Schwarcz. Electron spin resonance dating of the Pleistocene coral reef tract of Barbados, *Quat. Res.* **29**, 197-215 (1988).
4. S. Ikeda, M. Kasuya and M. Ikeya. ESR ages of Middle Pleistocene corals from the Ryukyu Islands, *Quat. Res.* **36**, 61-71 (1992).
5. A. Kai and T. Miki. Electron spin resonance of sulfite radicals in irradiated calcite and aragonite, *Radiat. Phys. Chem.* **40**, 469-476 (1992).
6. R. Debuyst, M. Bidiamambu and F. Dejehet. Diverse CO_2^- radicals in γ- and α-irradiated synthetic calcite, *Bull. Soc. Chim. Belg.* **99**, 535-541, (1990).
7. T. Miki and A. Kai. Rotating CO_2^- centers in coral and related materials, *Jpn. J. Appl. Phys.* **29**, 2191-2192 (1990).
8. S. Ikeda, D. Neil, M. Ikeya, A. Kai and T. Miki. Spatial distribution of CO_2^- and SO_3^- radicals in massive coral as environmental indicator, *Jpn. J. Appl. Phys.* **31**, L1644-1646 (1992).
9. P. J. Isdale. Fluorescent bands in massive corals record centuries of coastal rainfall, *Nature* **310**, 578-579 (1984).
10. M. Furusawa and M. Ikeya. Electron spin resonance imaging utilizing localized microwave magnetic field, *Jpn. J. Appl. Phys.* **29**, 270-276 (1990).
11. S. Mezaki (Ed.). *The coral reefs of Ishigaki Island.* World Wide Fund for Nature Japan, Tokyo (1991). (in Japanese).
12. A. Tokuyama, Z. Yonaha, N, Taira, T. Hirota and M. Ageta. Geochemical survey of water around the Shiraho Coral Reef, Ishigaki Island. In: *The coral reefs of Ishigaki Island.*S. Mezaki (Ed.). pp. 183-208, World Wide Fund for Nature Japan, Tokyo (1991). (in Japanese).
13. R. C. Highsmith. Coral growth rates and environmental control of density banding, *J. Exp. Mar. Biol. Ecol.* **37**, 105-125 (1979).
14. R. G. Currie. Solar cycle signal in surface air temperature, *J. Geophys. Res.* **79**, 5657-5660 (1974).

Late Quaternary paleoceanography of the Japan Sea; a tephrochronological and sedimentological study.

Ken Ikehara[1], Kiyoshi Kikkawa[1], Hajime Katayama[1] and Koji Seto[2]
1 Geological Survey of Japan, Higashi 1-1-3, Tsukuba, Ibaraki 305, Japan.
2 Shimane University, Nishikawazu-machi 1060, Matsue, Shimane 690, Japan.

Abstracts Tephrochronological, sedimentological and micropaleontological studies for late Quaternay sediments of the Japan Sea clarified that the paleoceanography of the Japan Sea was influenced by the global sea level change and paleoclimatic change on the Japanese Islands. At the last glacial maximum (about 13,600 - 23,000 years B.P.), deep water circulation was prevented by the low-salinity surface water cap related to the prevention of the inflow of oceanic water to the Japan Sea due to shallow entrances and the lower sea level. After the inflow of oceanic water, the sea bottom became oxidized conditions. During the Younger Dryas event (10,000 - 11,000 years B.P.), an anoxic bottom condition was formed. Sedimentation rates were lower during the last glacial maximum, and higher during the transgressional period. Because the age of the inflection point in sedimentation rates (about 13,000 years B.P.) coincides with the age of climatic change (increasing of snowfall), the change was caused by the change of sediment discharge from land related to the climatic change.

Key words: climate, dark band, Japan Sea, sea level change, tephra.

INTRODUCTION

The Japan Sea is a marginal sea between the Japanese Islands and Asian continent and is separated from the East China Sea by the Tsushima Strait (sill depth, 130 m), from the Pacific Ocean by the Tsugaru Strait (130 m) and from the Okhotsk Sea by the Soya (55 m) and Tatarskiy Straits (15 m). Because of these shallow entrances, global sea level fluctuation during the Quaternary strongly affected the paleoceanography of the Japan Sea [15, 21]. Detailed correlations between global events and paleoceanographic changes in the Japan Sea, however, have not been clarified, because of only a few time-controlling points.

On the other hand, there are many active volcanos around the Japan Sea. They provide useful time-planes, tephras. Furthermore, the deeper areas of the Japan Sea are areas of mud deposition [4]. These geological and sedimentological settings suggest that the Japan Sea is a suitable location for marine tephrochronology.

The Geological Survey of Japan collected many sediment cores from the southern and eastern Japan Sea floor during 1986 - 1990 (136 cores from the southern part, 134 from the eastern part). In these sediment sequences, several tephras which can be indentified to the age-known tephras, were intercalated. Using these tephrochronological data, we can discuss the detailed history of the paleoceanography of the Japan Sea.

RESULTS

Lithology   Deep-sea sediments of the late Quaternary were mainly composed of silt and clay. Sandy sediments were found only as gravity flow deposits (turbidites) or volcanogenic materials. Many turbidites were found at the fringe of the Yamato Basin, Toyama Trough, Mogami Trough and Japan Basin and on Toyama Deep-sea Fan. Tephra layers intercalated in muddy sediments were observed in 48 of collected 270 cores. Dark and light colored alternation, which was reported in previous studies [7,12,21], was well recognized. Parallel and fine lamination was well preserved in dark colored bands. After the deposition of Aira-Tn ash (AT; 24,500 years B.P. [11]), two dark bands were formed (Fig.1). The older one (TL2) is about 60 - 165 cm thick and has a high sulfur content, containing framboidal pyrites, and a low C/S ratio (C/S<2) [5]. Fine parallel laminations and some beds with abundant planktonic foraminiferal tests occur in TL2. The younger one (TL1) is thin (about 1.5 - 15 cm thick) and has foraminiferal thin laminations. Light colored sediments are massive and structure-less mud except of bioturbation.

Tephrochronology   Twelve tephra layers were identified using the chemical composition of volcanic glass shards, which were analyzed by using an inductively

Table 1   List of tephra layers collected.

Name	Source	Age(ky B.P.)	Ref.
Baegdusan-Tomakomai Ash (B-Tm)	Baegdusan Volcano (North Korea)	1	[16]
Ulreungdo-Oki Ash (U-Oki)	Ulreungdo Volcano	9.3	[10]
Towada-Hachinohe Pumice (Hpfl)	Towada Caldera (Tohoku)	12-13	[1]
Tsumagoi Fall Pumice (YPk)	Asama Volcano (north Kanto)	13.6	[2]
Kitanihon 2 Ash (NJ2)	unknown	15?	
Daisen Kusatanihara Pumice (KsP)	Daisen Volcano (San'in)	18	[13]
Aira-Tanzawa Ash (AT)	Aira Caldera (south Kyushu)	24.5	[11]
Toyamaoki Ash (To)	unknown	30-35	
San'in 1 Ash (SAN1)	unknown (probably San'in)	30-35	
Ulreungdo-Yamato Ash (U-Ym)	Ulreungdo Volcano	30-35	[17]
Aso-4 Ash (Aso-4)	Aso Volcano (middle Kyushu)	80	[18]
Toya Ash (Toya)	Toya Caldera (Hokkaido)	90-100 (103-134; 100-120)	[9] ([22]; [8])

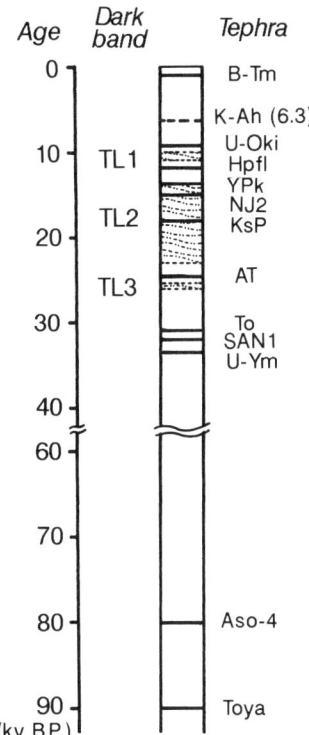

Figure 1 Generalized stratigraphy of Japan Sea sediments in the late Quaternary.

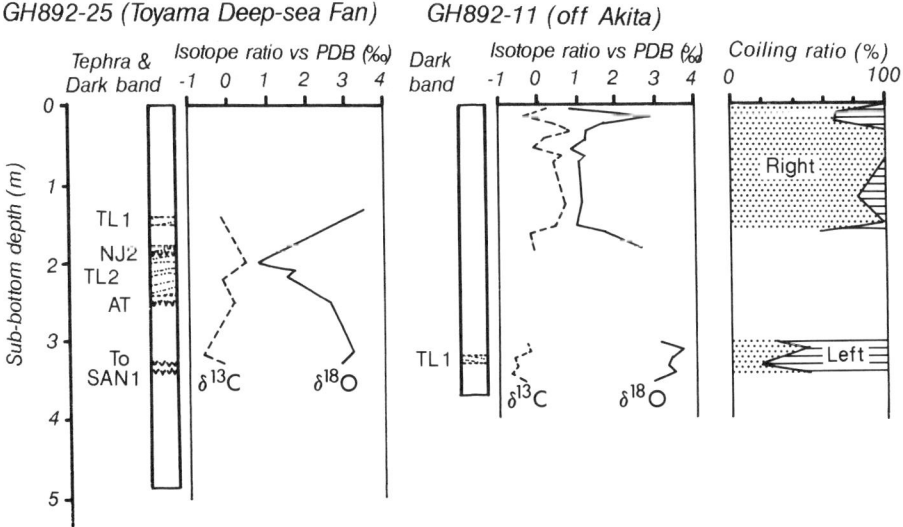

Figure 2 Isotope and micropaleontological results. Planktonic foraminifers are lack in the upper part of core GH892-25 and in the middle part of core GH892-11. *Neogloboquadrina pachyderma* was used for analyses.

coupled plasma emission spectrometry (ICP) (Fig.1). Some of them (B-Tm, U-Oki, Hpfl, YPk, KsP, AT, Aso-4 and Toya) are correlated to the late Quaternary tephras on land. The age of each tephra (Fig.1) was measured by several workers in on-land sequences using ^{14}C age determination (Table 1). Generally speaking, AT and Aso-4 ash (about 80,000 years B.P. [18]) have the widest distributions. U-Oki (about 9,300 years B.P. [10]) occurs only in the southern part and B-Tm (AD 800 - 1,000 [16]) only in the northern part (between 40 - 42 N). Distributions of tephras from San'in, Kanto and Tohoku district are limited to an area along Honshu.

Micropaleontology   Foraminifers in deep-sea sediments of the northern part show sporadic occurrences. No or very few planktonic and calcareous benthic foraminifers were observed in the upper sequence than TL1, reflecting shallow CCD in modern Japan Sea [3]. Then, ratio of the planktonic/benthic foraminifer is nearly zero and benthic foraminifer assemblages are composed by agglutinated ones in the upper sequences (Fig.2). The percentages of left coiling *Neogloboquadrina pachyderma* at TL1 horizon were higher than those after TL1 at off Akita. On the other hand, in the sequence lower than TL1, planktonic foraminifers were well-preserved (high P/B ratio) and benthic foraminifers with hyaline calcareous tests were contained (Fig.2). Benthic foraminiferal assemblages including *Bolivina pacifica* occurred in the TL1 and the dark band just beneath the AT ash.

Oxygen and carbon isotope   The oxygen isotope ratio of *Neogloboquadrina pachyderma* show almost the same pattern described by Oba et al. (1991) [15]; regarding the gentle decreasing in TL2 and abrupt increasing at the uppermost part of TL2, and highest peak at TL1 (Fig.2).

DISCUSSIONS

Sedimentation rates and climate

Sedimentation rates, newly redrawn using present tephrochronological data, in the last glacial maximum, were lower than those in the later stage, and especially than those in the transgressional phase (Fig.3). The inflection point is located at about 13,000 years B.P. In general, sedimentation rates in basins at the lower sea level are thought to be larger than those at the higher sea level, because the shelf was exposed and riverine materials were directly drained to the slope during the glacial time [19]. The difference in the temporal change of sedimentation rates between general model and the Japan Sea is caused by the change in the sediment discharge from land. In central Japan, the cold and dry climate was inferred from pollen analysis before 13,000 years B.P. [20]. Under the dry climate, rivers were not able to transport the sediments to the ocean. Sediments were

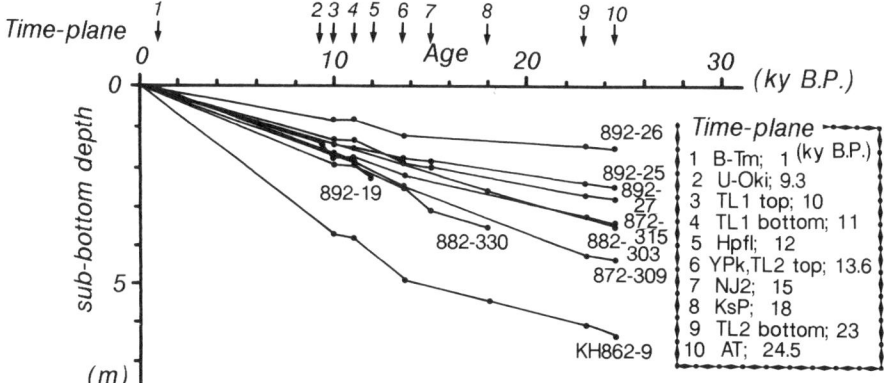

Figure 3 Late Quaternary sedimentation rates in the southern and eastern Japan Sea.

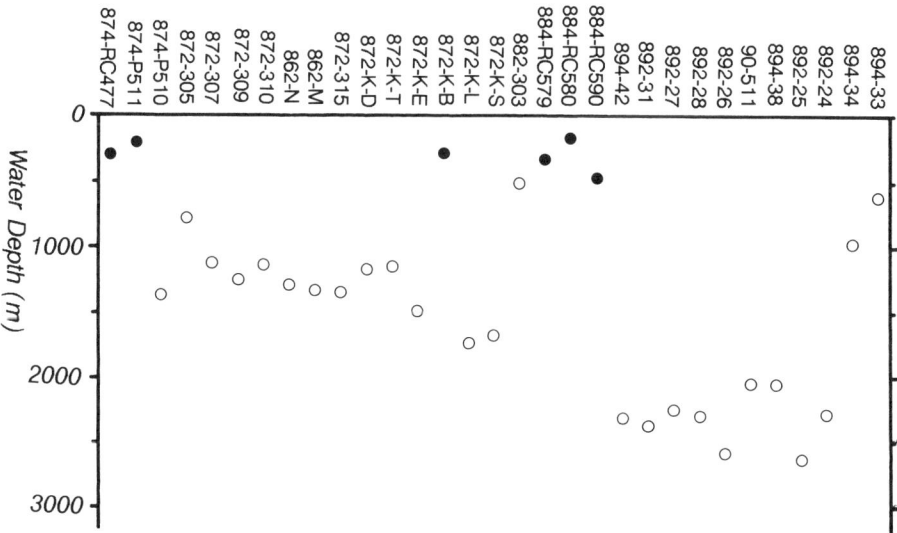

Figure 4 Occurrence of TL2 in the cores intercalating AT and/or KsP. Solid circles; not occurred, open circles; occurred.

transported to the ocean by melted snow, after 13,000 years B.P. depending on the amount of snowfall [20].

Paleoenvironments of dark bands

The stratigraphic relationship between the dark bands and tephra layers suggests that the they were formed at the same time throughout the Japan Sea.

The development of thin lamination, lack or poor occurrence of bioturbation, high sulfur content and occurrence of framboidal pyrite in TL2 suggests the anoxic bottom water condition. The decrease of the oxygen isotope ratio indicates the occurrence of a low-salinity surface water cap [14] which was formed in relation to the prevention of inflow of the oceanic water to the Japan Sea due to the lower sea level and shallow entrances. Under this condition, deep water circulation was prevented. On the other hand, TL2 did not occur in core sequences collected from places shallower than 500 m deep (Fig.4). This fact indicates that the bottom water over shallower basins was not under anoxic conditions even at the last glacial maximum.

Thin lamination, lack of bioturbation and the occurrence of *Bolivina pacifica* in TL1 also suggests the anoxic conditions. The oxygen isotope ratio indicates the inflow of the oceanic water, and the coiling ratio of *Neogloboquadrina pachyderma* suggests the influence of cold water. The age of TL1 (10,000 - 11,000 years B.P.) is correlated to the Younger Dryas event. Because the Younger Dryas event was reported from the southern offshore of Hokkaido [6], there is a possibility that TL1 is a figure of the Younger Dryas event in the Japan Sea.

SUMMARY

Based on the above new tephrochronological data, we have determined that the paleoceanography of the Japan Sea in the late Quaternary was controlled by a global sea level change and paleoclimate on the Japanese Islands. Using the other geochemical, micropaleontological and sedimentological data, the Quaternary paleoceanography of the Japan Sea and its relation to global change will be detailed in the near future.

Acknowledgements: We express our hearty thanks to Dr. K. Fujioka, Prof. H. Machida and Prof. K. Miura for tephra studies, and Prof. N. Niitsuma for isotopic study. We are grateful to the officer, crew and onboard scientists of the R/V Hakurei-Maru for kind help throughout the surveys.

REFERENCES

Y. Hayakawa. Pyroclastic geology of Towada volcano, *Bull. Earthq. Res. Inst., Univ. Tokyo*, **60**, 507-592 (1985).

Y. Hayakawa. Field excursion guide to volcanoes 1: Asama and Kusatsu Shirane, *Sci. Rep. Fac. Educ., Gunma Univ.*, **40**, 65-81 (1992). (in Japanese).

M. Ichikura and H. Ujiie. Lithology and planktonic foraminifera of the Sea of Japan piston cores, *Bull. Natl. Sci. Museum, Tokyo*, Ser. C (Geol. & Paleont.), **2**, 151-178 (1976).

K. Ikehara. Modern sedimentation off San'in district in the southern Japan Sea, In: *Oceanography of Asian Marginal Seas*, K.Takano (ed.), Elsevier, Amsterdam, 143-161 (1991).

T. Ito, N. Mita and H. Katayama. Carbon and sulfur content of the core, ST38 (P636) of GH89-4, In: *Preliminary Reports of Marine Geological Investigations on and around the Continental Shelf of the Eastern Margin of Central Japan Sea in FY1989*, Y. Okamura, M. Arita and Y. Okuda (eds), Geological Survey of Japan, Tsukuba, 166-172 (1990). (in Japanese).

N. Kallel, L.D. Labeyrie, M. Arnold, H. Okada, W.C. Dudley and J.-C. Duplessy. Evidence of cooling during the Younger Dryas in the western North Pacific, *Oceanol. Acta*, **11**, 369-375 (1988).

M. Kato. Japan Sea since the last glacial age; Sediments - mainly based on analysis of core KH-79-3, C-3, *The Earth Monthly (Gekkan Chikyu)*, **6**, 520-522 (1984). (in Japanese).

H. Machida and F. Arai. *Atlas of tephra in and around Japan*. Univ. Tokyo Press, Tokyo, 276p. (1992). (in Japanese).

H. Machida, F. Arai, T. Miyauchi and K. Okumura. Toya ash -a widespread late Quaternary time-maker in northern Japan-, *Quat. Res. (Daiyonki-kenkyu)*, **26**, 129-145 (1987). (in Japanese with English abstract).

H. Machida, F. Arai and H. Moriwaki. Tephras crossed over the Japan Sea, *Kagaku (Science)*, **51**, 562-569 (1981). (in Japanese).

E. Matsumoto, Y. Maeda, K. Takemura and S. Nishida. New radiocarbon age of Aira- Tn ash (AT), *Quat. Res. (Daiyonki-kenkyu)*, **26**, 79-83 (1987). (in Japanese with English abstract).

R. Matsumoto and Leg 128 shipboard scientific party. Dark/light rhythms of the Oki Ridge sediments, *J. Sed. Soc. Japan*, **32**, 1-3 (1990). (in Japanese).

K. Miura and M. Hayashi. Quaternary tephra studies in the Chugoku and Shikoku districts, *Quat. Res. (Daiyonki-kenkyu)*, **30**, 339-351 (1991). (in Japanese with English abstract).

T. Oba. Paleoenvironment of the Sea of Japan since the last glaciation, *The Earth Monthly (Gekkan Chikyu)*, **5**, 37-46 (1983). (in Japanese).

T. Oba, M. Kato, H. Kitazato, I. Koizumi, A. Omura, T. Sakai and T. Takayama. Paleoenvironmental changes in the Japan Sea during the last 85,000 years, *Paleoceanography*, **6**, 499-518 (1991).

S. Oike. Holocene tephrochronology in the eastern foothills of the Towada volcano, northeastern Honshu, Japan, *Quat. Res. (Daiyonki-kenkyu)*, **11**, 228-235 (1972). (in Japanese with English abstract).

A. Omura. Japan Sea since the last glacial age; uranium and thorium isotopes -mainly based on analysis of core KH-79-3, C-3, *The Earth Monthly (Gekkan Chikyu)*, **6**, 523-528 (1984). (in Japanese).

A. Omura, S. Kawai and S. Tamanyu. Dating of volcanic products by the radioactive disequilibrium system between 238U and 230Th, *Bull. Geol. Surv. Japan*, **39**, 559-572 (1988). (in Japanese with English abstract).

H.W. Posamentier and P.R. Vail. Eustatic controls on clastic deposition II -sequence and systems tract models, In: *Sea-level Changes: an Integrated Approach*, C.K. Wilgus, B.S. Hastings, H. Posamentier, J. Van Wagoner, C.A. Ross and C.G.St.C. Kendall (eds), SEPM Spec. Publ., 42, SEPM, Tulsa, 125-154 (1988).

Y. Sakaguchi. Climate changes in central Japan since 38,400 yBP, *Bull. Depart. Geogr., Univ. Tokyo*, **10**, 1-10 (1978).

R. Tada. Sedimentation rhythm and oceanographic changes found in the Japan Sea sediments, *The Earth Monthly (Gekkan Chikyu)*, **13**, 606-612 (1991). (in Japanese).

I. Takashima, T. Yamazaki, E. Nakata and K. Yukawa. TL age of the Quaternary volcanic rocks and pyroclastic flow deposits around the Lake Toya, Hokkaido, Japan, *Jour. Mineral. Petrol. Econ. Geol.*, **87**, 197-206 (1992).

A Pleistocene stratotype at Alzenau (Vorspessart, Germany)

E. JUVIGNE [1], R. GEERAERTS [2], F. GEISSERT [3], H.-J. GREGOR [4],
M. HOTTENROTT [5], J.J. HUS [2], G. SEIDENSCHWANN [6] and R.C. WALTER [7]

1 *Fonds national de la Recherche scientifique, Laboratoire de Géologie du Quaternaire, Place du XX Août, 7, 4000 Liège, Belgium*
2 *Centre de Physique du Globe de l'IRM, 6381 Dourbes, Belgium*
3 *Impasse des Mésanges, 1, 67770 Sessenheim, France*
4 *Naturmuseum, Im Thäle, 3, 8900 Augsburg, Germany*
5 *Hessisches Landesamt für Bodenforschung, Leberberg, 9, 6200 Wiesbaden, Germany*
6 *Brehmstrasse, 6, 6455 Erlensee, Germany*
7 *Institute of Human Origins, 2453 Ridge Road, Berkeley CA 94709, USA*

Abstract.- A multidisciplinary research of a middle terrace deposit of the river Kahl (Tk3) at Alzenau (Germany) was carried out. Sedimentological and paleontological results show that the sequence can be correlated with three successive Isotope Stages, alternatively cold, warm and cold. Five tephra layers occur in the section; their magmas are evolved to highly evolved, with compositions suggesting they have erupted from the East Eifel Volcanic Field. $^{40}Ar/^{39}Ar$ laser ages of single volcanic minerals, along with normal magnetic polarity of the unit suggest that the Tk3 terrace of Alzenau could be correlated with the Isotope Stage series either #14 (cold)- #13 (warm)- #12 (cold), or #16 (cold)- #15 (warm)- #14 (cold). A comparison with chronostratigraphical models of other regional deposits shows controversial results.

Keywords: Europe, Germany, Lower Franconia,Vorspessart, Eifel, Pleistocene, geomorphology, Kahl, terrace, mollusc, macroremnant, pollen, tephra, paleomagnetism, Ar^{40}/Ar^{39} dating.

INTRODUCTION

The brickyard pit Zeller, is located at Alzenau in Vorspessart/ Lower Franconia/ Germany (fig.1). The studied deposit is essentially a remnant of the Tk3 terrace of the river Kahl which is a tributary of the river Main (fig. 2). The profile was first studied by Seidenschwann [21]. Seidenschwann and Juvigné [23] described two tephras embedded in a fine grained body of the sequence. Recently, different aspects of the deposits were investigated separately: sedimentology [22], molluscs [11], macrofossils [12], paleomagnetism [14], pollen [13], tephras [16]. These and additional results enable an original joint discussion of the chronostratigraphical position of the profile.

GEOMORPHOLOGICAL SETTING AND SEDIMENTATION PROCESSES
(G. Seidenschwann)

The brickyard pit exposes a sequence of nearly continuously accumulated sediments with a total thickness of about 22 meters (fig. 3). In the last decade, the overhang of the retreating wall has been described in detail at three different positions [21, 22, 23]. A short description of this section is given in the legend of the figure 3.

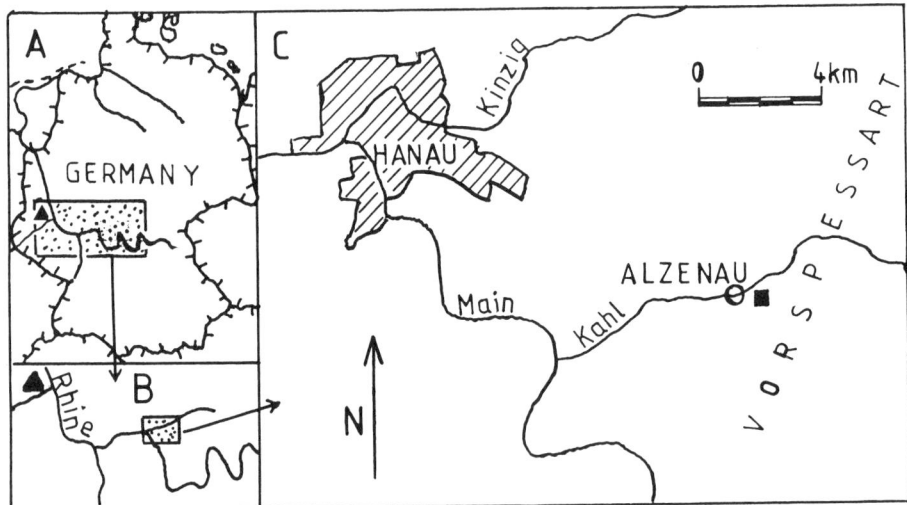

Figure 1.
Location map. In 'A' and 'B', black triangle= East Eifel Volcanic Field. In 'C', black square= brickyard pit Alzenau.

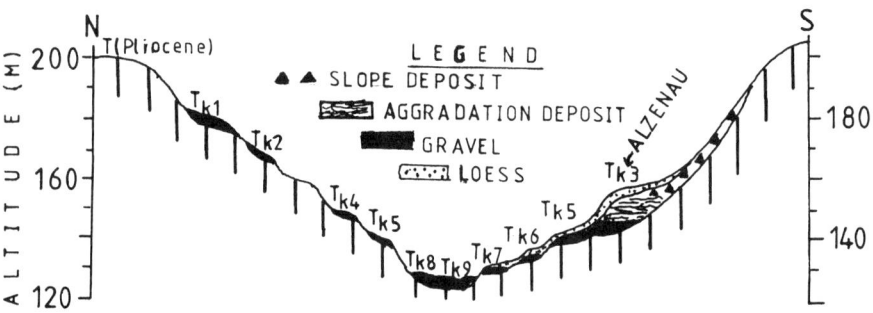

Figure 2.
Cross section of the Kahl Valley near Alzenau after Seidenschwann [22]. T(Pliocene)= Pliocene terrace; Tk1 to Tk9= Pleistocene terraces.

MOLLUSCS (F. Geissert)

Traces of molluscs were found in layers #10 to #13.
The most frequent species are listed below:
- *Columella col. columella* (BENZ) v.MARTENS, an arctic species
- *Vallonia tenuilabris* (Al. BRAUN), recently Centralasian, Sibirian, like previous species found in loess faunas and other sediments of cold climate character.
- *Pisidium (Odhneripisidium) stewarti* (PRESTON), a small mussel, which occurs in many Pleistocene and Early Holocene aquatic sediments, so far as it is cold enough. Recent spreading: Sibiria and central Asia.
- Furthermore to be counted to the previous species: *Semilax kochi* (ANDREAE), *Vertigo genesii* (GREDLER), Vertigo parcedentata (AL. BRAUN), also abundant occurrence of species of loess faunas, e.g. *Succinea* (Succinelle) *oblonga* and *Trichia hispida* (LINNÉ).

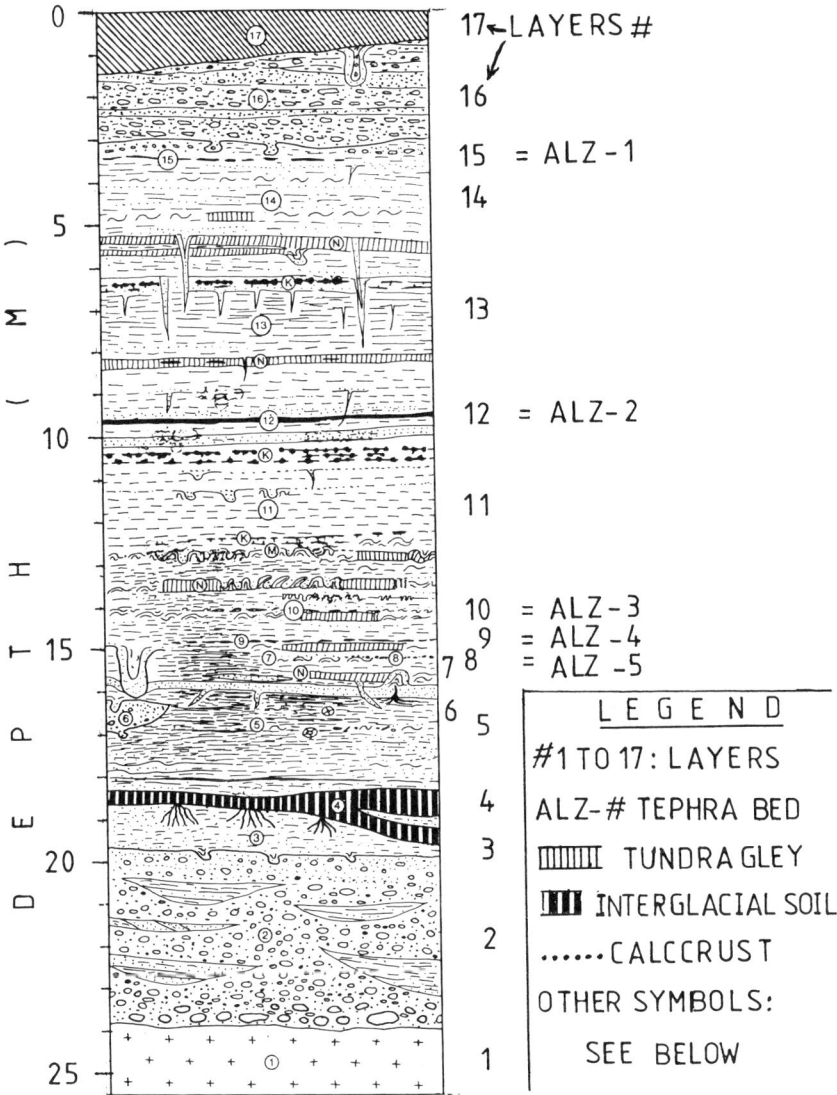

Figure 3.
Section of the Alzenau (state of wall in 1992).
Legend and explanation. The deposit begins with a basal bloc horizon overlain by coarse sandy gravel (#2), in its turn overlain by cryoturbated (cold period) sandy silt (#3). Layer #4 is fine grained sediment recording an intensively developed fossil gleysol with root horizon (interglacial soil). That soil is overlain by humic silt (#5). The series of gravelly sands (#6), loamy silts with thin sandy layers (#7, #11, #13, #14) give unmistakable proof of cold climate conditions, especially an extremely cold phase in the upper section. This cold phase is indicated by a continuous sequence including cryoturbations (fossil ice-wedges), tundra gleys (N) and mollusc faunas (M). Within this sequence there are: 1) five tephra layers (ALZ-1 to 5); 2) tundragley (N= Nassboden); 3) calcareous crust horizons (K).
Layer #16 consists of coarse sands, gravel, and debris building up a local fan, which overlies the aggradation sediments of the Kahl.
Unit #17 is a differentiated loess cover containing two interglacial paleosols overlain by Weichselian loess including the Eltville Tephra.

The above species are indicators of cold climate conditions.
Indicators of forest fauna, e.g. *Clausilia pumila* (C. PFEIFFER) and *Perforatella bidentata* (GMELIN), both of eastern origin, are also present in a wet soil in the uppermost part of the layer #10. Those species occur in more temperate periods of cold epochs.
Though chronostratigraphic evidence is not available, statements concerning the climate at the time when they were embedded are possible. Since the mollusc remnants are very fragile, they were not supplied from upwards by the river, and should lie in a primary position. Both the indicators of absolute cold, and the composition of the spectrum of species give unmistakable proof of cold climate conditions during the emplacement of the relevant unit.

PALEOECOLOGY (H.-J.Gregor)

Paleoecological investigations were done on fine grained humic sediments from layers #4 to #12 (fig. 3). Detailed results were published by Gregor [12]. The table 1 illustrates the occurrence of fossil fruits and seeds in the different basal layers.

Table 1.
Fossil content in layers #4 to #12.

Layers #4 to 12 after figure 3; (U) = Upper; (M) = Middle; (L)= Lower; += present; - = absent.
Macroremnants: 1 = Mosses (Calliergon divisp.), 2= Funghi (Cenococcum geophilum), 3= Ferns (Salvinia cf. natans). Gymnosperms: 4= Picea sp., 5= Pinus sp. Angiosperms: 6= Potamogeton natans, 7= Carex cf. pseudocyperus, 8= Carex cf. caespitosa, 9= Schoenoplectus lacustris, 10= Rumex sp., 11= Batrachium aquatilis, 12= Rubus sp., 13= Acer sp., 14= Hippuris vulgaris, 15= Menyanthes trifoliata, 16 = Twigs. Fauna: 17= Bones; 18= Gastropods; 19= Insects (beetles).

LAYERS #	1	2	3	4	5	6	7	8	9	10	11	12	13	14	15	16	17	18	19
12	+	-	-	-	-	-	-	-	-	-	-	-	-	-	-	-	-	-	-
11	-	-	-	-	-	+	-	-	-	-	-	+	-	-	-	-	-	+	-
10 (U)	-	-	-	-	+	+	+	+	-	+	-	-	-	+	-	-	-	-	+
10 (M)	+	-	-	-	-	+	+	-	+	-	+	-	-	-	-	-	-	-	-
10 (L)	-	+	-	-	-	-	-	-	-	-	-	-	-	-	-	+	-	-	-
7	-	-	-	-	-	-	-	-	-	-	-	-	-	-	-	-	+	-	-
5	+	+	-	-	+	+	+	+	+	+	+	+	-	+	+	+	+	-	+
4	+	-	+	+	+	+	+	+	+	+	+	-	-	+	+	+	-	-	+

The profile from Alzenau yielded poor fruit and seed flora results in different layers, especially in the basal horizons. There is no important break among the positon of the layers at the base (#4 and #5) and the top (#12). Upwards in the profile, no plant remains was found.
The main flora consists of water, reed and marsh plants such as Potamogetonaceae, Cyperaceae, Ranunculaceae, Hippuridaceae, Menyanthaceae and Polygonaceae. These observations suggest a wetland area as we have today around lakes, marshes, or wetland areas with high levels of ground water. The surrounding forest record is poor, and has only produced such fossils as *Acer, Rubus, Pinus and Picea*.
The macroremnants correspond to a mesoclimate (Cfb-climate, *sensu* Köppen) characterized by warm summer (f.e. *Schoenoplectus*) and mild winter. There is no sign of any cold period. Those remnants are likely preserved *in situ* within the paleosol of the horizon #4 in figure 3. In the overlying layers (#5 and above), the macroremnants could have been reworked along with the host material upwards in the basin, and they do not necessarily have a stratigraphical significance.
This palaeobotanical record provides no chronological information. Various European floras of Pleistocene age, whether older or younger, show similar spectra. The Cromerian flora (England) differs totally from the flora found in Alzenau, precluding biostratigraphic comparisons.

POLLEN (M. Hottenrott)

Pollen were only found in layers #5 to #10, from which 22 samples were analysed. It should be emphasized that pollen was not found in the interglacial soil of layer #4. The precise position of the samples and detailed results can be found in a previous paper [13].
Since 10 samples contain less than 50 pollen grains, and another 3 contains between 50 and 100 pollen grains, the relevant spectra have very low statistical significance, and therefore are very weak climate indicators. Only 5 samples have more than 250 grains.
After the processes of emplacement of the sediments (see above), only a very small part of the pollen content should be synchronous of the deposition of its host layer, and it is likely that an important part of the pollen content was previously embedded in sediments stored for different time ranges, upwards in the basin. Therefore, the vertical distribution of the taxa cannot be discussed as it is done after pollendiagrammes from peat-bogs or lacustrine deposits. The presence of high amounts of *Pediastrum*, especially in layers #5, #9, and #10 are likely an exception; that frequent taxa is in good agreement with the processes of sedimentation in humid area (see above).
The overall dominant tree pollen grains were produced by coniferous (*Pinus, Picea, Abies*). Since it is well known that those pollen grains can be transported over very long distances (hundreds of km), they do not necessarily indicate the regional presence of the relevant forest. This is in agreement with the rare occurrence of the relevant macroremnants in the deposit (see above: macroflora).
In any of the spectrum, the sum of the regional thermophile tree pollen (*Quercus+ Ulmus+ Tilia+* traces of other ones) do not exceed 5 %, and those taxa are absent in numerous samples. The dominant local trees should have been *Betula, Juniperus, Salix* and *Alnus*, along with high amounts of non-arboreal pollen as *Graminae, Polypodiaceae,* and *Artemisia*. Similar spectra have been described in Middle Europe only in cold period horizons [e.g. 1, 2, 28]. It has also to be pointed out that no typical taxa of Pliocene or Earliest Pleistocene were found.

TEPHRA LAYERS (E. Juvigné)

Five tephra layers occur in the fine grained sequence of the profile [16]; they are named respectively ALZ-1 to 5 (fig. 3).
Mineral determination.- Petrographic and microprobe analyses of single minerals (pyroxenes, amphiboles, olivine) were done.
Estimation of the magma composition.- Macroscopical observations in the field, previous petrographical investigations, and chemical analyses of bulk microsamples have showed, that: 1) each tephra is mixed with local material; 2) the glass has been strongly weathered into clay. Results of microprobe analyses are sumarized in the table 2.
The comparison with the composition of juvenile magmas of the Eifel Volcanic Field shows that all the tephras at Alzenau are strongly depleted in $CaO+ Na_2O+ K_2O$. This can be explained by alteration of glass, along with leaching of mobile elements. Since the Lower Paleozoic rocks (upper crust in the Eifel) contain about 4.8 % of '$CaO+ Na_2O+ K_2O$', a high xenolith content in the tephra can also deviate the composition of the magmatic component in the same way.
Due to the uncertainties on the chemical composition of the tephras at Alzenau, the juvenile magma composition cannot be established, but only estimated. Therefore it cannot be determined using current classifications of volcanic rocks.

ALZ-1 occurs as very scarcely thin greyish green lenses.
The magma should be basic to intermediate ($SiO_2<64.02\%$), and highly evolved ($MgO>0.98\%$, but not much more).
The mafic mineral suite consists of clinopyroxene/aluminian fassaïte (32-78%), olivine/forsterite Fo_{85-92} (0-21%), brown amphibole/pargasite (0-68%). The width of those minerals is up to 450 µm. Pyroxenes and olivine are strongly weathered.
ALZ-2 occurs as the thickest (up to 15 cm) tephra layer in the profile. Its material is also obviously reworked and mixed with adjacent material. The coulour of the tephra layer is very

light grey at the base, and becomes medium grey upwards; this could represent compositional zonation within the magma chamber. In the upper part of the layer, the material is banded with fine grained fluvial sediment (local reworking), as shown by mixed brown and grey deposits.

Table 2.
Chemical composition (major elements) of tephra layers, adjacent loesslike material, and Lower Paleozoic rocks.
Due to the absence of glass shards in tephra layers, 0.02 g of bulk sample from each layer was fused with so much Spectroflux to a homogeneous glass bead; similar method was applied to the adjacent loesslike material. Hence totals of analyses were close to 50%; values were normalized up to 100% in table 2. All analyses of tephra layers and adjacent material were undertaken with a CAMEBAX electron microprobe.
O.M.= Overlying Material; T.L.= Tephra Layer (mixture of pure tephra and local material); U.M.= Underlying Material; T.M.= trend of composition of Tephra Material (including xenoliths); T1, T2= Values of T.M. recalculate for extreme values of 'CaO+ Na$_2$O+ K$_2$O' in the Eifel magmas (respectively 13% and 25%); L.P.= averaged composition of about 300 samples from different beds of the Lower Paleozoic in Belgium after detailed data from Wilmart [27]; all analyses were done by XRF of whole rock.

Elements	A L Z- 1					A L Z- 2				
	O.M.	T.L.	U.M.	T.M.	T 1	T 2	O.M.	T.L.	U.M.	T.M.	T 1	T 2
Si O$_2$	83,33	70,94	81,94	<70,94	<64,02	<55,19	74,81	60,24	77,64	<60,24	<54,24	<46,75
Ti O$_2$	0,69	1,22	0,70	>1,22	>1,11	>0,95	1,33	1,87	0,73	>1,87	>1,68	>1,45
Al$_2$ O$_3$	8,67	19,33	10,48	>19,33	>17,44	>15,04	16,19	26,14	12,12	>26,14	>23,52	>20,29
Cr$_2$ O$_3$	0,00	0,09	0,00	>0,09	>0,08	>0,07	0,02	0,02	0,06	<0,02	<0,02	<0,02
Fe O$_t$	2,12	3,32	2,62	>3,32	>3,00	>2,58	2,76	6,59	3,61	>6,59	>5,93	>5,11
Mn O	0,06	0,19	0,01	>0,19	>0,17	>0,15	0,22	0,30	0,06	>0,30	>0,27	>0,23
Mg O	0,36	1,26	0,56	>1,26	>1,14	>0,98	0,73	1,50	0,83	>1,50	>1,35	>1,16
Ca O	0,20	0,18	0,31	<0,18⎤			0,30	0,15	0,30	<0,15⎤		
Na$_2$ O	1,24	0,95	0,75	≈0,95 ⎬ 13,00	25,00		0,71	0,54	1,21	<0,54 ⎬ 13,00	25,00	
K$_2$ O	3,26	2,44	2,56	<2,44⎦			2,86	2,58	3,37	<2,58⎦		

Elements	A L Z- 3					A L Z- 4				
	O.M.	T.L.	U.M.	T.M.	T 1	T 2	O.M.	T.L.	U.M.	T.M.	T 1	T 2
Si O$_2$	72,95	67,57	74,35	<67,57	<61,65	<53,15	77,44	62,47	70,94	<62,47	<57,24	<49,34
Ti O$_2$	1,07	1,15	1,06	>1,15	>1,05	>0,90	1,00	2,17	0,91	>2,17	>1,99	>1,71
Al$_2$ O$_3$	13,13	18,38	15,68	>18,38	>16,77	>14,46	12,84	18,83	12,95	>18,83	>17,27	>14,87
Cr$_2$ O$_3$	0,01	0,05	0,00	>0,05	>0,04	>0,04	0,01	0,13	0,00	>0,13	>0,12	>0,10
Fe O$_t$	7,55	6,71	3,75	≈6,71	≈6,12	≈5,28	2,85	8,00	9,78	≈8,00	≈7,33	≈6,32
Mn O	0,01	0,10	0,03	>0,10	>0,10	>0,08	0,07	0,08	0,14	≈0,08	≈0,07	≈0,06
Mg O	1,03	1,39	1,04	>1,39	>1,27	>1,09	0,93	3,27	0,97	>3,27	>3,00	>2,58
Ca O	0,44	0,67	0,26	>0,67⎤			0,42	1,96	0,52	>1,96⎤		
Na$_2$ O	1,25	1,18	0,61	≈1,18 ⎬ 13,00	25,00		1,35	0,54	1,14	<0,54 ⎬ 13,00	25,00	
K$_2$ O	2,50	2,73	3,14	≈2,73⎦			2,74	2,47	2,52	<2,47⎦		

Elements	A L Z- 5					LOWER PALEOZOIC (Ardenne-Eifel Massif)
	O.M.	T.L.	U.M.	T.M.	T 1	T 2	
Si O$_2$	77,10	63,95	79,29	<63,95	<56,76	<50,66	64,60
Ti O$_2$	1,22	2,03	1,24	>2,03	>1,86	>1,61	0,93
Al$_2$ O$_3$	13,25	17,10	12,29	>17,10	>15,71	>13,55	18,66
Cr$_2$ O$_3$	0,03	0,01	0,10	<0,01	<0,01	<0,01	NOT DET.
Fe O$_t$	4,03	7,18	3,00	>7,18	>6,60	>5,69	7,66
Mn O	0,01	0,11	0,02	>0,11	>0,10	>0,09	1,07
Mg O	0,84	4,30	0,69	>4,30	>3,95	>3,41	2,09
Ca O	0,44	2,86	0,39	>2,86⎤			0,52
Na$_2$ O	0,72	0,40	0,71	<0,40 ⎬ 13,00	25,00		0,71
K$_2$ O	2,27	1,96	2,18	<1,96⎦			3,57
TOTAL:							99,81

The magma should be basic to intermediate ($SiO_2 < 54.24\%$), and highly evolved ($MgO > 1.16\%$, but not much more).
The mafic mineral suite consists of clinopyroxene/aluminian subsilicic fassaïte (7-49%), olivine/forsterite Fo_{83-90} (5-36%), brown amphibole/ pargasite (13-69%), apatite (0-25%), and biotite (0-13 %). The width of those minerals is up to 280 µm. Most of the pyroxenes and all the olivines are strongly weathered.
ALZ-3 occurs as scarcely visible thin greyish green lenses.
The magma should be basic to intermediate ($SiO_2 < 61.65\%$), and highly evolved ($MgO > 1.09\%$, but not much more).
The mafic mineral suite consists of clinopyroxene/aluminian, subsilicic fassaïte and diopside Di_{62-86} (30-64%), olivine/forsterite Fo_{86-91} (2-12%), brown amphibole/magnesio-hastingsite and pargasite (3-13%). The width of those minerals is up to 440 µm. Pyroxenes and olivine are strongly weathered.
ALZ-4 occurs as frequent greyish green lenses up to 3 cm in thickness.
The magma should be basic to intermediate ($SiO_2 < 57.24\%$), and evolved ($MgO > 2.58\%$, but not much more).
The mafic mineral suite consists of clinopyroxene/aluminian, subsilicic fassaite and exceptionally diopside Di_{85} (76-88%), olivine/forsterite Fo_{85-89} (10-23%), brown amphibole/ magnesio-hastingsite and pargasite (0-4%). The width of those minerals is up to 170 µm. Most of the minerals show no weathering facies.
ALZ-5 occurs as light greyish green lenses up to 2 cm in thickness.
The magma should be basic to intermediate ($SiO_2 < 58.76\%$), and evolved ($MgO > 3.41\%$, but not much more).
The mafic mineral suite consists of clinopyroxene/aluminian fassaïte (88-95%), olivine/ forsterite Fo_{85-89} (0-2%), brown amphibole/magnesio-hastingsite and pargasite (6-12%). The width of those minerals is up to 130 µm. Only the olivines are weathered.

Apart from the uncertainties of the magma composition, it can be seen that the tephras are basic to intermediate, and evolved (ALZ-4,5) to highly evolved (ALZ-1,2,3), suggesting a correlation with the East Eifel Volcanic Field (EEVF). Since leucite has not been found the correlation with the Rieden Volcanic Phase is unlikely.

PALEOMAGNETISM (J.J. Hus and R. Geeraerts)

The present contribution deals with preliminary results of magnetostratigraphical investigation of the profile. During a previous investigation, 300 loose samples were taken with a interval of 5 cm, from the tephra layer ALZ-2, to the layer #2 (fig. 3), and a magnetic susceptibility profile was obtained [14].
Later, 140 oriented hand samples were retrieved to build up a magnetostratigraphy based on the polarity changes of the geomagnetic field during the past. These samples were taken mainly in the upper part between the tephra layers ALZ-1 and ALZ-2, and in the lower part from the tephra layer ALZ-4 to layer #2 (fig. 3), leaving a gap of about 5 m between ALZ-2 and ALZ-4. Before removal from the outcrop the hand samples were oriented towards the vertical and magnetic north, by shaping a horizontal plane on top of each sample with knife, on which the direction of magnetic north was marked.
The magnetic susceptibility of the loose samples, which was measured with a Kappabridge KLY-1 is given on a mass specific basis in figure 4A. The tephra layers can easily be recognized because of their high weak field response. No other pronounced susceptibility highs could be revealed except in the blueish grey clay below the humic horizon #4. The other soils, which are tundragleys, cannot be clearly differentiated on the basis of their magnetic susceptibility.
The oriented hand samples were trimmed into cubes of 4 cm, and their remanence measured with a cryogenic SCT magnetometer.
Several pilot samples, representing the main lithologies, were stepwise demagnetized in increasing alternating fields in order to isolate the most stable remanence component. Finally, all the samples were "cleaned" in an alternating field of 25 millitesla in order to remove the

viscous overprint.

Inclination, declination and latitude of the virtual paleomagnetic pole of the individual stable magnetization directions, are given in figure 4B. It can be noticed that all the samples examined possess a stable, normal magnetization component. Swings in inclination and declination point to the registration of field secular variation changes in the sediments. The average inclination in the upper part is about 5° higher than in the lower part, but in both, shallower than the inclination of about 67.20° corresponding to an axial dipole field for this area. A declination swing from east to west is recorded in the upper part, turning back to east near the tephra layer ALZ-2. In the lower part, the declination is strongly east near the tephra layer ALZ-4, but becomes slightly west below it. As normal magnetozones were found, no precise chronology can be proposed yet with the available information.

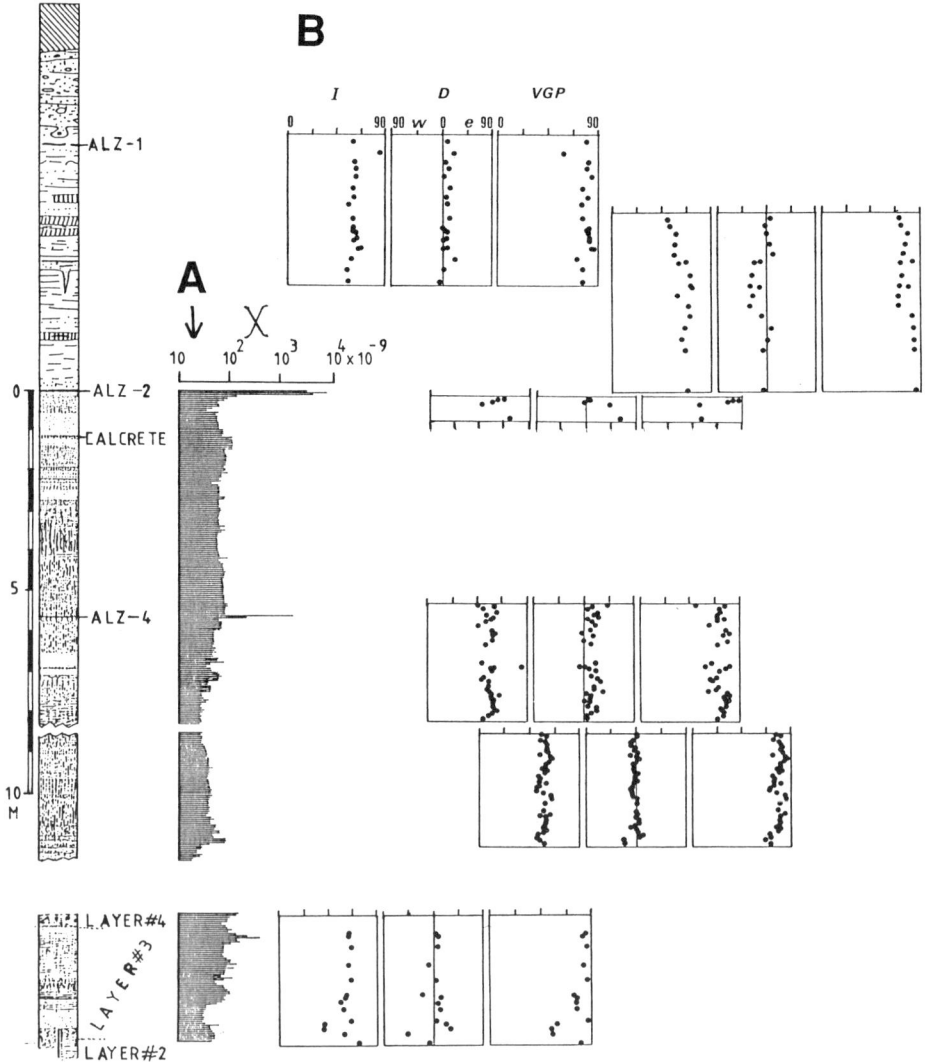

Figure 4.
Paleomagnetic measurements versus depth of the Alzenau sediment complex (see also figure 3). 'A', magnetic susceptibility profile. 'B', inclination, declination and latitude of virtual paleomagnetic pole.

40Ar/39Ar AGE OF TEPHRAS (R.C. Walter)

Detailed results of 12 analyses of sanidines and hornblendes from ALZ-2 and ALZ-3 are represented in table 3.

Table 3.
$^{40}Ar/^{39}Ar$ laser ages of sanidines and hornblendes from ALZ-2 and ALZ-3. E.= error; W.M.= weighted mean; W.U.= weighted uncertainty

LAB. #	TEPHRA	MATERIAL	Ca/K	% RAD.	AGE ± E. (ka)	W.M.±W.U. (ka)
5273-01	ALZ-2	Sanidine	0.0001	62.0	514 ± 52	
5273-02	ALZ-2	Sanidine	0.0001	32.5	781 ± 97	574 ± 121
5275-01	ALZ-3	Sanidine	0.0012	97.1	461 ± 6	
5275-02	ALZ-3	Sanidine	0.0025	78.0	441 ± 7	
5275-03	ALZ-3	Sanidine	0.0024	95.2	459 ± 6	
5275-04	ALZ-3	Sanidine	0.0016	95.5	443 ± 6	
5275-05	ALZ-3	Sanidine	0.0016	49.2	391 ± 9	446 ± 10
5276-01	ALZ-3	Hornblende	4.2325	48.0	339 ± 47	
5276-02	ALZ-3	Hornblende	4.9708	64.9	575 ± 73	
5276-03	ALZ-3	Hornblende	5.0496	70.5	607 ± 24	
5276-04	ALZ-3	Hornblende	4.6061	62.1	508 ± 41	543 ± 56
5276-05	ALZ-3	Hornblende	4.7473	62.3	874 ± 37	

ALZ-2.- Two sanidines were dated respectively at 514±52 ka, and 781±97 ka (weighted mean: 574±121 ka).
ALZ-3.- Five sanidines range from 391±9 ka to 461±6 ka; their weighted mean is 446±10 ka.
Five hornblendes range from 339±47 ka to 874±37 ka. The age of one of the hornblendes (5276-05: 874±37 ka) deviates largely from all the other ages, and it can be ignored. The weighted mean for the hornblendes (without 5276-05) is 543±56 ka.

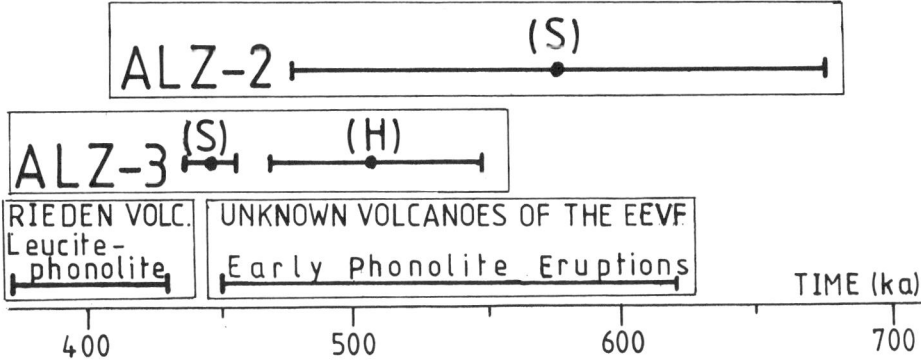

Figure 5.
$^{40}Ar/^{39}Ar$ laser ages of single grains from various tephras: ALZ-2 and ALZ-3, (S)= sanidines, (H)= hornblendes, bars= weighted means after values of table 2; Rieden volcano and 'early phonolite eruptions' in the EEVF, single grains from some phonolitic and mafic tephras after Bogaard et al. [4], Bogaard and Schmincke [5] and Schmincke [20].

Despite the fact that all the minerals of ALZ-3 erupted at the same time, the hornblendes give an age which is about 100 ka older than the sanidines. Moreover the sanidines of ALZ-2 and ALZ-3 respectively give controversial ages in comparison to the stratigraphical position of the tephra layers. Due to dispersion of all ages, ALZ-2 and ALZ-3 can only be roughly situated between about 450 ka and 600 ka, but it can be emphasized that most of the individual ages of sanidines (fig. 5) are concentrated around 450 ka.

Those values, compared with the age of Lower to Middle Pleistocene widespread tephra sheets of the EEVF obtained by the same method [4, 5], indicate that ALZ-2 and ALZ-3 are likely to be correlated with the recent part of the 'Early Phonolite Eruptions' in the East Eifel Field (fig. 5). From a chronological standpoint, the correlation with the early phase of the Rieden Volcano (about 435 ka) cannot be firmly ruled out, but the lack of leucite in the tephras at Alzenau allows us to put more weight on the correlation with the 'Early Phonolite Eruptions' (fig. 5).

CHRONOSTRATIGRAPHICAL POSITION OF THE SEQUENCE

In any of the paleontogical investigations (molluscs, pollen, macrofossils), indicators of Pliocene or even Lower Pleistocene material have been found in the sequence.

The fine grained layers #5 to #14 (layer #14 includes layer #15= ALZ-1) were deposited under stable sedimentation conditions involving a simultaneous high supply of loess; neither important climatic change, nor tectonic movement seem to have occurred during that period in the region. Hence the accumulation should have lasted the time range of one Isotope Stage. The molluscs, pollen, and macrofossils content, the occurrence of periglacial features, and tundragley formation indicate that the relevant period must be a cold Isotope Stage.

The high development of the paleosol recorded in the layer #4 (fig. 3), along with its macroremnant content suggest a correlation with a temperate to warm Isotope Stage.

Cryoturbations occurring within the flood sediments of layer #3 demonstrate that this layer was deposited under periglacial conditions (cold Isotope Stage).

Sedimentological and paleontological results indicate that the deposits of the Tk3 terrace of the Kahl can be correlated with three successive Isotope Stages which are alternatively cold (layers #2 and #3), warm (paleosol in layer #4), and finally cold (layers #5 to #14).

The geomorphological setting along with a well distributed rate of erosion/aggradation of the

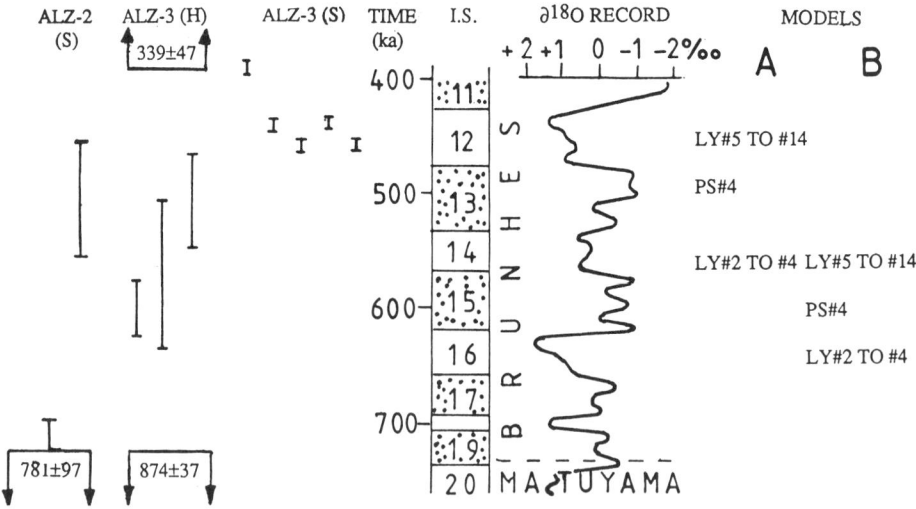

Figure 6.
Chronostratigraphical position of the Tk3 terrace of the Kahl. SPECMAP scale after Imbrie et al. [15]. Explanation.- ALZ-2,3= $^{40}Ar/^{39}Ar$ laser ages of sanidines (S) and hornblendes (H) of ALZ-2 and ALZ-3 (vertical bar= confidence interval after values of table 3); I.S.= Isotope Stages; LY#, PS#= layers, paleosol numbered after figure 3.

Kahl during the Pleistocene allow us to place the Tk3 terrace formation at about 800 ka [see fig. 2: Plio/Pleistocene boundary (2.400 Ma [9, 29]) approximately 190 m asl; Tk3 terrace approx. 150 m asl; flood plain approx. 130 m asl].
Due to the approximate chronostratigraphical position as defined above, the normal polarity enforces the sequence to be correlated either with the lower part of the Bruhnes Epoch (lower limit at 740 ka), or with the Jaramillo Event (920 ka to 1 Ma).
$^{40}Ar/^{39}Ar$ laser ages of volcanic sanidines and hornblendes indicate that ALZ-2 and ALZ-3 as well as the relevant sediment body can be correlated either with the Isotope Stage #12 (cold) or the I.S.#14. The longer and cooler I.S.#12 seems to be more appropriate than the I.S.#14 to correspond to the huge volume of aggradation sediments of the Tk3 terrace.
Taking all the results into account, the Tk3 terrace complex of the Kahl should correspond to one of the following Isotope Stage series (fig. 6):
Model A: I.S.14 (cold)- I.S.13 (warm)- I.S.12 (cold)
Model B: I.S.16 (cold)- I.S.15 (warm)- I.S.14 (cold)

COMPARISON WITH OTHER REGIONAL CHRONOSTRATIGRAPHICAL RESULTS (G. Seidenschwann)

The Tk3 terrace of the Kahl can be correlated with:
-the valley aggradation sediments of the Middle Main Valley [5, 8, 17],
-the upper part of the complex of the "Kelsterbacher Terrasse" and the "Mosbacher Sande" of the lower Main area [19, 21, 22, 24, 25],
-the TR4 terrace of the Middle Rhine Valley [3, 26].
According to Bibus [3], the existence of volcanic heavy minerals is also proved within a 15 m thick series of sandy sediments of the TR4 terrace of the Rhine.
$^{40}Ar/^{39}Ar$ laser ages of about 450 ka were obtained by Lippolt *et al.* [18] from sanidines of tephra layers on top of sediments of the TR8 terrace of the Rhine. A correlation with the early volcanic phase in the East Eifel Field as defined by Bogaard and Schmincke [5], was also proposed. On the one hand those $^{40}Ar/^{39}Ar$ laser ages [18] show a great discrepancy to stratigraphical and paleomagnetic investigations of terrace sediments. Brunnacker and Boenigk [7], Bibus [3] and Fromm [10] showed that in the Middle Rhine region, an aggradation terrace partly includes a reverse magnetic polarity (Matuyama Epoch) and is overlain by the oldest tephras of the EEVF as defined by Bogaard and Schmincke [5]. That aggradation terrace is supposed to be correlated with the aggradation terrace of the rivers Main and Kahl respectively. On the other hand, the TR8 terrace of the Rhine (about 450 ka) maycorrespond to the Tk6 terrace of the Kahl, which is stratiphically much younger than the Tk3 terrace of the Kahl (fig. 2).
Provisional paleomagnetic results obtained in Marktheidenfeld (Middle Main Region) show that the Brunhes/Matuyama boundary occurs at a position on top of the highest valley aggradation series.

CONCLUSION

A multidisciplinary research of the Tk3 terrace of the river Kahl at Alzenau shows that the sequence can be correlated with one of the following Isotope Stage series: 1) #14 (cold)- #13 (warm)- #12 (cold); 2) #16 (cold)- #15 (warm)- #14 (cold).
A comparison with chronostratigraphical interpretation of other regional deposits shows controversial results which suggest the following:
1) the Tk3 terrace of the Kahl could be older than it is proposed in the model developed in this paper,
2) the chronostratigraphical models of other regional deposits could be too old,
3) the correlations of the river terraces of the Kahl, Main, Rhine system could be partly wrong,
4) all the above models are wrong!!!

REFERENCES

1. B.Bastin. Recherche sur l'évolution du peuplement végétal en Belgique durant la glaciation de Würm, *Acta Geographica Lovaniensia*. **9**, pp. 136 p.(1971).
2. J.L.Beaulieu de and M. Reille. A long Upper Pleistocene pollen record from Les Echets, near Lyon, France, *Boreas* **13**, 111-132 (1984).
3. E. Bibus. Zur Relief-, Boden- und Sedimententwicklung am unteren Mittelrhein, *Frankf. geowiss. Arb., Ser. D*, **1**, 296 pp. (1980).
4. P. Bogaard, C.M. Hall, H.-U. Schmincke and D. York. $^{40}Ar/^{39}Ar$ laser dating of single grains; ages of quaternary tephra from the East Eifel Volcanic field, FRG, *Geophys. Res. Lett.* **14**, 1211-1214 (1987).
5. P. Bogaard and H.-U. Schmincke. Aschelagen als quartäre Zeitmarken in Mitteleuropa. *Geowissenschaften* **6**, 75-84, (1988).
6. K. Brunnacker. Über Ablauf und Altersstellung altquartärer Verschüttungen im Maintal und nächst dem Donautal bei Regensburg, *Eiszeitalter u. Gegenwart* **15**, 72-80 (1964).
7. K. Brunnacker and W. Boenigk. Über den Stand der paläomagnetischen Untersuchungen im Pliozän und Pleistozän der Bundesrepublik Deutschland, *Eiszeitalter u. Gegenwart* **27**, 1-17 (1976).
8. K.Dietz. Zur Reliefentwicklung im Main-Tauber-Bereich, *Rhein- Main. Forsch.* **93**, pp.123 (1981).
9. Ding Zhongli, N. Rutter, Liu Tungsheng, M.E. Evans and Wang Yuchun. Climatic correlation between Chinese loess and deep-sea cores: a structural approach. In: *Loess, environment and global change*. Liu Tungsheng (Ed.), pp. 168-186, Science Press, Beijing (1991).
10. K. Fromm. Paläomagnetische Bestimmungen zur Korrelierung altpleistozäner Terrassen des Mittelrheins, *Mainzer geowiss. Mitt.* **16**, 7-29 (1987).
11. F. Geissert. Mollusken aus dem Cromer-Profil von Alzenau i.Ufr., *Jber. wetterau. Ges. ges. Naturkunde* **140-141**, 97-108 (1989).
12. H.-J. Gregor. Vorläufige Mitteilung über makrofloristische Untersuchungen des Talverschüttungsprofils Alzenau i.Ufr. (Ziegeleigrube Zeller), *Jber. wetterau. Ges. ges. Naturkunde* **140-141**, 109-120 (1989).
13. M. Hottenrott. Zur Pollenführung der früh-mittelpleistozänen Sedimentfolge von Alzenau in Unterfranken (Ziegelei-grube Zeller), *Jber. wetterau. Ges. ges. Naturkunde* **140-141**, 133-141 (1989).
14. J.J. Hus and R. Geeraerts. Gesteinsmagnetische und paläomagnetische Untersuchungen an pleistozänen Ablagerungen der Ziegeleigrube Zeller, Alzenau i.Ufr., *Jber. wetterau. Ges. ges. Naturkunde* **140-141**, 121-132 (1989).
15. J. Imbrie, D. Hays, D.G. Martinson, A. McIntyre, A.C. Mix, J.J. Morley, N.G. Pisias, W.L. Prell and N.J. Shakleton. The orbital theory of Pleistocene climate: support from a revised chronology of the marine ^{18}O record. In: *Milankovich and Climate*. A. Berger et al. (Eds), pp. 269-306. Reidel Publ. Company, Dordrecht (1984).
16. E. Juvigné and G. Seidenschwann. Das Talverschüttungsprofil von Alzenau i.Ufr.(Ziegeleigrube Zeller) eine Typlokalität früh- mittelpleistozäner Tephren, *Jber. wetterau. Ges. ges. Naturkunde* **140-141**, 143-172 (1989).
17. H. Körber. Die Entwicklung des Maintals. *Würzb. Geogr. Arb.* **10**, 170 pp.(1962).
18. H.J. Lippoldt, U. Fuhrmann and H. Hradetzky. $^{40}Ar/^{39}Ar$ age determinations on saninidines of the Eifel volcanic field (Federal Republic of Germany): constraints on age and duration of a middle Pleistocene cold period. *Chem. Geol. (Isot. Geosci. Sect.)* **59**, 187-204 (1986).
19. H.-D. Scheer. Die pleistozänen Flußterrassen in der östlichen Mainebene, *Geol. Jb. Hessen* **104**, 61-86 (1976).
20. H.-U. Schmincke, P. Bogaard v.d. and A. Freundt. Quaternary Eifel Volcanism, Excursion 1A1, International volcanological Congress, 188 pp. Mainz (1990).
21. G.Seidenschwann. Zur pleistozänen Entwicklung des Main- Kinzig- Kahl- Gebietes, *Rhein-Main. Forsch.* **91**, 194 pp. (1980).
22. G.Seidenschwann. Die pleistozäne Talverschüttung im Kahl- und Kinziggebiet, ihre Gliederung und geomorphologisch- stratigraphische Stellung innerhalb der Terrassenfolgen von Kahl und Kinzig, *Jber. wetterau. Ges. ges. Naturkunde* **140-141**, 71-96 (1989).
23. G. Seidenschwann and E. Juvigné. Fundstellen mittelpleistozäner Tephralagen im Randbereich des Kristallinen Vorspessarts, *Zeit. deut. geol. Ges.* **137**, 625-655 (1986).
24. A. Semmel. Area between the Scandinavian and the Alpine Glaciation. 1. Periglacial Sediments and their Stratigraphy, *Eiszeitalter u. Gegenwart* **23/24**, 293-305 (1973).
25. A. Semmel. Der Stand der Eiszeitforschung im Rhein-Main-Gebiet, *Rhein-Main. Forsch.* **78**, 9-56 (1974).
26. A. Semmel. Geomorphologie der Bundesrepublik Deutschland, *Geogr. Z. Erdkdl. Wiss.* **30**, 149 pp. Wiesbaden (1984).

27. E. Wilmart. Modèle géochimique des sédiments paléozoïques du Sud de la Belgique. Méthodologie pour l'étude des sédiments pélitiques. *Rapport final au F.R.F.C., Programme 112, Université de Liège, Laboratoires associés de Géologie, Pétrologie et Géochimie*, 318 pp., Liège (1984).
28. G. Woillard. Grande Pile peat-bog: a continuous pollen record for the last 140,000 years, *Quaternary Research* **9**, 1-21 (1978).
29. W.H. Zagwijn and H. Hager. Correlations of continental and marine Neogene deposits in the South-Eastern Netherlands and the lower Rhine district. *Meded. Werkgr. Tert. Kwart. Geol.* **24**: 59-78 (1987).

Analysis of paleo-environments based on
electric conductivity of the STICS-water
————On five drill cores in Kanto-plain,
central Japan.

M.Koarai[1] and Research Group for Alluvium Deposit[2]

[1] Seacoast Division, River Bureau, Ministry of Construction,
Kasumigaseki 2-1-3, Chiyoda Ward, Tokyo 100, Japan

Abstract Electric conductivity of the STICS-water was measured for presumption of the sedimentary environments on five drill cores in Tsukuba upland and Tokyo lowland. From measurement rusults, electric conductivity method is more effective and convenient in analyzing the sedimentary environments in detail, for supplement of other paleo-environment analyses such as diatom analysis.

1. Introduction

It is very important to clarify the sedimentary environments and paleogeography, to consider geomorphic or geological history. Micro-fossil analyses such as diatom analysis have been conventional method for the presumption of the sedimentary environments; however, such methods need the existence of micro fossil in sediments, and the identification of fossil requires a specialized proficiency.

On the other hand, there have been reports which suggest to study sedimentary environments through chemical analysis of sulfur concentration. The total sulfur measurement method (Koma et al;1985) and the sulfur analysis of the sediment by the H_2O_2 treatment - turbidimetric method (Sato;1989) are the examples. Such methods all take note of the fact that marine sediments contain a large amount of SO_4 ion.

---

[2]Member of the Research Group for Alluvium Deposit:Kazuo Ando, Yoshiaki Hirano, Shigeharu Hochigai, Mamoru Koarai, Masao Maeda, Toshiyuki Matsumoto, Toshio Nakayama, Ichiro Noguchi, Masahide Shimamura.

Electric conductivity measurement is a well known method for determining the amount of anion in the solution such as SO_4, Cl and NO_3 ions. Higher electric conductivity value indicates higher content of anions in the solution. As it is suggested that the amount of anion included in the clayey sediments reflects the difference of marine and non-marine paleo-environments, the measurements of electric conductivity and pH of stirred clayey sediments into water (STICS-water) is effective for analysis of paleo-environments (Yokoyama and Sato;1987). Such method is easy and comprehensive, and is able to be carried out regardless of the existence of micro-fossil in sediments. The condition of samples does not affect the result either. Thus, the electric conductivity measurement provides simple installations and minimal errors.

Yokoyama and Sato (1987) reported that the results of paleo-environments analysis using electric conductivity corresponded well with diatom assemblages in Osaka Group. This method is expected to be more effective in analyzing the transgression and regression in precise detail compared to the diatom analysis.

In this article, we introduce two examinations of electric conductivity method in Kanto-district, central Japan. One is the measurement of electric conductivity and pH of STICS-water from three drill cores samples of Pleistocene in Tsukuba upland, Ibaraki Prefecture (one drill core by Geographical Survey Institute(GSI); two by Geological Survey of Japan(GSJ)), and we compared these measurement results with diatom assemblages by GSI and GSJ. Another is the measurement of electric conductivity and pH from two drill cores samples of Holocene in Tokyo lowland conducted by the Research Group for Alluvium Deposit. This group consists of applied geologists from goverment, muricipal offices and consultant companies, and has worked on analyses of drill cores collected in Tokyo area to obtain detailed geological profile of Tokyo lowland.

2.Analysis method of paleo-environments based on electric conductivity

The principle of electric conductivity method is that the amount of anion contained in the clayey sediments reflects the difference of marine and non-marine paleo-environments, as the amount of anion in marine water is more than that is found in fresh water. Therefore, marine sediments indicate high electric conductivity and low pH, whereas fresh water sediments indicate low electric conductivity and high pH.

Next, we introduce measurement method by Yokoyama and Sato (1987).
(1) Preparation of sample.
Dry the sample at 110 °C for 48 hours. Crush the sample, and weigh 10g for the measurement sample.
(2) Measurement of electric conductivity and pH
Add 120cc pure water to the sample and stir for 3 minites. Measure the electric conductivity and pH after an hour. Measure them again after five days. If the difference of the measurement values between after an hour and after five days is small, the later is adopted for measurement result.

(a) Tsukuba upland

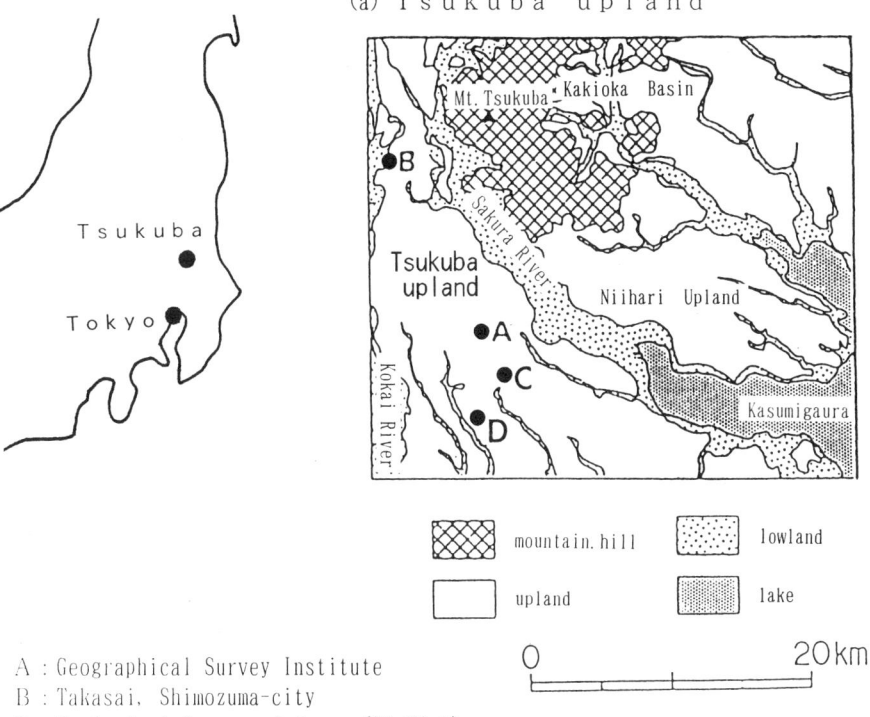

A : Geographical Survey Institute
B : Takasai, Shimozuma-city
C : Geological Survey of Japan (GS-TS-1)
D : Geological Survey of Japan (GS-TS-2)
E : Hibiya, Chiyoda Ward
F : Kiba, Kohto Ward

(b) Tokyo lowland

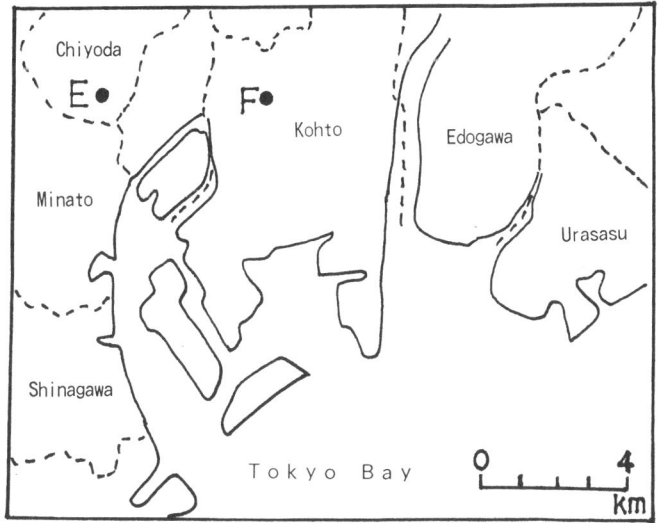

Figure 1 ; Location maps of studied area

3. Result of the analysis of drill cores in the Tsukuba platform

Tsukuba upland is located in the surthern part of Ibaraki Prefecture. It is situated in the central part of the Kanto Plain, which is 60km from Tokyo Metropolis. The area is developing for the largest science city in Japan and one of satelite towns in Tokyo Metropolis.

The geologic units at the area are as follows in ascending order by Unosawa et al (1985); Pre-Quaternary basement which forms the Tsukuba mountainous land, the Shimousa Group of Middle and Late Pleistocene which forms Tsukuba and Niihari upland, the Younger Kanto Loam Formation of Late Pleistocene which covers upland and terraces. The Simousa Group consists of six formations; the Jizodo Formation, Yabu Formation, Kamiizumi Formation, Kamiiwahashi Formation, Kioroshi Formation and Jyoso Formation in ascending order.

Verification of the effectiveness of electric conductivity method in Tsukuba upland was carried out using the samples taken from the three drill cores bored by the government research institute and an outcrop (Figure 1).

Figure 2A shows the columnar section of 190m boring at the site of GSI. Pollen analysis and diatom analysis had been carried out for following five horizons (GSI; 1981 MS); No.1 (29.24～29.47m), No.2 (69.43～69.68m), No.3 (69.95～70.20m), No.4 (117.16～117.40m), No.5 (169.25～169.49m). The result of the diatom analysis is shown in Figure 3. From this result, the sedimentary environment of each sample is presumed as follows (GSI; 1981 MS). No.1 is the marine sediment on coastal zone with an inflow of fresh water. No.2 is the fresh water sediment. No.3 is also the fresh water sediment. No.4 is the marine sediment on coastal zone. No.5 is the fresh water sediment in stable area. From the result of the pollen analysis, No.2, No.3 and No.5 indicate cold paleo-climate, and No.1 and No.4 indicate warm paleo-climate. As it is presumed that the transgression had progressed during the warm paleo-climate period, the result of the pollen analysis correspond with the result of the diatom analysis (GSI; 1981 MS).

Electric conductivity and pH were measured for the following 8 horizons near the 5 horizons where the diatom analysis had been carried out earlier; a (29.24～29.27m), b (29.45～29.47m), c (69.55～69.60m), d (69.78～69.80m), e (70.50～70.70m), f (117.16～117.25m), g (117.30～117.40m), h (169.50～169.53m). From the measurement results(Table 1(a)), electric conductivity values are divided roughly into two groups, a group of c, d, e, h of which electric conductivity ranges 0.27～0.45, and a group of a, b, f, g of which electric conductivity ranges 0.62 ～0.84. The values given by the former group correspond with the values of electric conductivity of fresh water sediments measured by Yokoyama and Sato (1987). They also correspond with the results of the diatom analysis (No.2, 3, 5). The latter group indicates the values given by electric conductivity of blackish water sediments also measured by Yokoyama and Sato (1987). On the other hand, the latter group corresponds with the results of the diatom analysis (No.1, 4) in regard to having a higher electric conductivity and being under a strong influence of marine water. However, taking a marine

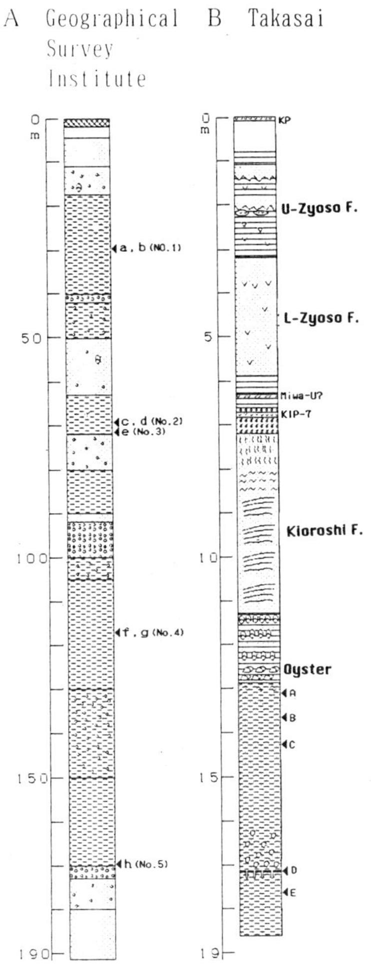

255

A Geographical Survey Institute
B Takasai

Table 1 ; Measurement results of electric conductivity and pH of STICS-water in the Tsukuba upland.

(a) Geographical Survey Institute

Sample	Cond (mS/cm)	pH
a	0.692	6.58
b	0.840	6.51
c	0.357	6.88
d	0.449	6.95
e	0.328	7.04
f	0.620	7.20
g	0.704	7.25
h	0.274	7.55

(b) Takasai

Sample	Cond (mS/cm)	pH
A	0.611	7.29
B	0.779	7.33
C	0.777	5.05
D	1.421	3.55
E	0.384	6.45

(c) Geological Survey of Japan (GS-TS-1, GS-TS-2)

Sample	Depth (m)	Cond (mS/cm)	pH
1	1.40	0.124	6.97
2	3.10	0.151	6.91
3	5.40	0.305	6.80
4	10.13	0.150	6.98
5	14.52	0.352	6.91
6	20.10	0.597	6.96
7	42.65	0.417	7.15
8	43.50	0.353	7.21
9	44.90	0.560	7.13
10	45.35	1.258	6.80
11	52.15	2.060	6.58
12	57.00	3.200	4.40
13	67.35	1.676	4.54
21	2.30	0.143	6.42
22	2.70	0.154	6.60
23	17.30	0.706	6.50
21	20.35	1.327	6.67
22	23.68	3.290	5.72
23	27.18	2.600	5.10

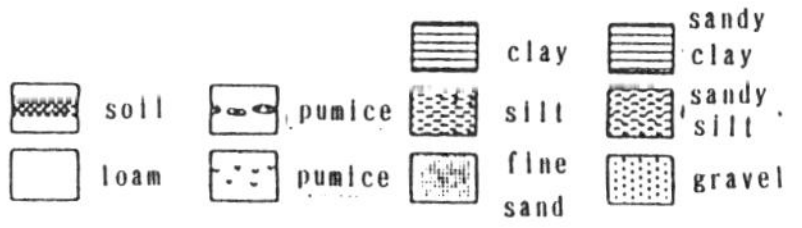

Figure 2 ; Columnar sections of 190m boring at GSI and a outcrop at Takasai.

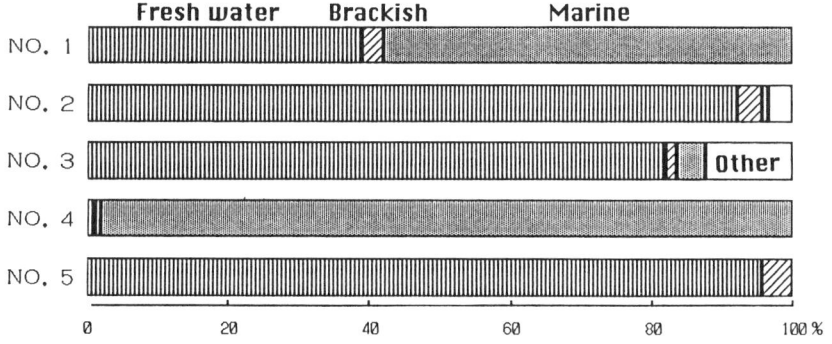

Figure 3 ; Result of daitom analysis by Geographical Survey Institute.

water influence into consideration, the marine sediments show over 0.6mS/cm in electric conductivity, whereas the non-marine sediments show under 0.5mS/cm. It was therefore difficult to distinguigh between marine and non-marine sediments from their pH values since every sample indicated a neutral pH ranging 6 ～7.

Figure 2B shows the columnar section of a outcrop at Takasai, Shimozuma City, Ibaraki Prefecture. It represents the northern most margin of the Paleo-Tokyo Bay (Masuda et al;1987). Electric conductivity and pH were measured for five samples. From the result of the measurement(Table 1(b)), the radical change of electric conductivity value which correspond to the environmental change from fresh water to marine was recognized between sample E and D. After that, as the regression progressed gradually, marine environment turned into blackich environment.

Figure 4 shows the columnar sections of drill cores (GS-TS-1, GS-TS-2) by GSJ, to make "The environmental geologic map of the Tsukuba Science City and its surroundings." Diatom analysis of these drill cores had been carried out by GSJ (Unosawa et al;1988). Electric conductivity and pH for 13 samples of GS-TS-1 and 6 samples of GS-TS-2 were measured.

From the measurement results(Table 1(c) and Figure 4), electric conductivity values of marine sediments show over 0.35mS/cm excepting sample 4. And electric conductivity values of non-marine sediments show under 0.35mS/cm excepting sample 9,10,12,13,26. Some samples of fresh water sediments presumed by micro-fossil analysis show higher electric conductivity. That reason is presumed that electric conductivity is higher by influence of humus acid, as these samples are humus.

4. Result of the analysis of drill cores in Tokyo lowland

Two drill cores in Tokyo lowland were measured. One is drill core at Hibiya, Chiyoda Ward (Hb sample), another is at Kiba, Kohto Ward (Kb sample) (Figure 1). The result of the analysis of Hb sample is shown in Table 2(a) and Figure 5, and that of Kb sample is shown in Table 2(b) and Figure 6. These diatom assemblages were analyzed by Mr. Kazuo Ando.

These samples were taken from Holocene deposit which named the Yurakucho Formation. The result of diatom analysis shows that these samples are marine sediments.

The result of the analysis of Hb sample suggests that electric conductivity and pH are in the reverse correlation. Electric conductivity measurement corresponds well with the diatom assemblage; for example, electric conductivity value drops as the marine diatom near the level 0m decreases. Electric conductivity values are divided roughly into three groups; under 0.2mS/cm, 0.5～0.8mS/cm and over 0.9mS/cm. It is assumed that each group indicates the sedimentary environment of fresh, blackish and marine water, respectively.

The result of the analysis of Kb sample shows that electric

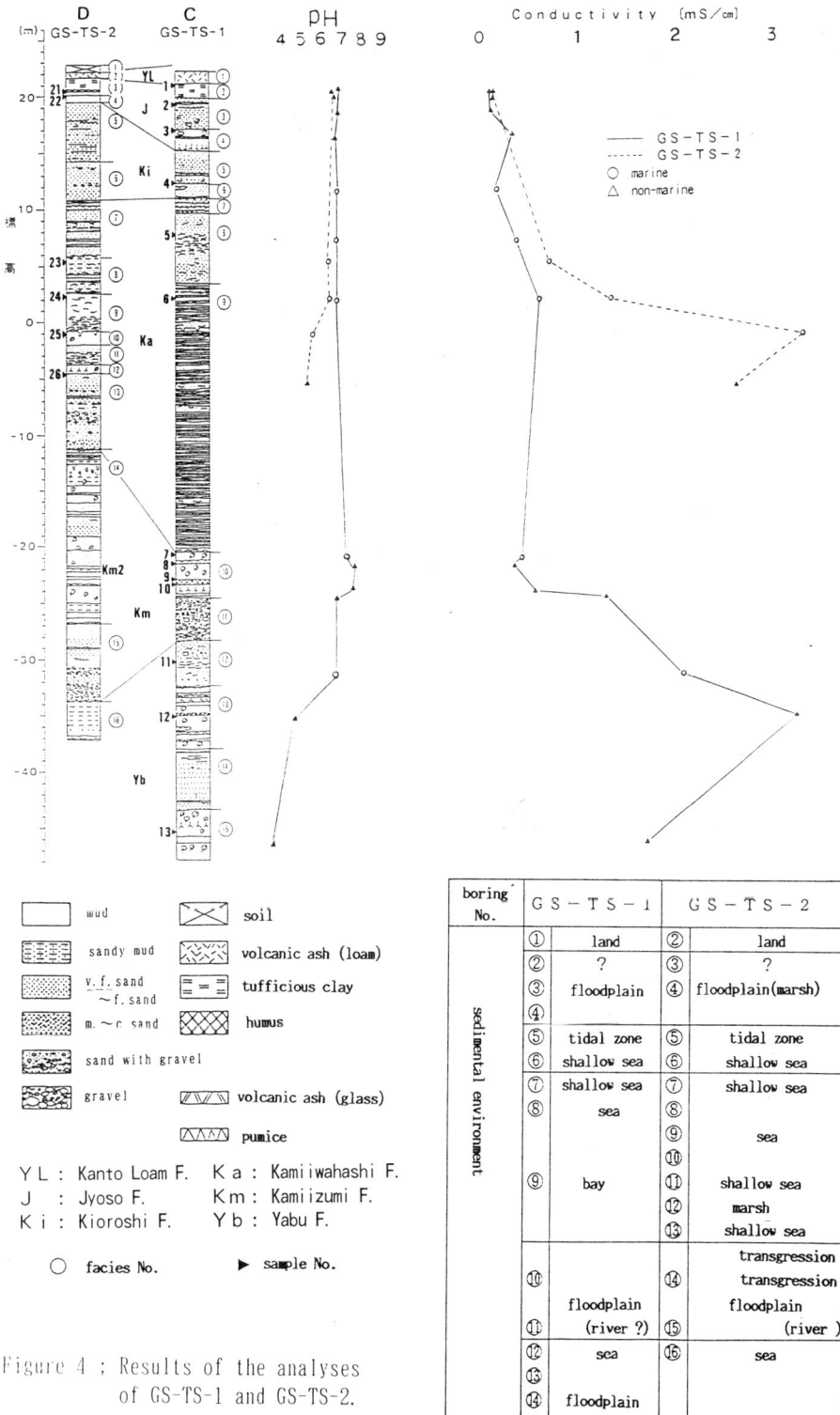

Figure 4 ; Results of the analyses of GS-TS-1 and GS-TS-2.

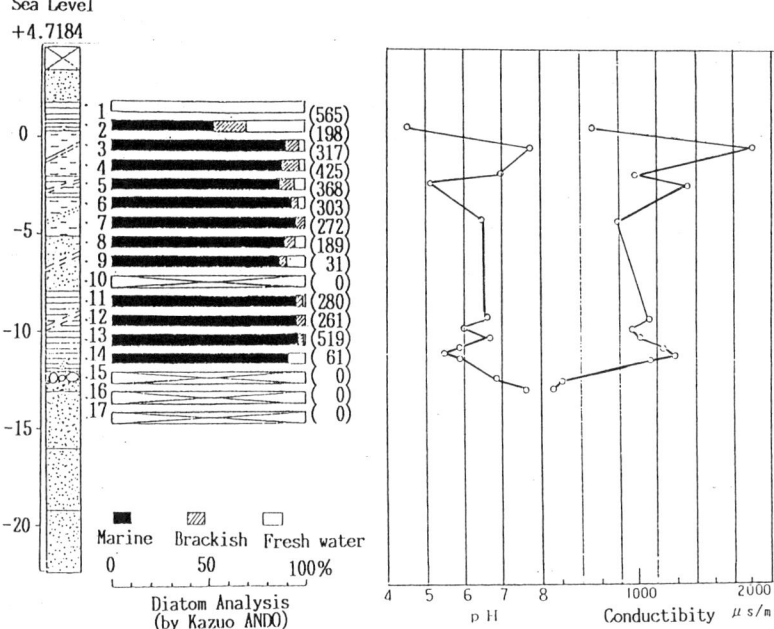

Figure 5 ; Result of the analysis of Hibiya drill core.

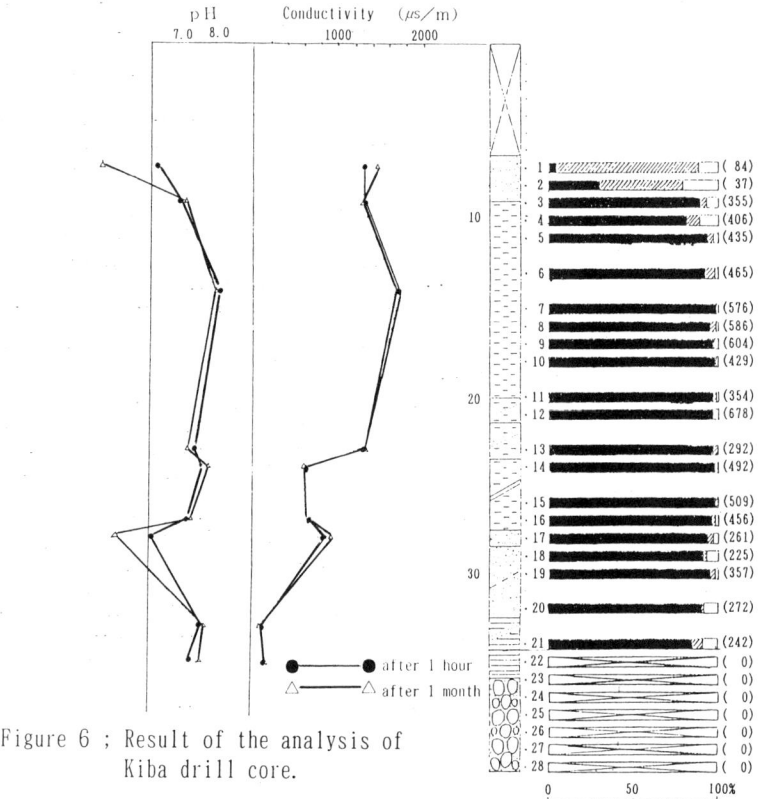

Figure 6 ; Result of the analysis of Kiba drill core.

Table 2 ; Measurement results of electric conductivity and pH of STICS-water in Tokyo lowland.

(a) Hibiya drill Core

Sample No.	Depth GL-(m)	Soil	after 1 hour	
			pH	Cond.
263(15)	3.95	clayey sand	4.5	0.59
436(1)	4.95	clayey silt	7.7	2.20
436	6.30	clayey silt	6.9	0.98
460(1)	6.80	clayey silt	5.1	1.35
453(12)	8.70	very fine sand	6.5	0.79
11	13.66	silty clay	6.6	1.13
460	14.20	silty clay	6.0	0.96
A	14.70	silty clay	6.7	1.01
485	15.20	silty clay	5.9	1.27
14	15.48	silty clay	5.5	1.38
459(1)	15.80	silty clay	5.9	1.14
459	16.48	sandy clay	6.9	0.21
16	16.55	sandy clay	7.6	0.13

(b) Kiba drill core

Sample No.	Depth GL-(m)	Soil	after 1 hour		after 1 month	
			pH	Cond.	pH	Cond.
436	7.0	fine sand	6.2	1.26	4.6	1.46
485	9.0	fine sand	6.9	1.34	7.0	1.32
11	14.0	silt	8.0	1.66	7.8	1.65
16	23.0	silt with sand	7.3	1.33	7.1	1.33
12	24.0	sandy silt	7.5	0.65	7.7	0.62
14	27.0	silty clay	7.1	0.69	7.2	0.70
459	28.0	silty clay	6.1	0.88	5.0	0.94
15	33.0	clayey fine sand	7.5	0.11	7.6	0.12
460	35.0	clay	7.2	0.19	7.5	0.19

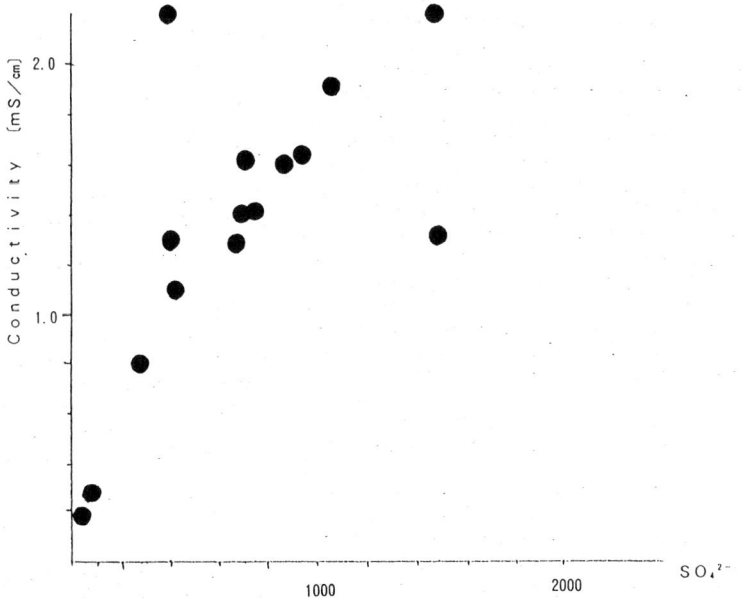

Figure 7 ; Relationship between amount of SO_4^{2-} and electric conductivity of STICS-water.

conductivity values are stable very well and the difference of the measurement values between after an hour and after a month is minimal. The change of diatom assemblage was not found with respect to the increase of electric conductivity value near the depth 27 ~28m (between the samples 12 and 16). The decrease of electric conductivity value was not found with respect to the increase of blackish diatom near the depth at 7~8m (between the samples 485 and 436). The measurement of electric conductivity was not effective at the depth of 7~8m, probably due to the samples containing too much sand.

Figure 7 shows the results of chemical analysis of Hb sample using ion chromatograph. Electric conductivity value is directly propotional to sulfate anion concentration. Electric conductivity found similar relationship with chlorine anion.

5. Conclusion

The results of the analyses stated above clarified the difference of values of electric conductivity of STICS-water between marine sediments and non-marine sediments. Thus, such method is effective for the distinguishing marine or non-marine environments of Pleistocene and Holocene sediments on Kanto-district. However, as the boundary value of electric conductivity between marine and non-marine sediments may vary depending upon districts and horizons, it is not possible to give the estimation of paleo-sedimentary environments using the measurement value of one sample only. It is required to measure electric conductivity successionaly and comprehensively estimate paleo-environments.

This method is very simple and easy, because it does not require specialized analytic skill (i.e. micro-fossil analysis), long measurement time and expensive measurement installations. Therefore, the supplementary use of this method such as pre-investigation of diatom analysis, allows a very effective presumption of paleo-environments and geological history. For example, the time and the expenses spent on analysis could be more efficient through carrying out the diatom analysis of drill samples focusing on the changes resulted from successive analysis of electric conductivity.

But, the results of electric conductivity measurement does not correspond with the result of diatom analysis, for the samples containing too much sand or humus. For this reason, this method does not necessarily apply to all samples. Therefore, a further consideration should be made regarding the analysis of sandy samples and the evaluation of the influence of underground water or humus.

The Research Group for Alluvium Deposit has been conducting actively the collection of drill cores in Tokyo lowland, the measurement of electric conductivity and diatom analysis. In the future, the group is planning to increase the number of comparison between electric conductivity values and diatom assemblages. By using the results of the comparison, it is examined whether electric conductivity method could be used for non-fossil sediments insted of diatom analysis.

Acknowledgements

On the study of Tsukuba drill cores, we extend our gratitude to; Prof. Takuo Yokoyama of Doshisha University for helping us to measure electric conductivity and pH; Dr Ichiyo Isobe of Geological Survey of Japan and Mr. Shinzo Ohi of Geographical Survey Institute for obtaining the samples.

REFERENCE

Goegraphical Survey Institute. Report of pollen and diatom analyses of drill samples from a ground subsidence observation well in Tsukuba Science City (in Japanese). 16p. (1981).

T. Koma, T. Sakamoto and A. Ando. Total sulfur content and sedimentary environment of the late Cenozoic formations in Ibaraki Prefecture, Japan, Bull. Geol. Surv. Japan. 34, 279-293 (1983).

F. Masuda, M. Ishibashi and M. Ito. A New Locality of Marine Molluscan Fossils from the late Pleistocene on the Nortern Margin of the Paleo-Tokyo Bay: Takasai, Shimozuma City, Ibaraki, Environmental study in Tsukuba. 10, 79-89 (1987).

H. Sato. Surfur Analysis of the Sediment by the H_2O_2 Treatment-Turbidimetric Method: A Simple Method for Studying the Paleoenvironment, The Quaternary Research. 28, 35-40 (1989).

M. Sato and T. Yokoyama. Analysis of paleo-environments based on electric conductivity of the STICS-water -On two drilled cores at Osaka Bay, Japan-, Jour. Geol. Soc. Japan. 98, 825-839 (1992).

A. Unosawa, I. Isobe, H. Endo, Y. Tagutschi, S. Nagai, T. Ishii, T. Aihara and S. Oka. Explanantory text of the environmental geologic map of the Tsukuba Science City and its surroundings, scale 1:25,000. Miscellaneous Map Series (23-2), Geol. Surv. Japan, 139p. (1988).

T. Yokoyama and M. Sato. Analysis of paleo-environments based on electric conductivity of stirred clayey sediments into water -On two drill cores of Osaka Group in Senriyama Hills and Biwako Group at bottom of Lake Biwa, Japan-, Jour. Geol. Soc. Japan. 93, 667-679 (1987).

Holocene dinoflagellate cyst assemblage in lagoonal lakes along the east coast of Japan

N. KOJIMA

Department of Biology, Faculty of Science, Osaka City University, Sugimoto 3-3-138, Sumiyolshi-ku, Osaka 558, JAPAN

Abstract The change of Holocene dinoflagellate cyst assemblage is analyzed in the core sample borings from inland-sea lakes, Lake Hamana and Lake Kasumigaura, on the east coast of Japan. The fossil assemblage fluctuations are caused by the transition of water properties accompanying the development of a barrier-system during sea-level changes. The cysts appearing in the two lagoonal regions have many common species. Heterotrophs (e.g. *Brigantedinium* and *Selenopemphix*), in particular, have high occupation ratios in the two lakes. The alteration of occupation percentage in many species does not clearly indicate their motional factor. The change of an aggregate fossil cyst, however, is distinctly related to the influence of salinity change. The analysis also revealed that the possession ratio of a specific species, *Polykrikos schwartzii*, is affected by a particular water property (perhaps salinity). The index may also indicate small- scale sea-level changes during the Late Holocene from the viewpoint of the change of inflow sea water in the lakes.

KEYWARDS: dinoflagellate cyst, Holocene, inland-sea lake, salinity , sea-level changes,

INTRODUCTION

The species composition of dinoflagellate cyst fossils does not change remarkably during the Quaternary age. This is because the first appearance and extinction of dinoflagellate cysts are rare in Quaternary sections due to the limited amount of time available during that period for the evolutionary process[1]. This fact shows that the dinoflagellate cyst doesn't contribute to the biostratigraphy for the Quaternary age.
The distribution of modern cyst assemblages connected directly to Quaternary cysts indicates clearly different distribution patterns based on environmental divergency[2]. Moreover, many Quaternary cyst fossil assemblages are well preserved and obtained from large quantities of modern thanatocoenothis-like assemblages in comparison with poor cyst data for the pre-Quaternary period. Hence, they may provide information about environmental change at that time.
For this paper, inland-sea lakes develop a barrier-system with a recent history of environmental change were chosen for the study field. The author then shows the fluctuation characteristics of Holocene dinoflagellate cyst assemblages and discusses their significance.

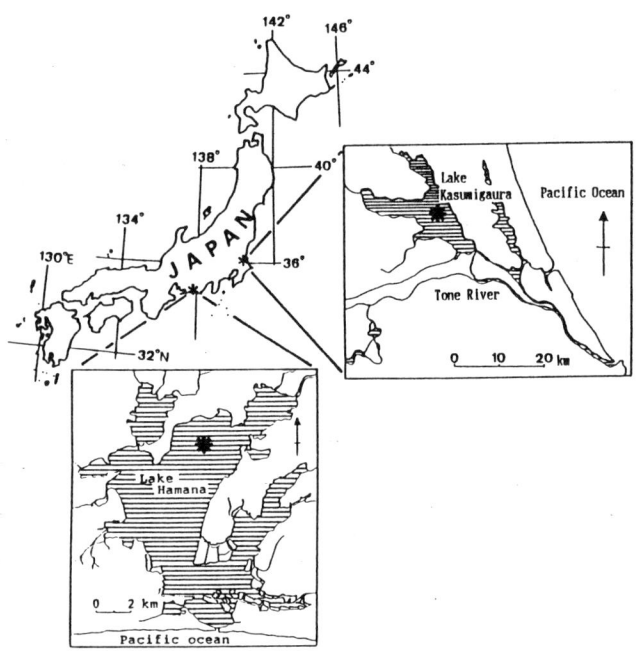

Figure 1. Location of study sites.

MATERIALS AND METHODS

The samples were obtained from Lake Kasumigaura and Lake Hamana situated on the east coast of Japan (Fig.1). The boring samples were mainly composed of pelitic sediments and did not show bioturbation by the benthos because of the existence of an intensely reducing environment. Many tephra also were recognized in them which act as a good time marker [3,4]. The carbon-14 ages of the lowest part of the two sites are from 7000 to 10000 years. The dates are older than ca. 6000 yBP, which was the period of highest sea-level around the Japanese coast. The bottom sample of Lake Kasumigaura includes peaty sediments[4].

The collected samples were treated by a palynological method[5] as follows. 1) An approximately 10% solution of hydrochloric acid is added to the sample to dissolve the calcium carbonate. 2) The remains are treated with an approximately 50% solution of hydrofluoric acid to remove silicate particles. 3) The organic remnants passed through a mesh with 200 μm opening to remove large impurities. 4) The useful sized remnants were retained continuously on a 20 μm opening mesh sieve. The remainder were enclosed on slide glasses and observed by a light microscope. The data were based on identifying 300 cyst individuals per sample and the total per unit volume ($1cm^3$).

RESULTS

The total number cysts was small near the bottom part of the cores in the two lakes but was an almost fixed quantity above the lower samples (Fig.2). In Lake Kasumigaura, however, the

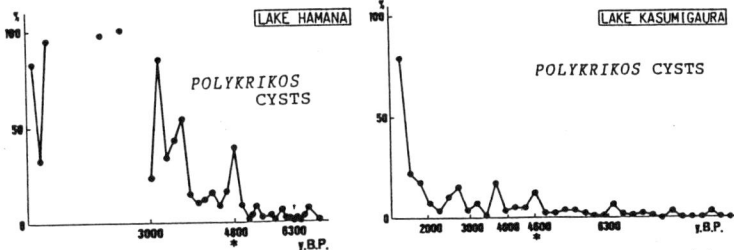

Figure 2. Quantity fluctuation of dinoflagellate cysts and chrolophyceae cells of the two lakes.
(Lake Hamana's fig.: after Kojima, 1989)

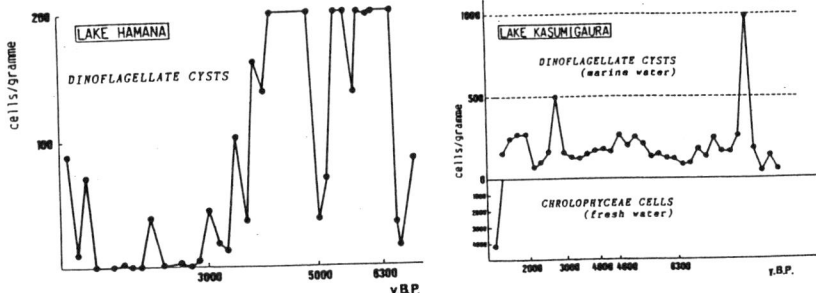

Figure 3. Percentage fluctuations of *Polykrikos* cysts of the two lakes.
(Lake Hmana's fig.: after Kojima, 1989)

marine dinoflagellate cysts changed to fresh water algae (mainly *Pediastrum*) (Fig.2) approximately 300 years ago. In Lake Hamana, the cyst almost disappeared approximately 3000 years ago. The condition of sporadic appearance of cysts continued for 2500 years after that, and then returned to marine dinoflagellate cysts again 500 years ago (Fig.2).

The composition of the cysts in the two lakes is dominated by heterotrophic species (Fig.3). Speceis of *Brigantedinium* were most abundant and *Selenopemphix* cysts were subdominant in the assemblage among others. Each species has its specific fluctuation of the appearance ratio, and their percentage curves are not always in parallel. The occupancy proportion of *Polykrikos schwartzii* clearly increases with regression of the salinity level in the lakes (Fig.4). In autotrophic species, *Pheopolykrikos hartmanii* occasionally occupied more than 70% in the assemblage. This phenomenon is not frequent but remarkable.

DISCUSSION

Saito[6] revealed that Holocene sea-level changes influenced the environment of inland-sea lakes having barrier-systems in Japan. To synthesize the result and this tendency of cyst appearance we are led to the next interpretation. That is to say, the initial cyst-increasing period corresponds to the age of invading sea water with rising sea level in the valley formed in the last glacial age. Consequently, abundant cyst assemblages are obtained continuously from the sediments accumulated during the higher sea-level stage centered around 6000 yBP. The following age begins with a lowering of sea level, and a decrease ultimately in aggregate cysts with the emergence of sandy sediments at the bay mouth (Fig.5). The different timing of the

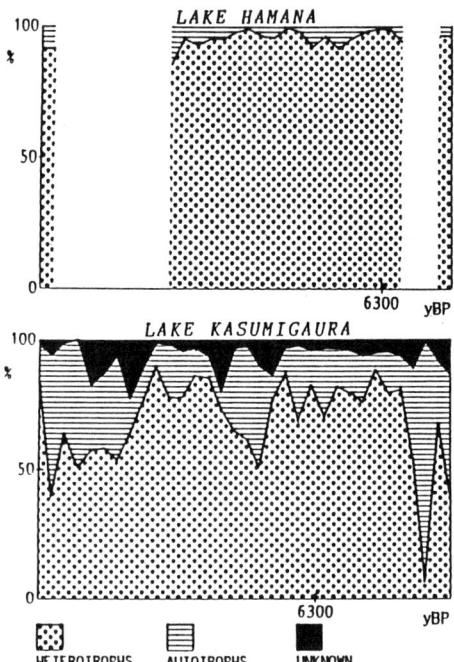

Figure 4. Occupation ratio of heterotrophic and autotrophic dinoflagellate cysts of the two lakes.

change from a marine to brackish environment between the two lakes shows that the two lakes have different factors which cause the barrier-system to form. Namely, the formation of the barrier systems occurred in early times in Lake Hamana, but at a later date in Lake Kasumigaura because of its large bay mouth and the disparate sedimentary environment. The time lag of the barrier formation caused the difference on the low salinity period. The composition of cyst species of the lakes is similar to that of Lake Saroma, Akkeshi bay[7] and off Honjo, off Oga[8] in comprison with the assemblage of the bottom sediments around Japan. The assemblage resembles the northern group around the Japanese Isles. The two lakes have almost the same species; however, the appearance ratio in Lake Hamana shows an intense fluctuation. Lake Hamana's assemblage is presumably more sensitive to salinity change than Lake Kasumigaura's, because Lake Hamana has more changeable barrier at the lake mouth than Lake Kasumigaura.

P. schwartzii appeared characteristically in the two lakes. As previously indicated regarding *P. schwartzii*, the species becomes an index of the progression from a marine to a brackish environment[5]. The index species had maintained its percentage at a low level untill approximately 5000 yBP. The percentages, however, rose suddenly after 5000 yBP, and increased steadily after that. This age, 5000 yBP, agrees with the so-called "Middle Jomon little regression", which has an abrupt sea-level lowering. Consequently, the regression accelerates the emergence of the sandy sediments around the bay mouth area, and speeds up the decline of salinity level in the lakes (Lake Hamana's temporary cyst reduction at ca. 5000 yBP may also stem from this event.). In Lake Hamana, though the cyst appearance reduces unexpectedly at approximately 3000 yBP, the incident may result from the so-called "Yayoi little regression". This regression is not recognized in the cyst data of Lake Kasumigaura.

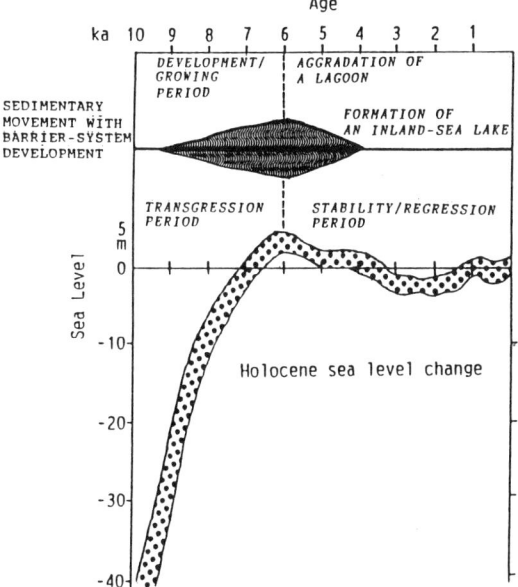

Figure 5. The relation between Holocene sea-level changes and the development of sedimentary body. (adapted from Saito, 1987)

It has been found that the Holocene dinoflagellate cyst assemblage of inland-sea lakes developing a barrier-system is strongly influenced by sea-level changes. The initial stage of a transgression and the last stage of regression, in particular, have a dominant effect on the total cyst number. The cyst alteration between the two stages, however, cannot be interpreted. The comparative evident fluctuational curve of the possession ratio of each cyst has then many obscure points. The relationships between many cysts, physical and chemical information, and data for other lakes should be further studied in detail.

ACKNOWLEDGEMENTS

I am much indebted to Prof. K. Matsuoka for giving an opportunity to study of the samples of Lake Hamana. I also thank to Dr. Y. Inouchi for providing some samples of Lake Kasumigaura.

REFERENCES

1. D. K. Goodman. Dinoflagellate cysts in ancient and modern sediments. In: *The biology of dinoflagellates*. F.J.R. Taylor (Ed.). pp.649-722. Chap. 15. Blackwell, London (1987).
2. F. J. R. Taylor. Ecology of dinoflagellate. A. General and marine ecosystems. In: *The biology of dinoflagellates*. F.J.R. Taylor (Ed.). pp.398-502. Chap. 11. Blackwell, London (1987).
3. N. Ikeya, H. Wada and M. Ohmori. On the boring core sediments from Haman Lake. *Geosci. Rep. Shizuoka Univ.* **13**, 67-111 (1987). (in Japanese)
4. Y. Saito, Y. Inouchi and S. Yokota. Coastal lagoon evolution influenced by Holocene sea-level changes Lake Kasumigaura, central Japan. In: *The memoirs of the geological society of Japan*. Y. Inouchi, T. Tokuoka,

K. Takayasu, K. Anma, Y. Makino and H. Nirei (Eds). No. 36, pp.103-118. Geological society of Japan, Tokyo (1990). (in Japanese)

5. N. Kojima. Dinoflagellate cyst analysis of Holocene sediments. *Trans. Proc. Palaeont. Soc. Japan, N. S.* No. 155, 197-211 (1989).

6. Y. Saito. Formation model of marin alluvial sediment controled through sea-level changes. *The Earth* (Chikyu) 9, 533-541 (1987). (in Japanese)

7. K. Matsuoka. Organic-walled dinoflagellate cysts from surface sediments of Akkeshi Bay and Lake Saroma, north Japan. *Bull. Fac. Lib. Arts. Nagasaki Univ. Nat. Sci.* 28, no. 1, 35-123 (1987).

8. K. Matsuoka. Organic-walled dinoflagellate cysts from surface sediments of Nagasak Bay and Senzaki Bay, West Japan. *Bull. Fac. Lib. Arts Nagasaki Univ. Nat. Sci.* 25, no. 2, 21-115 (1985).

The Late Quaternary Environment around Lake Nojiri in Central Japan

Nojiri-ko Excavation Research Group
Secretariat: c/o Department of Geology, Faculty of Science, Shinshu University, Asahi 3-1-1, Matsumoto 390, JAPAN

Abstract

The Upper Pleistocene sediments distributed around Lake Nojiri are Kamiyama Loam and Kannoki Formations, Ikejiri-gawa Mud-flow Deposits, Nojiri Loam and Nojiri-ko Formations, and Kashiwabara Black Volcanic Ash and J-retsu Formations in ascending order. The Nojiri-ko and Nojiri Loam Formations are divided into three Members, respectively: Lower, Middle and Upper, with marked unconformities.

Fossil pollen assemblages of the upper part of the Kannoki Formation (Pinaceae zone) were dominated by such coniferous pollen as *Picea*, *Tsuga*, *Abies* which show a coldest climate during the Last Glacial Maximum. The occurrence of *Oiceoptoma thoracicum* from the lowest silt bed of the Lower Nojiri-ko Member III suggests a paleoclimate ranging from upper cool temperate to subfrigid. Most vertebrate fossils collected in these excavations are *Palaeoloxodon naumanni* and *Sinomegaceros yabei*. Many foot prints of elephants artiodactyl were also found on the upper surface of the ash layer (Konaazuki) in the Lower Nojiri-ko Member.

Artefact clusters were found in the Middle Nojiri-ko Member I, with a bone cleaver, some refitted bone flakes and chips, rib bones of Nauman's elephant and stone flake tools. This case not only implies a circumstancial evidence for the kill-site, but also suggests that the elephant hunters made bone tools in kill-buchering processes.

1 Introduction

Lake Nojiri is a beautifull lake in Shinano-machi, Nagano Prefecture, central Japan, 645 meters above sea level with a maximum depth of 38.5m. It is surrounded by four Pleistocene stratovolcanoes, Myoko, Kurohime, Iizuna and Madarao (Fig. 1). Many Palaeolithic artefacts are distributed around the lake [1, 2].

The Nojiri-ko Excavation has been carried out since 1962 to restore natural environment around Lake Nojiri where Palaeolithic men lived in the Last Glacial Age. Main excavations were carried out at Tategahana on the west shore of the lake and the others were also carried out at Kannoki, Mukoshinden and Nakamachi Hill (Fig. 2).

The Nojiri-ko Excavation Research Group is composed of both specialists and non-specialists groups. In each excavation total participants were about 2.000 people. As the excavation proceeded, they organized an excavation team themselves in a network system, with twenty-seven local circles (Friends of Nojiri-ko Society) distributed all over Japan and eleven research groups of various fields such as geology, tephrochronology, paleomagnetism, diatom study, palynology, paleobotany, conchology, entomology, mammalogy, trace fossil study and archaeology.

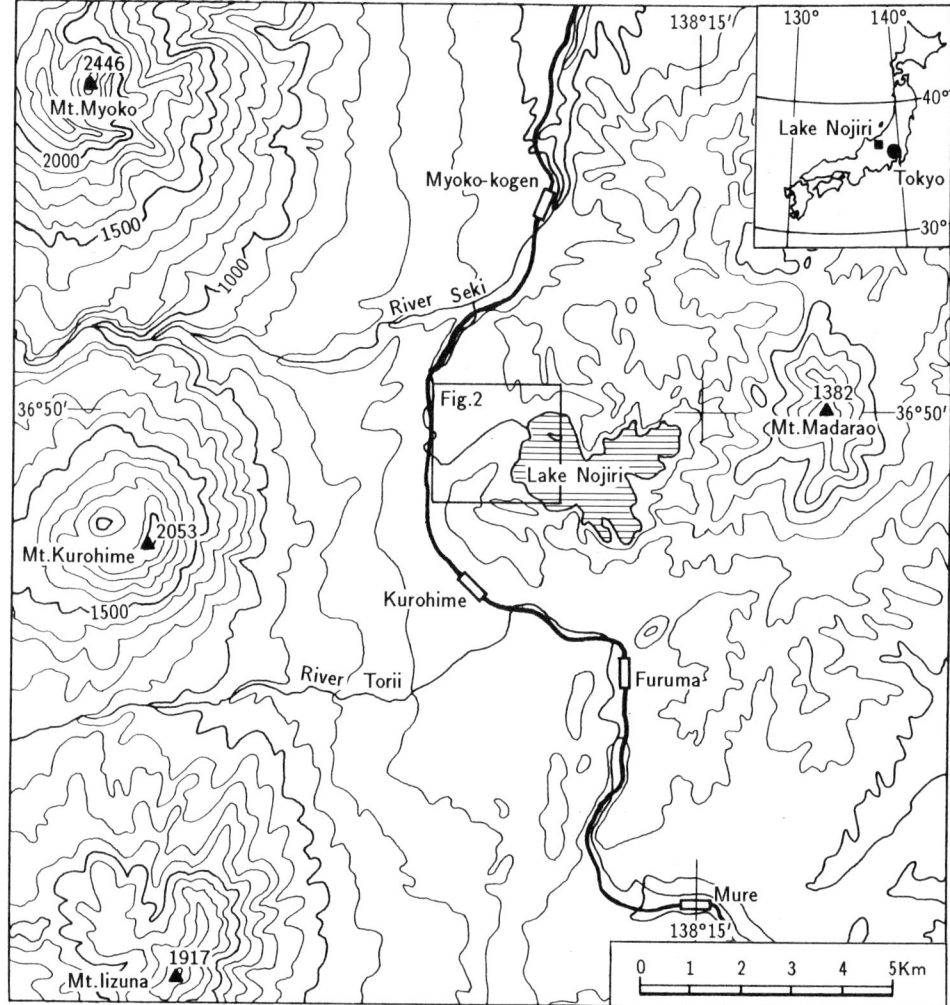

Figure 1. Location of Lake Nojiri.

2 Geology

The Upper Pleistocene sediments distributed around Lake Nojiri are Kamiyama Loam and Kannoki Formations, Ikejiri-gawa Mud-flow Deposits, Nojiri Loam and Nojiri-ko Formations, and Kashiwabara Black Volcanic Ash and J-retsu Formations in ascending order (Table 1 and Fig. 3). The Nojiri-ko Formation is characterized by the presence of plentiful fossils and artefacts.

The Nojiri-ko and Nojiri Loam Formations are divided into three Members, respectively: Lower, Middle and Upper, with marked unconformities.

Furthermore, three or four subdivisions (Sections) can be recognized in each Member of those Formations [4]. The Nojiri Loam Formation is composed of many volcanic ash layers which make it possible to correlate precisely between aeolian deposits on the hill and subaqueous beds in the lake considered from the mineral compositions [5, 6]. Around the shore of Lake Nojiri, the Nojiri-ko Formation lies conformably on the Kannoki Formation in the southern part of the excavation site, while in the northern part it lies on the

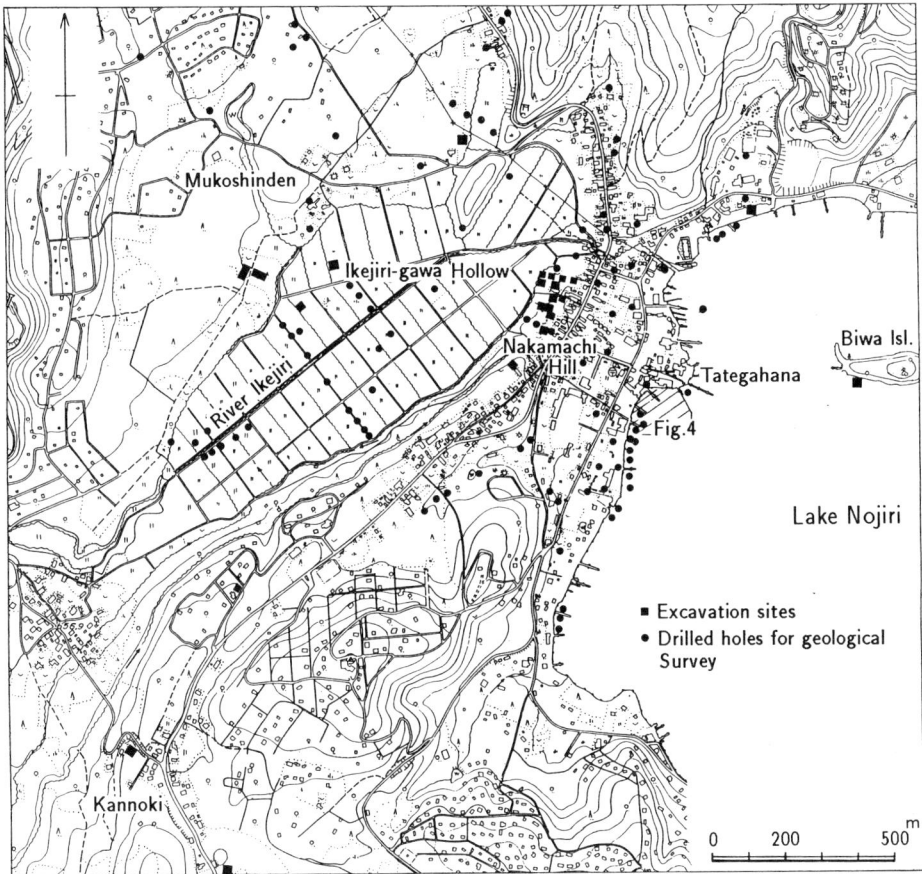

Figure 2. Excavation sites around Lake Nojiri.

basement rocks unconformably without the Kannoki Formation.

A core sample of 45.1 meters length obtained from the bottom of Lake Nojiri shows that Lake Nojiri was formed immediately after the deposition of the Upper Kamiyama Loam I [7].

3 Fossil occurrence and paleoenvironment

The strata of the Kannoki Formation can be divided into such two pollen assemblage zones as the *Cryptomeria* and the Pinaceae zones. The latter zone indicates the coldest climate during the Last Glacial Maximum.

The fossil pollen assemblages of Nojiri-ko Formation are composed of the *Larix-Betula*, the *Picea-Fagus*, the *Tsuga-Picea*, the *Quercus-Fagus* and the *Picea-Abies-Tsuga* zones in ascending order. Pollen assemblage of Section II of Upper Nojiri-ko Member are characterized by abundant coniferous pollen as *Picea*, *Abies* and *Tsuga*.

The Holocene J-retsu Formation has the cool-temperate elements as *Fagus* and *Quercus* [8].

Molluscan fossils of the Nojiri-ko Formation are *Inversidens japanensis*, *Anodonta* sp., *Semisulcospira* sp., *Lymnaea* sp. and *Viviparus?* sp. [9]. Among them, *I. japanensis* is abundant, while the other species are found rarely. *I. japanensis* was yielded from

Section III of the Lower Nojiri-ko Member to Section I of the Upper Nojiri-ko Member of the Nojiri-ko Formation, especially with abundant occurrence in Section IIIB of Lower Nojiri-ko Member (fig. 4), Section I and II of the Middle Nojiri-ko Member and Section I of the Upper Nojiri-ko Member. There are eight types of the mode of occurrence of *I. japanensis* in sediments. Among them three types are composed of intact valve specimens which are considered to be authochtonous. A greater part of the shells are detached valves, suggesting that they were transported before deposition.

Insect fossils of the Nojiri-ko Formation were found from the Lower Nojiri-ko Member I to the Upper Nojiri-ko Member III. It is significant that many fossils of nearly ten species of coprophagous beetles such as *Copris pecuarius*, *Aphodius elegans* etc. were found from the Lower Nojiri-ko Member I to the Upper Nojiri-ko Member I, suggesting

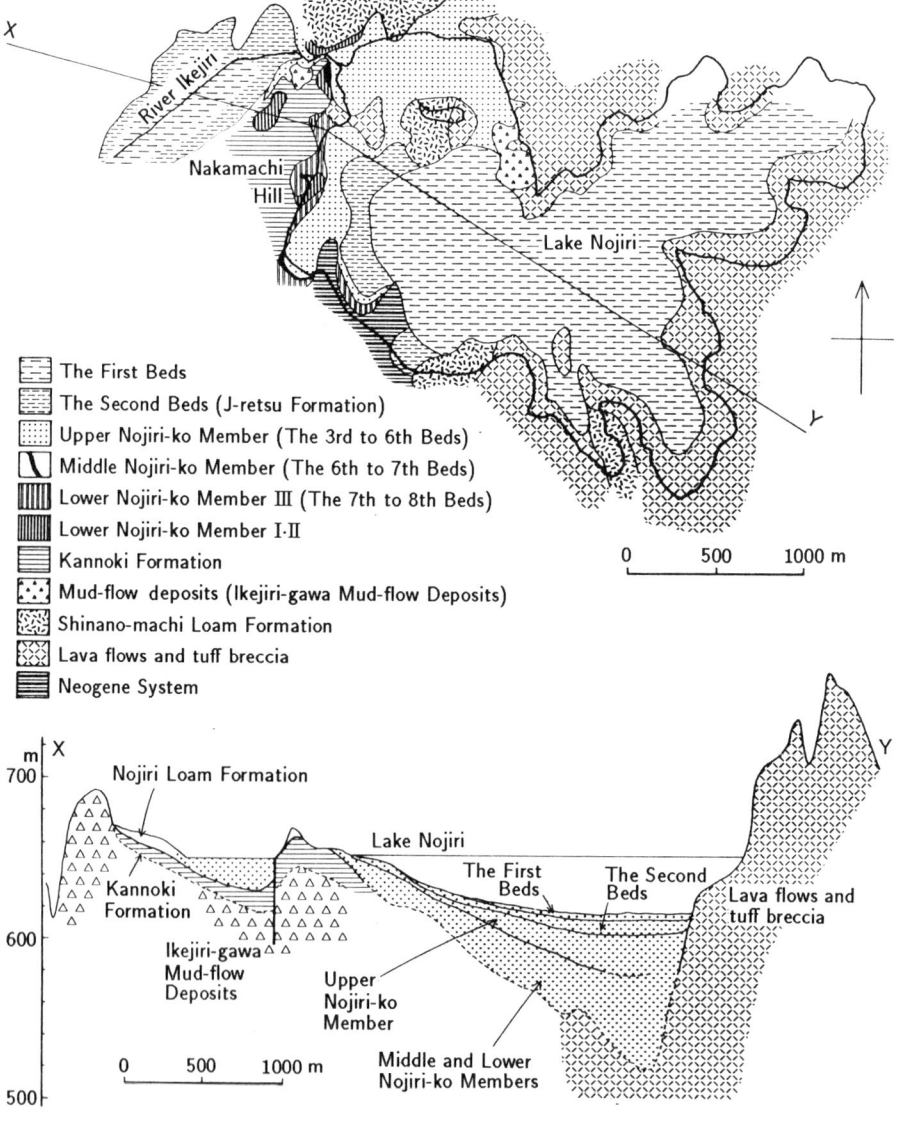

Figure 3. Geological map around Lake Nojiri and its section along the line X–Y [3].

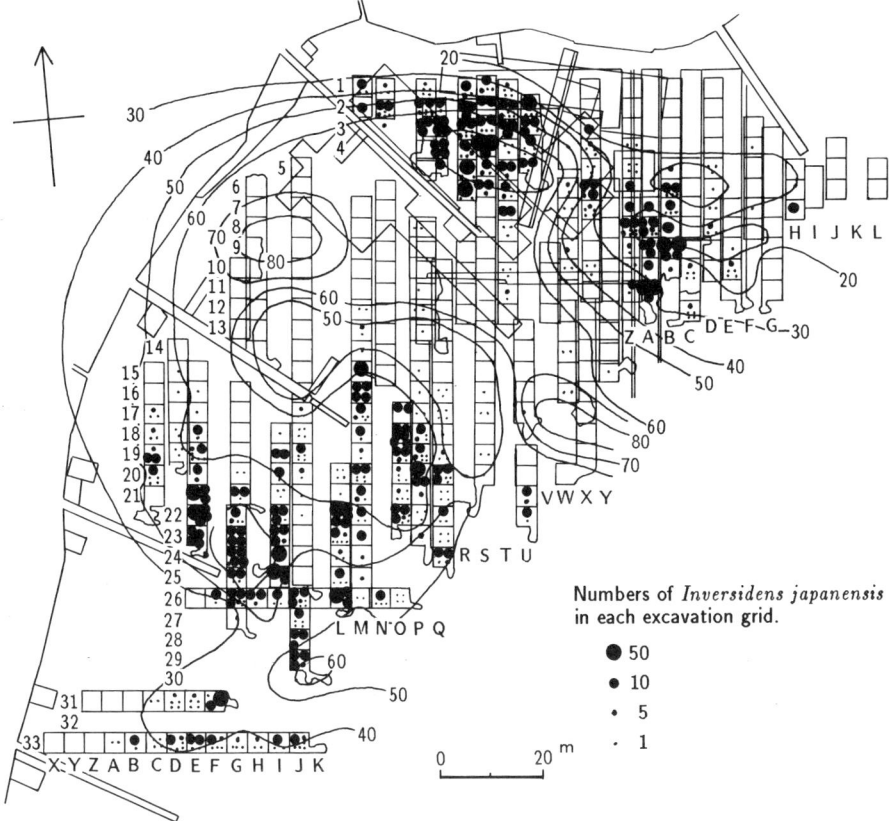

Figure 4. Distribution of *Inversidens japanensis*, yielded from Section III B of the Lower Nojiri-ko Member[9]. Curved lines in the figure show the isopach one from 20~80cm of the Section III B. The letters A~Z and numbers 1~33 show the rows and lines of the excavation grids, respectively.

continuous inhavitation of herbivorous animals of medium to large size. The occurrence of *Oiceoptoma thoracicum* from the lowest greenish grey silt bed of the Lower Nojiri-ko Member III suggests a paleoclimate ranging from upper cool temperate to subfrigid [10]. Similarly, *Donacia gracilipes* found from the Upper Nojiri-ko Member II and III at the Ikejiri-gawa Hollow suggests nearly the same climatic condition [11].

Most insect fossils were discovered in thick peat or peaty silt beds at the Ikejiri-gawa Hollow. It is remarkable that many fossils of pupae of moths and flies, and cocoons of small wasps were yielded from most part of the beds. They provide us important information on paleoenvironmental analysis, though not identified to species level, because they usually pupate in rather dry land conditions. It is interesting that both of those Donaciinae beetles, *Donacia hiurai* and *Plateumaris constricticollis*, obtained from the Upper Nojiri-ko Member II and III are distinctly smaller in size of pronota and elytra than those of the modern specimens collected in and around Nagano and Niigata Prefectures [10].

Most vertebrate fossils collected in these excavations are *Palaeoloxodon naumanni* and *Sinomegaceros yabei*. Well preserved molars, axial skeletons and forelimb bones in *Palaeoloxodon naumanni* were yielded from the Middle Nojiri-ko Member I. These bones possibly belong to an individual specimen. From the Upper Nojiri-ko Member I well preserved hindlimb bones were yielded, which are probably belonging to two individuals. More than 100 tusk fragments of *Palaeoloxodon naumanni* from the Lower Nojiri-ko Member

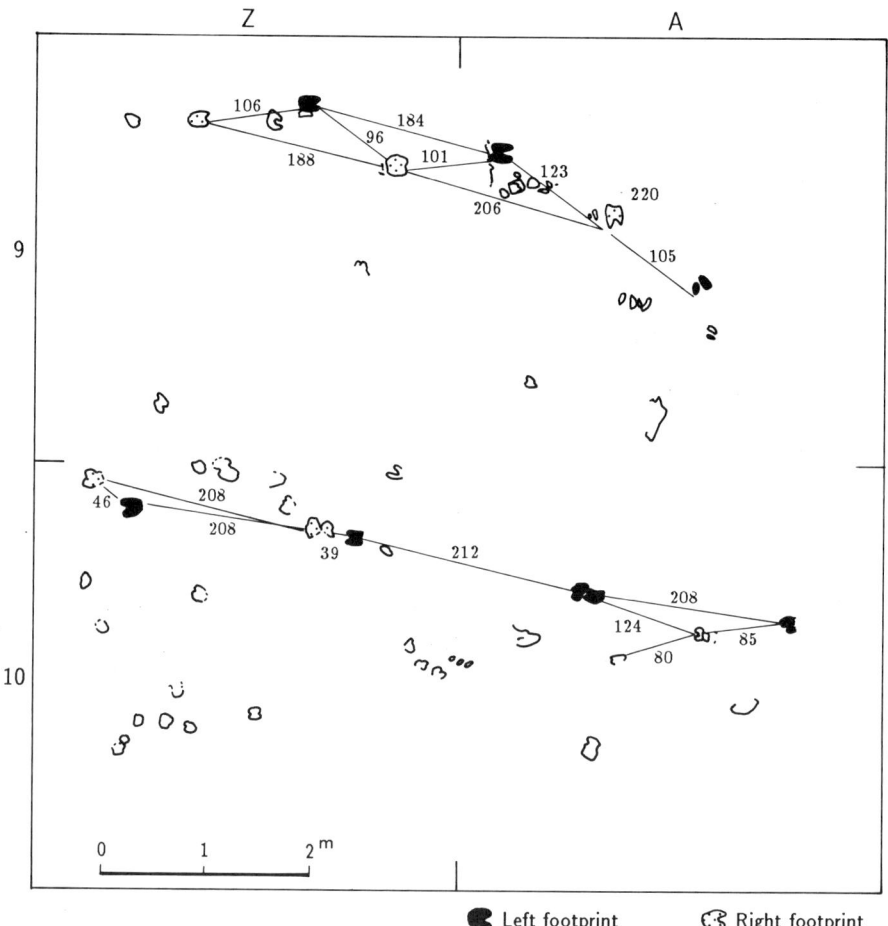

Figure 5. Distribution of artiodactyl footprints. Letter A, Z and number 9, 10 outside of the square show the grid number same to ones in the Fig. 4. Numbers in the square show the distance between each footprint by cm[13].

IIIB were found within 5-6 meters away from original tusk specimenn. Such modes of occurrence provides us valuable information on fossilization process [12].

The most interesting facts are two lines of artiodactyl footprints eight meters long preserved clearly on the upper surface of the ash layer (Konaazuki) where trace fossils are abundant [13](Fig. 5). Many foot prints of elephants were also found on the same layer in the Lower Nojiri-ko Member during the 11th excavation in 1990. During the Nakamachi Hill Excavation in 1991 a jaw and a molar of Naumann's elephant were found from the Kannoki Formation, thus the existence of elephant in this area dated back probably more than 70,000 years.

An interesting mode of artefact cluster was found in two different stratigraphic horizons [14]. One is in the Lower Nojiri-ko Member IIIB where bone flakes, bone spiral flakes and boulder gravels were yielded. The other is in the Middle Nojiri-ko Member I, with a bone cleaver, some refitted bone flakes and chips, rib bones of Naumann's elephant and stone flake tools. The latter case not only implies a circumstancial evidence for the kill-site, but also suggests that the elephant hunters made bone tools in kill-buchering processes.

Table 1a.
Synthetic diagram of the results by Nojiri-ko Excavations.

	Aeolian Sediments		Subaquatic Sediments		^{14}C date	Culture
		Marker bed (cm)		Marker bed (cm)	$\times 10^4$ y	
Holocene	Kashiwabara Volcanic Ash Formation	20, 100 Kibidango I	Recent / J-retsu Formation	25, 20, 5	9530±190 / 10720±260	Neolithic / Jomon Culture
Upper Pleistocene (Quaternary) — Upper Nojiri Loam Formation II		20, 80 Ajishio	Upper Nojiri-ko Formation III	45	13400±300 / 14340±320	Late Palaeolithic / Micro-lithic Industry
Upper Nojiri Loam Formation I		Nuka	Upper II	70		Backed blade Industry
		40, 80 Joichi Pink		70 Joichi Pink	34497±668	
Mid Nojiri Loam II, I		30 Akasuko / Chuni pink / 50 Higesuko	Mid Nojiri-ko III, II, IV B	25, 20, 40 Akasuko / Higesuko	31920±700, 38819±1579, 42540±1420 / 41700±1260, 41516±1023 / 45812±1289, 45100±1191	3 / Nojiri-ko Culture / Lithic Industry with longitudinal flakes
Lower Nojiri Loam III, II, I		100 Doraikarei Konaazuki Breccia Zone / 40~80, 30, 50 Santen Set Kigoma Nomi	Lower Nojiri-ko III, II, I	35, 35 / 180 Doraikarei Konaazuki Breccia Zone / 70, 100 Santen Set Kigoma Nomi	43520±1340, 43307±1200, 46230±2430	4 / Middle Palaeolithic / Industry with bone tool and small lithic flake tools
Kamiyama Loam Formation		Haizara	Kannoki Formation	110 Haizara	5	
Ikejiri-gawa -Mud-flow-		Kibiokoshi	Ikejiri-gawa -Mud-flow-	Kibiokoshi	6	

Legend:
- Weathered fine volcanic ash
- Fine volcanic ash
- Coarse volcanic ash
- Pumiceous lapilli
- Scoriaceous lapilli
- Two pyroxene andesitic lapilli
- Hornblende andesitic lapilli
- Black volcanic ash

4 Conclusion

The results of the studies on our Nojiri-ko Excavations are summarized shown in Table 1a, 1b.

The coldest climate during the Last Glacial Age corresponds to the Pinaceae zone of the Kannoki Formation.

Most vertebrate fossils collected in these excavations are *Palaeoloxodon naumanni* and *Sinomegaceros yabei*. Many artiodactyl and elephants foot prints were found on the upper surface of the ash layer in the Lower Nojiri-ko Member.

A bone cleaver, some refitted bone flakes and chips, rib bones of Naumann's elephant and stone flake tools were yielded in the Middle Nojiri-ko Member I. This case not only implies a circumstancial evidence for the kill-site, but also suggests that the elephant hunters made bone tools in kill-buchering processes.

Table 1b.
Synthetic diagram of the results by Nojiri-ko Excavations.

		Artefacts	Mammals	Insects	Pollen Zone	Vegetation	Climate C — W
Recent					Pinus-Cryptomeria	Secondary forest	
J-retsu Formation				Chrysolina aurichalcea	Fagus-Quercus	Deciduous broad-leaved forest	
Nojiri-ko Formation — Upper	III	Linear relief Pottery	Palaeoloxodon naumanni / Sinomegaceros yabei / Ursus arctos	Donacia gracilipes	Picea-Abies-Tsuga	Mixed forest of conifer and d.b.l. trees	
	II	Backed blade		Donacia gracilipes		Sub-arctic coniferous forest	
	I				Quercus-Fagus	Deciduous broad-leaved forest	
Nojiri-ko Formation — Mid.	III	Bone spiral flake		Geotrupes auratus / Pachysternum haemorrhoum	Tsuga-Picea	Mixed forest of conifer and deciduous broad-leaved trees	
	II / I / IV	Bone cleaver Wooden spear					
Nojiri-ko Formation — Lower	B / A / III	Bone scraper Bone spiral flake Bone point Graver, Scraper	Foot prints of Naumann's elephant	Geotrupes auratus Aphodius elegans Aphodius brachysomus Oiceoptoma thoracicum Copris pecuarius Geotrupes auratus	Picea-Fagus		
	II / I				Larix-Betula	Northern cool-temperate coniferous forest	
Kannoki Formation					Pinaceae	Sub-arctic coniferous forest	
					Cryptomeria	Cool-temperate coniferous forest	
Ikejirigawa -Mud-flow-							

Legend: Dark band / Mud-flow sediments / Crack zone / Clay / Silt / Sand / Gravel / Sand and gravel

By the systematic analysis on the fossils and artefacts combining with geological, tephrochronological, paleogeographical, paleomagnetic, paleoclimatological and paleoecological studies, we will be able to clarified the environmental changes from 70,000 y B.P. to the Recent.

Acknowledgment.

The writer wishes to state thanks to various support of the Town-office of Shinano-machi, Kami-minochi-gun, Nagano Prefecture, and especially to many inhabitants in the excavation sites. Financial support of our research works had been depending on the Grand-in-Aid for Scientific Research from the Ministry of Educaion, Science and Culture in Japan.

References

[1] M. Minato (Ed.). *Japan and its Nature*. Heibonsha, Tokyo (1977). (in English)

[2] Anthropology and Archaeology Research Group. Study on the Paleolithic sites around Lake Nojiri. In: *The Paleolithic Site and Paleoenvironment in and around Lake Nojiri*. T. Utashiro (Ed). pp. 215-249. Geological Society of Japan, Tokyo (1980). (in Japanese).

[3] Acoustic Research Subgroup for Nojiri-ko Excavation. Acoustic stratigraphy in Lake Nojiri, *Monograph Assoc. Geol. Collab. Japan*, 32, 23-36 (1987). (in Japanese).

[4] Utashiro, T. (Ed.). *The Paleolithic site and Paleoenvironment in and around Lake Nojiri*. Geological Society of Japan, Tokyo (1980). (in Japanese).

[5] Geology Research Group for Nojiri-ko Excavation. Geology of the excavation site at Lake Nojiri and its surrounding area, Ser. 6 (1986-1988). *Monograph Assoc. Geol. Collab. Japan*, 37, 1-13 (1990). (in Japanese).

[6] Volcanic Ash Research Group for Nojiri-ko Excavation. Grain constitution of the Quaternary sediments in and around the Nojiri-ko Excavaton sites, Ser. 3. *Monograph Assoc. Geol. Collab. Japan*, 37, 29-38 (1990). (in Japanese).

[7] Geology Research Group for Nojiri-ko Excavation. Stratigraphical study of a 40m drilling core with its implication to the history of Lake Nojiri, *Monograph Assoc. Geol. Collab. Japan*, 37, 15-20 (1990). (in Japanese).

[8] Palynologycal Research Group for Nojiri-ko Excavation. Fossil pollen assemblages of the Kannoki Formation and the Lower Nojiri-ko Member, and the change of palaeoenvironment during the period after the deposition of the Ajishio Volcanic Ash Layer, *Monograph Assoc. Geol. Collab. Japan*, 37, 61-76 (1990). (in Japanese).

[9] Fosill Mollusc Research Group for the Nojiri-ko Excavation. Freshwater molluscs from the Pleistocene Nojiri-ko Formation (part 5), *Monograph Assoc. Geol. Collab. Japan*, 37, 85-92 (1990). (in Japanese).

[10] Fossil Insect Research Group for Nojiri-ko Excavation. Fossil Insects obtained from the Nojiri-ko Formation during the 10th Nojiri-ko and the 5th Hill Site Excavations. *Monograph Assoc. Geol. Collab. Japan*, 37, 93-110 (1990). (in Japanese).

[11] Fossil Insect Research Group for Nojiri-ko Excavation. Fossil Insects obtained from the Nojiri-ko Formation in the 10th Nojiri-ko Excavations and the 4th Hill Site Excavation, *Monograph Assoc. Geol. Collab. Japan*, 32, 117-136 (1987). (in Japanese).

[12] Fossil Mammal Research Group for Nojiri-ko Excavation. Vertebrate fossils from the Nojiri-ko Formation during the Lake Nojiri Excavation (1987-1989), *Monograph Assoc. Geol. Collab. Japan*, 37, 111-134 (1990). (in Japanese).

[13] Trace Fossil Research Group for Nojiri-ko Excavation. Trace fossils from the Nojiri-ko Formation found at the 10th Nojiri-ko Excavation, *Monograph Assoc. Geol. Collab. Japan*, 37, 145-160 (1990). (in Japanese).

[14] Anthropology and Archaeology Research Group. Paleolithic and Jomon Culture of the Nojiri-ko Site (1987-1989), *Monograph Assoc. Geol. Collab. Japan*, 37, 135-144 (1990). (in Japanese).

Vegetation and climate since Last Glacial Maximum in Darjeeling (Mirik Lake), Eastern Himalaya

CHHAYA SHARMA and M.S. CHAUHAN

Birbal Sahni Institute of Palaeobotany,
Lucknow-226007,INDIA

Abstract

Pollen analysis of a sedimentary profile from Mirik Lake, situated in the temperate zone of Darjeeling Himalaya has unravelled the vegetation history to infer the corresponding palaeoclimate back to the Last Glacial Maximum. This mountain region where the above lake lies had mainly open grasslands around 20,000 years B.P. and besides the preponderance of grasses and sedges, herbaceous Ranunculaceae and Caryophyllaceae were other major elements which contributed to the gregarious ground vegetation cover. However, the existence of many arboreals such as *Quercus, Alnus, Pinus, Cupressus, Tsuga*, etc., recorded for the region exhibit their sporadic distribution.

The overall vegetation composition or grassland-phase reflects to the prevalence of cold and dry climatic conditions during the Last Glacial Maximum which was subsequently replaced by mixed broad-leaved forests during the mid-Holocene, dominated chiefly by arboreals such as *Quercus, Alnus, Betula, Carpinus*, etc. This indicates that from cold and dry climate, it changed to warm and moist climate. Later on, around 2,000 years B.P. marked decline in broad-leaved elements with a simultaneous increase in Poaceae, Rosaceae, Cyperaceae, Caryophyllaceae, Asteraceae is witnessed which demonstrates slight deterioration in the climate. This recent change can infact be attributed to the impact of the new settlement or human activities in the region resulting into the induced change in the emerged vegetation picture.

Key words Quaternary, Palaeovegetation, Palaeoclimate, Palynostratigraphy, Mirik Lake, Eastern Himalaya.

INTRODUCTION

Darjeeling, situated in the temperate belt of Eastern Himalaya has a number of lakes viz., Mirik, Sindhap, Sinchar, Jore-Pokhri, etc. In the present paper lacustrine sediments from Mirik Lake are pollen analysed and the investigations happen to be the maiden attempt on the palaeovegetation and palaeoclimatic oscillations for this region since the Last Glacial Maximum.

Mirik Lake is situated, at an elevation of 1,700 metre a.s.l. about 50 km south-west of Kurseong between 26° 31' & 21° 18' latitude and 87° 59' & 98° 53' longitude (Fig.1). The lake is quite large and measures about 500 m in length and 100 m at its widest, demarcated in the north by Darjeeling, south-wards by Siliguri, east-wards by Kurseong and on its west by Nepal. Originally, this lake was a big swampy expansion but recently aquatic weeds have been cleared and it is now a well developed tourist resort owing to the scenic grandeur of this himalayan region. But still the area towards right of the lake remains a thick swamp overgrown by *Typha* sp. and gregarious growth of other aquatics.

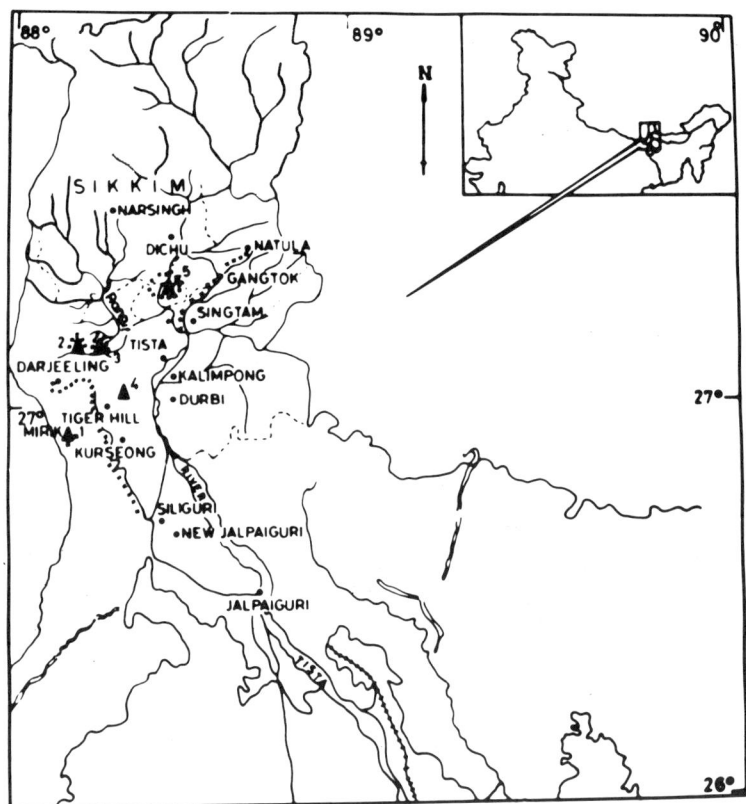

MAP OF DARJEELING AND SIKKIM AREA TO SHOW LOCATION OF LAKE SITES AND SURFACE SAMPLES

INDEX
▲ Lake site - 1 Mirik 2 Sindnap 3 Sinchar 4 Jore-pokhri 5 Khechipir
▪▪ Surface sample
• Township

Fig. - 1

GEOLOGY

The southward hills of mountaneous Darjeeling are mostly sedimentary rocks, whereas the rest are largely metamorphic representing different grades. However, in the southern hills there are regions where the younger sediments are overlaid by the older ones and this can be attributed to the local landslides or the erosion by fast water currents resulting into the present condition. The region where the present investigated lake is situated, is composed of mainly gneiss and granites of different ages.

Soil composition is of red and yellow clay and is generally covered with a thick layer of humus; felspar and mica schist are often exposed underneath the clay. Due to heavy rainfall in the region the soil is quite susceptable to erosion, but is checked due to the presence of broad-leaved forest cover in the region.

CLIMATE

Darjeeling has a humid climate with a mean maximum summer temperature of 22.45°C and average minimum temperature 15.3°C, but the mercury is as low as 8.3°C during the months of December and January. July and September are the main rainy-season months with a mean rainfall of 2595 mm. The cold season is from October to February, characterised by severe fog during the extreme cold in the region.

VEGETATION

The broad-leaved forests in the area are dominated chiefly by oaks viz., *Quercus pachyphylla*, *Q. lamellosa* and *Q. lineata*. Other taxa such as *Castanopsis hystrix*, *Machilus edulis*, *Nyssa sessilifolia*, *Magnolia* sp., *Alnus nepalensis*, *Eurya japonica*, *Acer campbelli*, *Elaeocarpus* sp., *Symplocos thiefolia*, *Rhododendron campanulatum* and *Arundinaria* are other close associates of oaks. The shrubby components of these forests providing a thick undergrowth comprises chiefly *Viburnum cotonifolium*, *V. erubescens*, *Berberis asiatica*, *Strobilanthes* spp., *Zanthoxylum acanthopodium*, *Daphne cannabina*, *Gaultheria griffithiana*, *Rosa moschata*, *R. serica*, *Rubus ellipticus*, *Crataegus cronulata* and *Cotoneaster* sp.

Also thick patches of *Cryptomeria japonica* forest stands are seen towards the higher slopes in the lake region and these are devoid of any marked undergrowth. On some hill slopes, particularly around Darjeeling township, plantations of *Pinus petula* have been raised, initiated by Britishers long back.

The herbaceous montane flora is quite rich, exhibiting luxuriance particularly underneath the broad-leaved forests and in moist situations. Most commonly encountered species are- *Viola biflora*, *Hypericum reptens*, *Ranunculus diffusa*, *Polygonum nepalensis*, *Rumex nepalense*, *Artemisia indica*, *Oxalis corniculata*, *Geranium nepalense*, *Gentiana pedicellata*, *Saxifraga diversifolia*, *Spirea micrantha*, *Anemone vitifolia*, *Potentilla fruticosa*, etc. However, in the marshy or swampy habitat along the streams or around the lakes *Polygonum plebeium*, *P. pterocarpium*, *Ammania baccifera*, *Rotala rotundifolia*, *R. indica*, *Cyperus rotundifolia*, *Scirpus* sp., *Hydrocotyle sibthorpioides*, *Ocimum americanum*, etc., are some of the species which grow profusely. Likewise, the ferns grow quite luxuriantly throughout the region under humid temperate climate and excessive precipitation. Commonly seen ferns are *Oleandra wallichi*, *Hymenophylla simosianum*, *Phymatodes malacodon*, *Asplenium ensiformae*, *Adiantum peruvianum*, *Alsophylla andersoni*, *A. spinulosa* and *Lygodium flexuosum*, etc.

STRATIGRAPHY AND RADIOCARBON DATES

The soil profile investigated here is only 1.3 metre deep, dug out from the swampy margin of Mirik Lake through Hillers peat auger provided with 50 cm long coring chamber. Beyond this depth it could not be possible to bore hole deeper profiles from the accessible spot owing to the hard stratum underneath. The 1.3 metre core yielded 26 samples, each at 5 cm interval.

Lithologically the profile from top to bottom is divisible into three distinct lithozones. The uppermost stratum constitutes the thinnest zone and is mainly composed of organic-mud mixed with clay. The subsequent or middle zone is of clay only, whereas the bottom one, which is the thickest lithozone is composed of silty-clay.

Stratigraphical details of the lithocolumn

Depth	Lithology
0-25 cm	Organic-mud with clay
25-75 cm	Clay
75-130 cm	Silty clay

Radiocarbon dates determined for the profile at three different horizons

S.No.	Nature of sediment	Depth	Laboratory No.	^{14}C dates
1.	Organic-mud with clay	10-30 cm	BS-949	470±90 years B.P.
2.	Clay	50-80 cm	BS-950	2,400±170 years B.P.
3.	Silty clay	100-125 cm	BS-936	17,900±600 years B.P.

POLLEN ANALYSIS

The usual technique of pollen analysis (Erdtman[1]) through the use of 10% KOH, HF and acetolysing mixture was adopted to librate the pollen and spores from the sediments. Both modern as well as fossil pollen spectra constructed here are based on a pollen sum of all terrestrial plants varying in number from 200 to 500 in each sample. However, some samples turned out to be palynologically barren. Two well marked pollen barren sedimentary phases are seen in the pollen diagram.

Modern Pollen/Vegetation Relationship

For the modern pollen/vegetatiuon relationship studies, a total of 22 surface samples, comprising moss-cushions as well as mud samples, were collected from different forest stands along a transect while ascending from Siliguri foot-hills having thick sub-tropical to heights of the temperate forests in Darjeeling, covering altitudinal range from 900 to 2700 metre a.s.l.(Table-1). Study of the modern pollen depositional pattern is indispensable for profile analysis as it provides the required comparative database for the interpretation of the pollen diagram in terms of past vegetation and climate in the region.

Table-1
Provenance of surface samples from Darjeeling Area

S.NO.	Locality	Details of surface samples
1-4.	Sukana	Moss-cushions from sal forest.
5&6.	Chunbati	Moss-cushions from sal-teak forest.
7&8.	Tindhara	Moss-cushions from sal-teak forest.
9&10.	Ghaiya Bari	Moss-cushions from mixed forest.
11&12.	Mahanadi	Moss-cushions from *Cryptomeria* forest.
13&14.	Gidhya Pahar	Moss-cushions from *Cryptomeria* forest.
15&16.	Tong	Moss-cushions from mixed broad-leaved forest.

17.	Sonada	Moss-cushions from mixed broad-leaved & conifer forest.
18.	Sonada	Soil sample from mixed broad-leaved & conifer forest
19&20	Ghoom	Moss-cushions from *Cryptomeria* forest.
21&22.	Jore-Pokhri	Moss-cushions from *Cryptomeria* forest.

The pollen spectra constructed (Fig.2) from the subtropical belt of sal-teak stands at lower elevations infact, exhibit dominance of non-arboreals over arboreals. The non-arboreals are characterised by their good representation which include mainly the grasses, Asteraceae followed by Cyperaceae, Cheno/Ams and *Polygonum*, whereas the arboreals viz., *Betula*, *Alnus*, *Pinus* and *Tsuga* are recorded in low frequencies.

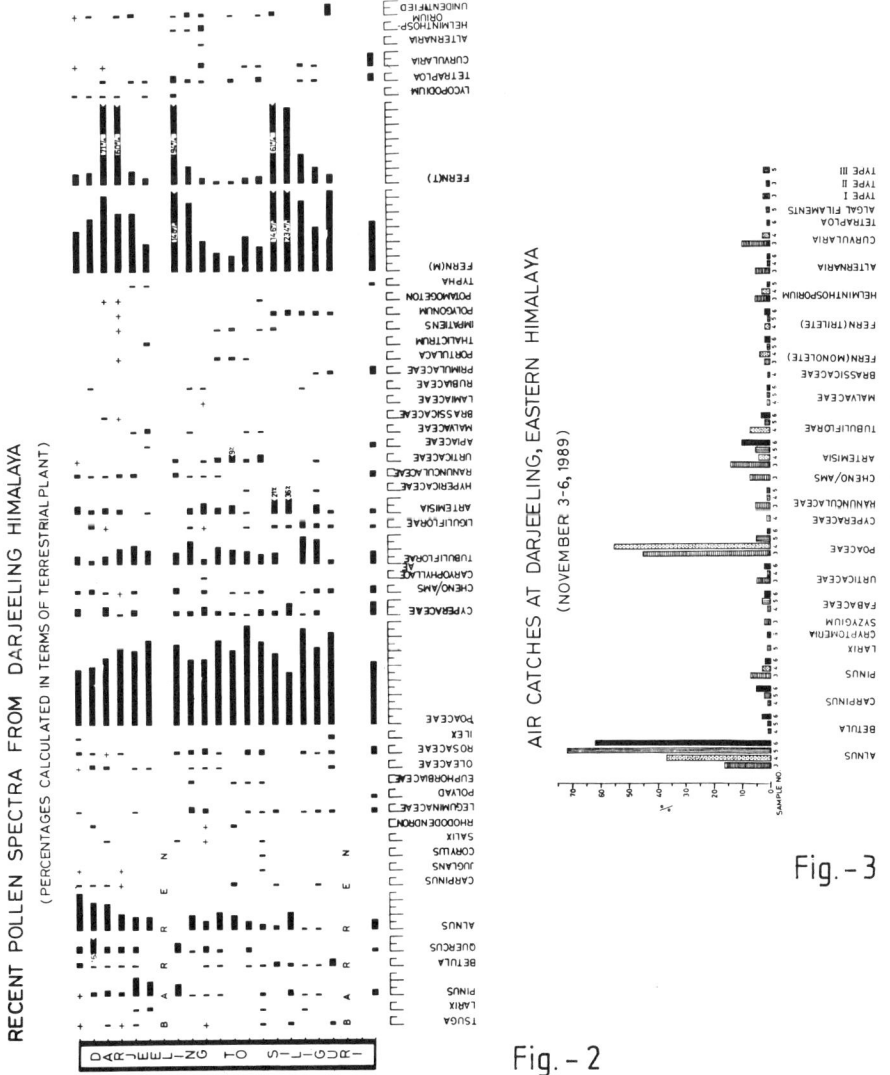

Fig. - 2

Fig. - 3

Also Fabaceae and Rosaceae have a poor representation. Interestingly, both *Shorea robusta* as well as *Tectona grandis* which are though the chief constituents of the forest, but remain un-represented in the pollen spectra.

The pollen spectra constructed from the mixed broad-leaved forests depict the dominance of *Quercus, Alnus, Betula* followed by *Carpinus, Juglans* and *Pinus*. In pollen spectra constructed from conifer dominated forests, *Pinus* is faithfully represented and attains the highest frequency, but showing pseudo-presence of *Tsuga* and *Larix* in the region which are infact high altitude taxa. In the mixed broad-leaved forests too, the ground coverage is dominated by grasses and sedges, followed by Asteraceae, Cheno/Ams and Ranunculaceae. *Lycopodium* spores are also recovered but in low frequencies and only in the pollen spectra constructed from mixed broad-leaved-conifer forests. *Tetraploa* and *Curvularia* fungal spores are better represented in contrast to *Helminthosporium* and *Alternaria* in all the above pollen spectra. The aeropalynological investigations (Fig.3) carried out for Darjeeling portrays excessively high values of *Alnus*, whereas other taxa viz., *Pinus*, *Carpinus* and *Betula* are encountered in good frequencies compared to the representation of *Larix*, *Cryptomeria* and *Syzygium*. The two families- Fabaceae and Urticaceae are also represented though sporadically. Among the non-arboreals, as usual the Poaceae attains the maximum frequency alongwith *Artemisia*, Tubuliflorae, Chenopodiaceae and Ranunculaceae as prominent constituents of modern pollen rain compared to Brassicaceae, Malvaceae and Cyperaceae together with fern spores which too are represented in the air-catches. The fungal spores of *Helminthosporium*, *Alternaria*, *Tetraploa* and *Curvularia* are encountered in moderate values.

The overall vegetational picture emerging from the investigated surface samples (moss-cushions) depicts more or less a true picture of the extant vegetation in the area. The representation of majority of broad-leaved taxa viz., *Quercus, Alnus, Betula, Carpinus*, etc. and *Pinus* correspond well with the existing composition of these taxa in their respective forest communities. However, in contrast to the above, the pollen spectra constructed from sal-teak subtropical forest and *Cryptomeria* stands fail to do so and depict a completely distored picture as expressed by the total absence of the pollen of *Shorea robusta, Tectona grandis* and *Cryptomeria* in the surface samples. Absence of these three prominent forest elements in the surface samples may be attributed to the poor preservation as well as microbial degradation of pollen irrespective to their being high pollen producers.

In the recent pollen spectra as well as air-catches, broadly *Alnus, Pinus, Betula* and *Carpinus* alongwith prominent herbage such as grasses, sedges, Cheno/Ams, Asteraceae and *Artemisia* show more or less coherence in their representation. *Quercus* is one taxon which was totally absent in the air-catches.

The comparative study of modern pollen spectra and aerospora of the region show that fern spores always remain sporadic in the air-catches despite their marked dominance in the recent spectra and it is probably due to their localised occurrence under peculiar habitat or much humid local situations which do not permit the suspension of spores in the air. However, *Helminthosporium, Tetraploa, Curvularia* are quite frequently represented in moderate values in the pollen spectra as well as in the air-catches.

Description of pollen diagram (Fig.4)
In the pollen diagram, the whole sequence is divided into six pollen assemblage zones I-VI numbered from oldest to the modern and aimed to precisely interpretation for better understanding of the vegetation succession.

Pollen Zone I (130-120cm) Grasses-Sedges-Rosaceae Assemblage:
This pollen zone is characterised by the exceedingly high values of grasses, followed by sedges, Rosaceae, Cheno/Ams, Hypericaceae, Tubuliflorae and Ranunculaceae. Arboreals

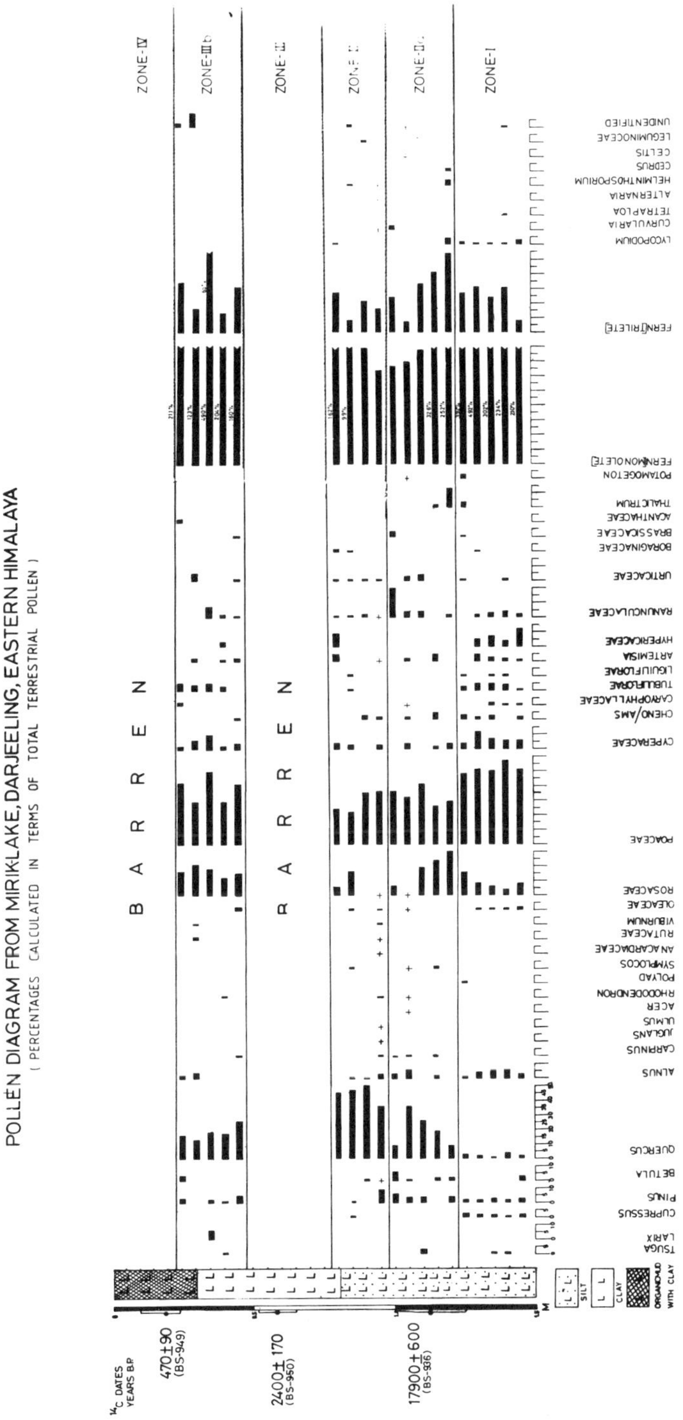
POLLEN DIAGRAM FROM MIRIK LAKE, DARJEELING, EASTERN HIMALAYA
(PERCENTAGES CALCULATED IN TERMS OF TOTAL TERRESTRIAL POLLEN)
ZONE-IV
ZONE-I
B A R R E N
B A R R E N
TSUGA
LARIX
CUPRESSUS
PINUS
BETULA
QUERCUS
ALNUS
CARPINUS
JUGLANS
ULMUS
ACER
RHODODENDRON
POLYAD
SYMPLOCOS
ANACARDIACEAE
RUTACEAE
VIBURNUM
OLEACEAE
ROSACEAE
POACEAE
CYPERACEAE
CHENO/AMS
CARYOPHYLLACEAE
TUBULIFLORAE
LIGULIFLORAE
ARTEMISIA
HYPERICACEAE
RANUNCULACEAE
URTICACEAE
BORAGINACEAE
BRASSICACEAE
ACANTHACEAE
THALICTRUM
POTAMOGETON
FERN(MONOLETE)
FERN(TRILETE)
LYCOPODIUM
CURVULARIA
TETRAPLOA
ALTERNARIA
HELMINTHOSPORIUM
CEDRUS
CELTIS
LEGUMINOCEAE
UNIDENTIFIED
14C DATES YEARS B.P.
470±90 (BS-949)
2400±170 (BS-950)
17900±600 (BS-936)
SILT
CLAY
ORGANIC MUD WITH CLAY

Fig. - 4

are not many and are represented by the low values of *Alnus, Quercus, Pinus, Cupressus* and *Tsuga* alongwith shrubby elements of Rosaceae and Oleaceae. Fern spores are recorded in extremely high frequencies throughout this zone.

Pollen Zone II (120-95 cm) Oak-Pine-Grasses-Rosaceae Assemblage:
This pollen zone is dated to 17,900±600 years B.P. showing consistent increase in *Quercus* but for an abrupt decline in its values at the top. Compared to the previous zone, *Betula* and *Pinus* also show improvement but Rosaceae after its initial improvement in its values register a decline towards the top. *Carpinus* has its appearance in this zone though scantily. Among herbs, grasses and sedges exhibit reduced values than witnessed before. Ferns though dominate throughout in this zone but decline considerably in the upper half.

Pollen Zone III (95-75 cm) Oak-Grasses-Sedges Assemblage:
This zone is characterised by much higher values of *Quercus* and its consistent representation as compared to above two pollen zones. Other tree taxa such as *Alnus, Betula, Pinus* are lowly represented, and *Juglans, Ulmus, Viburnum*, Anacardiaceae, Rutaceae turn up for the first time in this zone. Rosaceae remain sporadic in the beginning but show slight improvement in the upper part. Grasses exhibit static values at the beginning of the zone but decline considerably towards the upper half. Sedges do not show any appreciable change. Ranunculaceae and Urticaceae inspite of their consistent representation show reduced values than witnessed earlier.

Pollen Zone IV (75-45 cm) Barren zone:
Though, this zone was found palynologically barren but it could be ^{14}C dated to 2,400±170 years B.P.

Pollen Zone V (45-25 cm) Oak-Pine-Rosaceae-Grasses-Sedges Assemblage:
In this zone is witnessed an overall decline in the values of arboreals such as *Quercus, Pinus, Alnus* and *Betula* and a simultaneous improvement in the frequencies of grasses followed by Rosaceae, sedges, Tubuliflorae, *Artemisia*, Urticaceae and Ranunculaceae. Fern spores maintain much increased values throughout this pollen zone.

Pollen Zone VI (25-0 cm) Barren zone:
This zone too, is found palynologically barren though it is ^{14}C dated to 470±90 years B.P.

HISTORY OF VEGETATION AND CLIMATE

Enough data has been generated from the pollen analytical investigations of Quaternary lacustrine profiles from the subtropical, temperate and alpine lakes from North-West Himalaya and Western (Vishnu-Mittre et al.[2], Sharma & Singh,[3,4] Bhattacharyya,[5,6] Dodia[7]) as well as Central Himalaya (Vishnu-Mittre & Sharma,[8]) and as a consequence of which a very comprehensive sequence of past vegetation and climatic alterations has been reconstructed. However, Eastern Himalaya which possesses a diversified vegetation mosaic and fluctuations in climatic patterns did not get much attention on different Quaternary aspects, except for some sporadic information (Bhattacharya & Chanda,[9] D'Costa & Mukherjee,[10]). Therefore, in the present investigation in Mirik Lake which lies in Darjeeling hills an attempt has been made to trace out the palaeovegetation succession as well as to build up the contemporary climatic phases covering a span of about 20,000 years B.P. The pollen diagram constructed from Mirik Lake has an absolute maximum date of 17,900±600 years B.P. However, on the basis of date extrapolated from the rate of sedimentation it could be possible to stretch the vegetation succession and climatic oscillations in the region back to about 20,000 years which is considered to be the period of Last Glacial Maximum (LGM). The vegetation scenario in the region had commenced with open grasslands dominated chiefly by grasses and sedges, together with some other prominent herbaceous constituents such as Hypericaceae, Tubuliflorae, *Artemisia*, Cheno/Ams and Ranunculaceae. The arboreal elements were few and scantily represented by *Quercus*, *Alnus*, *Betula*, *Pinus* and Rosaceae. Their stray occurrence could be viewed either due to the confined distribution in the pockets or these might had been randomly distributed in the grasslands. The consistent representation of high altitude element *Cupressus*, together with *Tsuga* could be the result of the descedence of these taxa to lower elevations under the impact of cold climate.

Subsequently between 18,000-11,000 years B.P., there was a gradual increase in *Quercus* and *Pinus*, *Betula* and Rosaceae were also better represented than before, whereas grasses, sedges and other arboreals declined simultaneously during this period. This change in vegetation pattern signifies the replacement of grasslands by the forests in response to the amelioration of climate, mainly mixed oak-pine forests as evidenced by high frequencies and consistent representation of *Quercus* and *Pinus*. However, towards the termination of this phase i.e. 12,000-11,000 years B.P. the abrupt decline in oak frequency and other broad-leaved taxa coupled with and corresponding improvement in grasses and associated non-arboreals is inferred to a period of short-term deterioration of climate, changing from warm temperate to cool temperate.

Pinus existed during the late glacial period in Darjeeling, but today it is the introduced species *Pinus petula* which now grows in the region.

Between 10,000-4,000 years B.P., the significant event is witnessed and it is the further expansion of oak which attained consistent maximum frequencies and the first appearance of some thermophilous broad-leaved temperate elements viz., *Juglans*, *Rhododendron* and *Ulmus*, though scantily represented. On the other hand, the conifer taxa such as *Pinus*, and *Tsuga* alongwith grasses and other components of ground flora declined considerably. The marked expansion of broad-leaved forests with the beginning of this phase reflect the onset of a period of optimal climate which continued until 4,000 years B.P.

The ultimate phase, covering a time period of 2,000-500 years B.P. also portrays the continuance of oak forests in the region, though the major constituents of these forests such as *Quercus* as well as *Betula* dwindle sharply, whereas *Pinus* shows improved frequencies from the beginning, thereby reflecting a gradual deterioration of climate . The good representation of grasses alongwith Cheno/Ams, Asteraceae, Caryophyllaceae and *Artemisia* demonstrates the intensive agricultural practice and biotic degradation of forest, indicating prominent anthropogenic activities in the region.

Acknowledgements

The present work was carried out under a project sponsored by the Department of Science and Technology, New Delhi. The authors are grateful to Dr K.P. Jain, Acting Director of Birbal Sahni Institute of Palaeobotany, for encouragement and to Dr. Rajagopalan for providing radiocarbon assay of the samples.

References

1. G. Erdtman. *An Introduction to Pollen Analysis*, Waltham (1943).
2. Vishnu-Mittre, H.P. Gupta and R.D. Robert. Studies of the late Quaternary vegetation of Kumaon Himalaya, *Curr. Sci.* **36** (20),539-540 (1967).
3. C. Sharma and G. Singh. Studies in Late-Quaternary vegetational history in Himachal Pradesh-1. Khajiar Lake, *Palaeobotanist* **21** (2), 144-162 (1974a).
4. C. Sharma and G. Singh. Studies in Late-Quaternary vegetational history in Himachal Pradesh-2. Rewalsar Lake, *Palaeobotanist* **21** (3), 321-338 (1974b).
5. A. Bhattacharyya. Vegetation and climate during post glacial period in the vicinity of Rohtang Pass, Great Himalayan Range, *Pollen Spores* **30** (3-4), 417-427 (1988).
6. A. Bhattacharyya. Vegetation and climate during the last 30,000 years in Ladakh, *Palaeogeogr. Palaeoclimatol. Palaeoecol.* **73**, 25-38 (1989).

7. R. Dodia. Climate of Kashmir during last 700,000 years: The BalTal pollen profile, *Proc.Indian natn. Sci. Acad.* **54A** (3),481-489 (1988).
8. Vishnu-Mittre and C. Sharma. Vegetation and climate during the Last glaciation in the Kathmandu Valley, Nepal, *Pollen Spores* **1**,69-94 (1984).
9. K. Bhattacharya and S. Chanda. Quaternary pollen analysis of a peat sample from Gangtok (Sikkim), *Sci. Cult.* **52** (4), 139-140 (1986).
10. M. D'Costa and B.B. Mukherjee. Holocene history of ferns in Darjeeling hills, Eastern Himalaya, *Phytomorphology* **36**(1,2), 151-163 (1986).

Comparative study of the investigated profiles from Mirik Lake (Darjeeling), Estern Himalaya and Tsokar Lake (Ladakh), North-west Himalaya

Mirik Lake (Darjeeling)-1,7000 m a.s.l.			Tsokar Lake (Ladakh)-4,572 m a.s.1.		
500-2,000 yrs. B.P.	Oak-pine forest	Warm temperate with gradual detarioration of climate as well as evidences of agriculture			
2,000-4,000 yrs. B.P.	Barren zone				
4,000-10,000 yrs. B.P.	Oak forest	Amelioration of climate (climatic optimum)	10,000 yrs. B.P.	Rise in *Juniperus* and decline in steppe elements	Amelioration of climate (warm climate)
Around 11,000 yrs. B.P.	Increase in grasses and decline in oak	Cool oscillation	11,800 yrs. B.P.	Decline in *Juniperus* and rise in steppe elements	Cool oscillation
12,000-18000 yrs. B.P.	Oak-pine forest	Amelioration of climate (warm climate)	11,800-18,000 yrs. B.P.	Rise in *Juniperus* and decline in steppe elements	Amelioration of climate (warm climate)
Around 20,000 yrs. B.P.	Open grassland with scattered birch	Cold temperate	18,000-21,000 yrs. B.P.	Expansion of *Juniperus*	Amelioration of climate (warm cold)
			18,375 yrs. B.P.	Decline in *Juniperus* and corresponding increase in steppe elements	Deterioration of climate (cold climate)
			21,000-28,000 yrs. B.P.	Decline in *Juniperus* and increase in steppe elements	Deterioration of climate (cold climate)
			28,000-30,000 yrs. B.P.	Substantial rise in *Juniperus*	Amelioration of climate (warm climate)
			Prior to 30,000 yrs. B.P.	Alpine desert	Cold climate

Quaternary environmental changes in the northeastern Japan

S. TAKEUTI[1] and K. MANABE[2]
[1] Saito Ho-on Kai Museum of Natural History, Honcho 2-20-2, Aoba-ku, Sendai 980, JAPAN
[2] Department of Earth Science, Faculty of Education, Fukushima University, Matsukawa Machi, Fukushima, 960-12, JAPAN

Abstract– Most of the inland basins in Northeast Honshu, Japan, are distributed in lowland zones which run parallel with the island arc. Geochronological, palynological and paleomagnetic studies have been made on the sedimentary drill cores taken from six inland basins in Northeast Honshu Japan. One (E1) and three (E2, E3 and E4) local geomagnetic excursions recognized are dated to be between 14,000 and 17,000 yr. B.P. and between 35,000 and 50,000 yr. B.P., respectively. The Blake and Biwa I events are also recognized. From the results of pollen analysis, charactristic environmental changes are recognized, and the following pollen zones are established during the period from late Middle Pleistocene to Holocene. Zones NP8, NP6, NP4 and NP2 are characterized by mainly boreal coniferous forest under relatively cold and dry climatic conditions, corresponding to one of the glacial events in the late Middle Pleistocene, the first, second and third stadials in the Last Glaciation, respectively. Zones NP7, NP5, NP3 and NP1 are characterized by a temperate or cool temperate forest under a relatively warm and moist climate, corresponding to the Last Interglaciation, the first interstadial in the Last Glaciation, the second small interstadial and the Holocene, respectively.

INTRODUCTION

In Northeast Honshu, Japan, many inland basins of various scale are developed. Most of these basins are distributed in lowland zones which run parallel with the island arc (Fig.1). The Neogene System of the northeast Japan is widely distributed along the general trend of the island arc, unconformably overlying such basement rocks as Pre- Tertiary sedimentary and granitic rocks. The Neogene System can be divided into two types of facies, namely, the Neogene developed in the inner and outer belts of the island arc. One is characterized by the predominance of volcanic products and the other being poor in such materials. Quaternary volcanoes are distributed along the inner belt of the island arc. Most of the inland basins are also developed in the inner belt, while the beginning of deposition in these basins differ from one another, reflecting their forming processes. Quaternary strata in these basins are generally characterized by non-marine sequences including fluvial and lacustrine deposits which interbed with volcanic products and peat seams.

Three sedimentary drill cores, 300 m, 200 m and 50 m long, were taken from the Yamagata Basin. Pollen analyses, paleomagnetic measurements and radiocarbon datings were

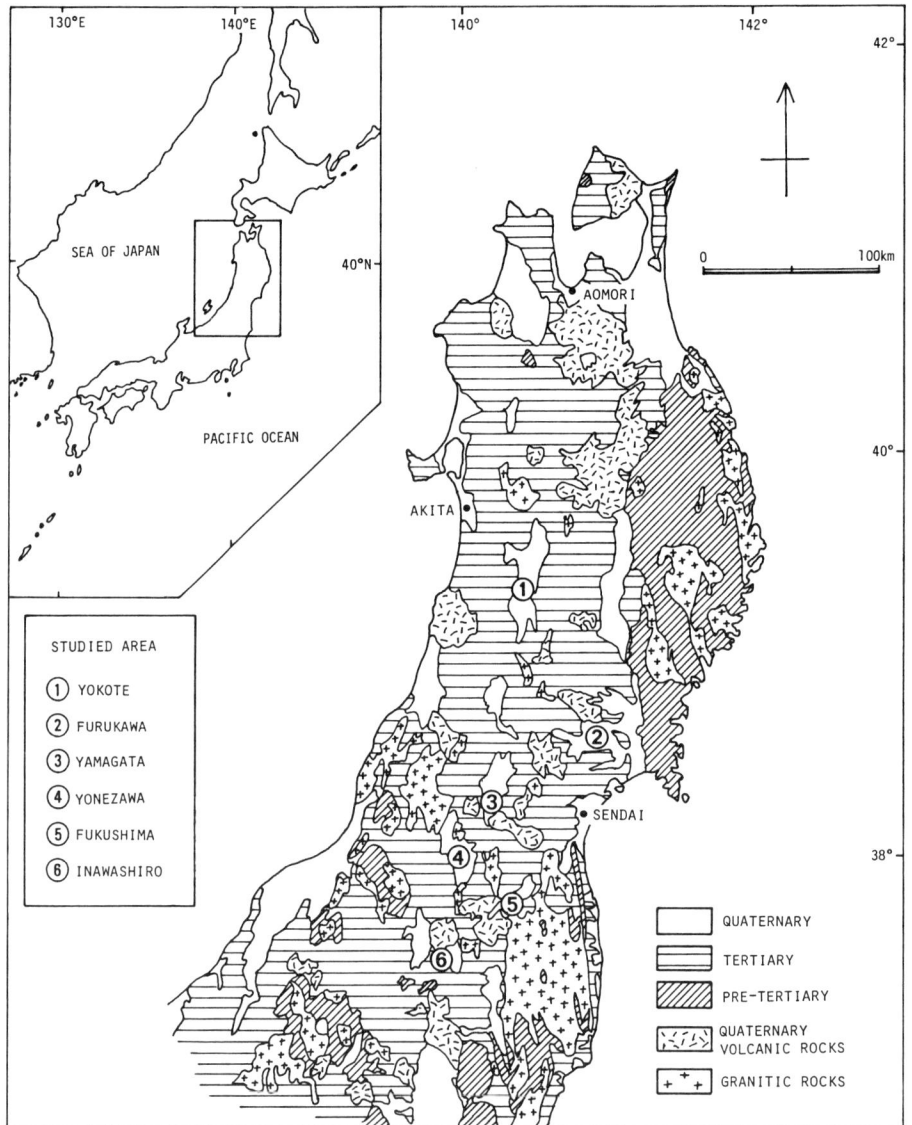

Figure 1. Geological map of the northeastern Japan showing the studied areas.

performed on sediment samples taken from these drill cores [1]. In the Fukushima Basin, a 96-m-long drill core was obtained. The relationship between vegetational changes and paleomagnetic stratigraphy was examined [2]. Late Pleistocene sedimentary cores were taken at marshy lands in the Inawashiro Basin, and the relationship between the changes of pollen assemblages and magnetostratigraphy was examined [3]. In other three basins, Yokote [4], Furukawa and Yonezawa [5] Basins, sedimentary cores were collected, and vegetational changes during the Pleistocene and Holocene were investigated by means of

pollen analyses.
The purpose of the present study is to summarize these results and to reconstruct environmental changes in the northeastern Japan during a period ranging from the Middle Pleistocene to the Holocene.

MAGNETOSTRATIGRAPHY

Comparison with the geomagnetic polarity time scale is useful for determining the age of a sedimentary sequence. Recently, some short geomagnetic polarity events and excursions within the Brunhes Chron have been discovered from various sections of the world. For example, the Blake event (ca 0.11 Ma) has been observed in the North Atlantic deep-sea cores by Smith and Foster [6], and the Biwa I event (ca 0.17 Ma) was discovered from Lake Biwa core in Japan [7]. Sedimentary section including reliable records of geomagnetic excursion as well as polarity reversal furnishes useful key to detailed correlation and chronology of the Quaternary sequences. However, some geomagnetic excursions seem to be difficult to trace globally; they are probably local geomagnetic events. In order to establish a correlation between sedimentary sequences of the inland basins in the northeastern Japan, magnetostratigraphic studies have been attempted.

The detrital remnant magnetization of cored sediments were measured. As a result of paleomagnetic measurements, four local geomagnetic excursions within the late Brunhes Chron were recognized, and were named E1, E2, E3 and E4 excursions in descending order. In the Inawashiro Basin, extrapolated radiocarbon dates indicate that the youngest E1 excursion occurred between 14,000 and 17,000 years B.P., and the older E2, E3 and E4 excursions occurred between about 35,000 and 50,000 years B.P. [8]. Two short reversed events were also recognized in cored sequences from Yamagata and Fukushima Basins [1,2]. These reversed events are situated below the E4 excursion. Although there is no radiometric dating available for correlating with the polarity time scale, they were correlated with the Blake event and the Biwa I event, on account of their pollen-stratigraphic positions.

In the following discussions, the above-mentioned paleomagnetic evidence and radiocarbon dates are adopted for establishing the chronology and making a correlation of sedimentary sequences of the studied basins.

POLLEN ASSEMBLAGE ZONES

Pollen analyses have been made on the cored sediments taken from Yokote, Furukawa, Yamagata, Yonezawa, Fukushima and Inawashiro Basins (Fig. 1).

Yokote Basin: Based on the pollen record from a 100-m-long drill core, Tohoku Regional Agricultural Administration Bureau [4] recognized five pollen assemblage zones, Y2-I, Y2-II, Y2-III, Y2-IV and Y2-V. In the present paper, zone Y2-IV is further subdivided into two zones. Six pollen zones are called AS1, AS2, AS4, AS5, AS6 and AS7 in descending order (Fig. 4).

Furukawa Basin: Pollen analysis has been made on a 77.5-m-long drill core. Pollen diagram is divided into five pollen assemblage zones, MF1, MF4, MF5, MF6 and MF7 in descending order (Figs. 2, 4).

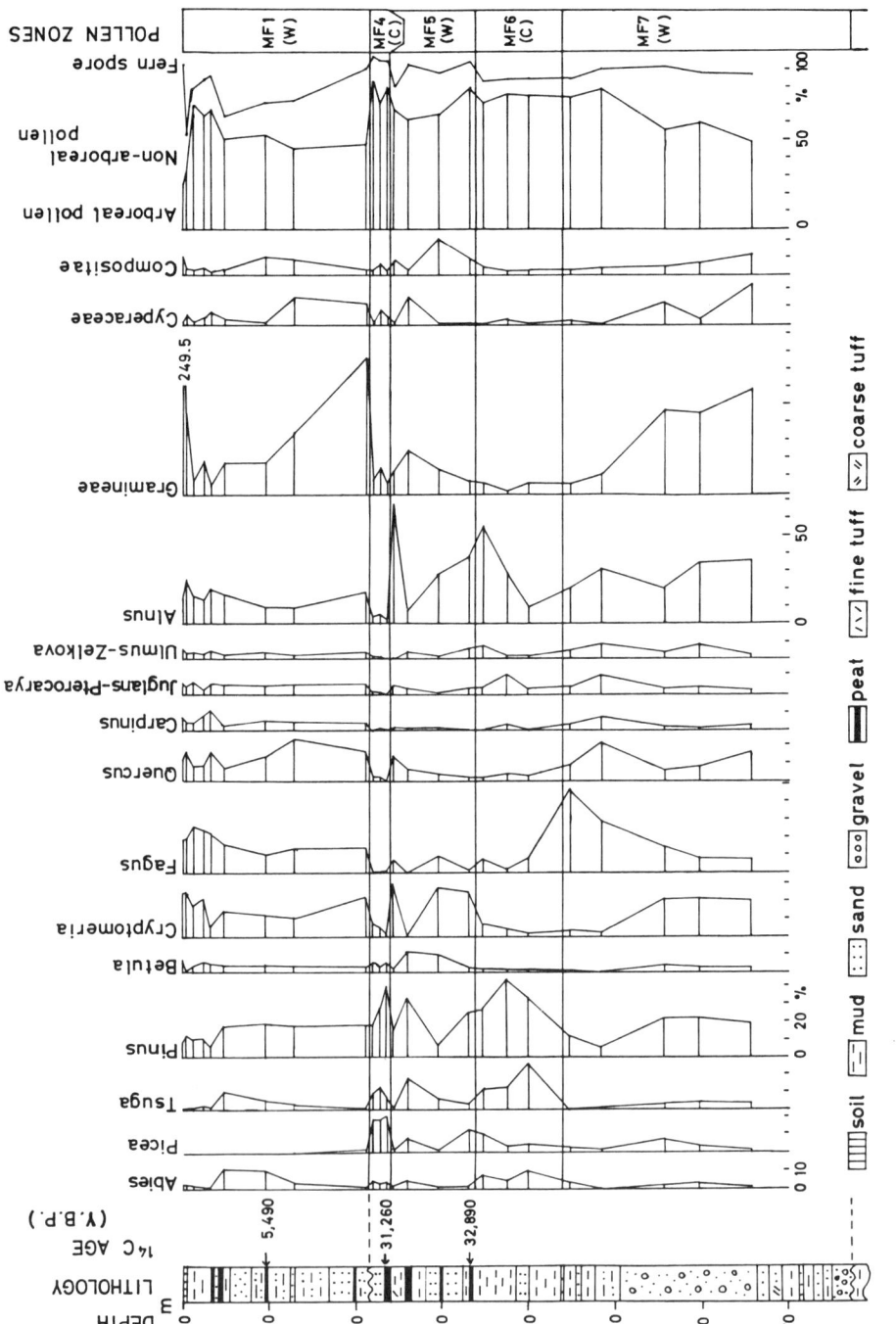

Figure 2. Pollen diagram showing selected taxa from Furukawa. Percentages are of the total tree pollen except *Alnus*.

Yamagata Basin: Among the pollen analysis data available from the three drill cores [1], this paper presents the result from a 260-m-long sequence in a 300-m-long drill core. Pollen diagram is divided into eight pollen assemblage zones, namely YN1, YN2, YN3, YN4, YN5, YN6, YN7 and YN8, in descending order (Figs. 3, 4).

Yonezawa Basin: Nagae et al. [5] have divided the drill core sequence into five pollen assemblage zones, Y2A, Y2B, Y2C, Y2D and Y2E. In this paper, zone Y2D is subdivided into two zones. Six pollen zones are called YF1, YF2, YF4, YF5, YF6 and YF7 in descending order (Fig. 4).

Fukushima Basin: Manabe et al. [2] have divided the drill core sequence into seven pollen assemblage zones, Fu1, Fu2, Fu3, Fu4, Fu5, Fu6 and Fu7. In this paper, zone Fu2 is subdivided into two zones. Eight pollen zones are called FF1, FF3, FF4, FF5, FF6, FF7 and FFS, in descending order (Fig. 4).

Inawashiro Basin: The drill core sequence from Hoshojiri in the basin has been divided into three pollen assemblage zones, H-I, H-II and H-III, [9]. In this paper, H-II is subdivided into two units. Four pollen zones are named FH1, FH2, FH4 and FH5 in descending order (Fig. 4).

CORRELATION OF SIX DRILL CORES

Judging from the tendency of each pollen assemblage zone, ^{14}C date and paleomagnetic evidence as mentioned above, each pollen zone in the six basins can be correlated one another as follows (Fig. 4).

No zone correlatable with FFS has been recognized. FFS is coeval with the late Early Pleistocene.

YN8 is correlative with FF8, and both of them are dated to the late Middle Pleistocene, called here zone NP8.

AS7, MF7, YN7, YF7 and FF7 are correlated one another and are datable as the Last Interglaciation, called here zone NP7.

AS6, MF6, YN6, YF6 and FF6 are mutually correlatable and are dated to the first stadial of the Last Glaciation, called here zone NP6.

AS5, MF5, YN5, YF5, FF5 and FH5 are mutually correlatable and are equated with the first interstadial of the Last Glaciation, called here zone NP5.

AS4, MF4, YN4, YF4, FF4 and FH4 are mutually correlatable and are equatable with the second stadial of the Last Glaciation, called here zone NP4.

YN3 is correlated with FF3, and both are coeval with the second interstadial of the Last Glaciation, called here zone NP3.

AS2, YN2, YF2 and FH2 are mutually correlatable and all are equated with the third stadial of the Last Glaciation.

AS1, MF1, YN1, YF1, FF1 and FH1 are mutually correlatable and all are of Holocene age, called here zone NP1.

VEGETATIONAL HISTORY AND CLIMATIC CHANGES

The vegetational succession in Northeast Honshu, Japan, during a period from late Middle Pleistocene to Holocene is discussed in the following lines based on the pollen data.

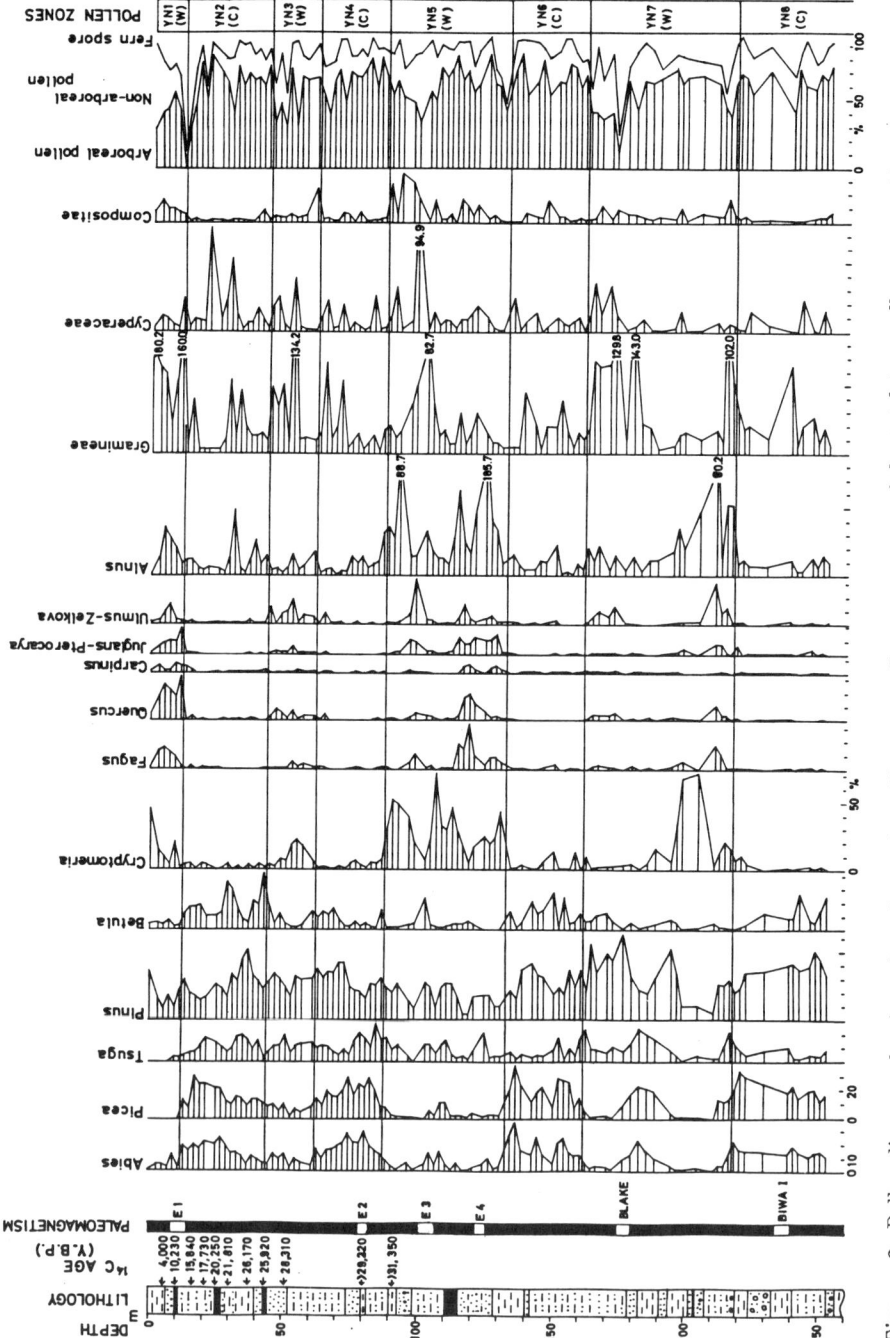

Figure 3. Pollen diagram showing selected taxa from Yamagata. Percentages are of the total tree pollen except Alnus. After Matsuoka and others [1] with some revision. Legend : see Figure 2.

Period of NP8: The late Middle Pleistocene, about 200,000 - 150,000 y.B.P. Lowland areas are roughly characterized by a boreal forest mainly composed of such conifer trees as *Pinus*, *Picea* which are accompanied by *Abies* and *Tsuga* and intermingled with *Betula* and *Alnus*. The vegetation indicates cold and dry climatic conditions.

Period of NP7: The Last Interglaciation, about 150,000 - 70,000 y.B.P. Various opinions have been expressed concerning the termination of this period [10]. However, this paper basically follows that of Japan Association for Quaternary Research [11]. The temperate forest, consisting dominantly of *Cryptomeria*, *Alnus* and *Pinus* and accompanied by deciduous broad leaved trees, replaced the preceding boreal coniferous forest. There is some evidence that certain vegetational changes occurred during this period. The boreal coniferous trees increased somewhat or the temperate trees decreased or both changes occurred around the middle of the period, around 110,000 y.B.P. The forest is inferred to have flourished under a warm and moist climatic condition. A relatively cool climatic condition temporarily appeared at approximately the same age as the Blake event which occurred around 110,000 y.B.P. The change of vegetation from 120,000 y.B.P. to 90,000 y.B.P. in the Kawadoi Basin, Yamagata Prefecture [12] is inferred to reflect the climatic change as mentioned above. This climatic deterioration is recognized in the pollen records also in the other places in Japan [13].

Period of NP6: The first stadial in the Last Glaciation, about 70,000 - 50,000 y.B.P. The boreal forest appeared again, though dominant taxa differed in each place. It is assumed that the climatic condition became cold and dry.

Period of NP5: The first interstadial in the Last Glaciation, about 50,000 - 35,000 y.B.P. The forest dominated by *Cryptomeria* and associated with *Fagus* and *Quercus* expanded, considerably at the expense of coniferous forest. *Alnus* is inferred to have flourished in swamp areas. The climate during this period is estimated to be relatively warmer and more moist than during the preceding period.

Period of NP4: The second stadial in the Last Glaciation, about 35,000 - 28,000 y.B.P. The mixed conifer and broad leaved boreal forest dominated by *Pinus*, *Picea*, *Abies* and *Tsuga* replaced the preceding temperate forest. It is assumed that this vegetational change was caused by a climatic detrioration, becoming cold and dry.

Period of NP3: The second interstadial in the Last Glaciation, about 28,000 - 24,000 y.B.P. In the Yamagata and Fukushima Basins, the conifer and broad leaved boreal trees have slightly declined. On the other hand, *Cryptomeria* and temperate broad leaved trees have increased. It is inferred that a slight climatic amelioration occurred, though it was temporary. In Kawadoi [12], Yonezawa [5] and Inawashiro [9] Basins, this change is not recognized. This may be due to the difference in latitude and altitude. In these areas, vegetations have not changed even though climatic conditions became slightly ameliorated.

Period of NP2: The third stadial in the Last Glaciation, about 24,000 - 10,000 y.B.P. The coniferous and broad leaved boreal forest expanded again. Such temperate trees as *Cryptomeria*, *Fagus* and *Quercus* reduced their areas of distribution but did not become extinct. The climate was coldest within the Last Glaciation.

Period of NP1: Holocene, about after 10,000 y.B.P. This period is characterized by the deciduous broad leaved and coniferous temperate forest mainly composed of *Fagus*, *Quercus*, *Cryptomeria* and *Pinus*, intermingled with *Ulmus-Zelkova*, *Juglans-Pterocarya* and *Carpinus*. Under warm and moist climatic conditions, *Alnus* and Gramineaceae are inferred to have flourished in and around swamps and marshy areas.

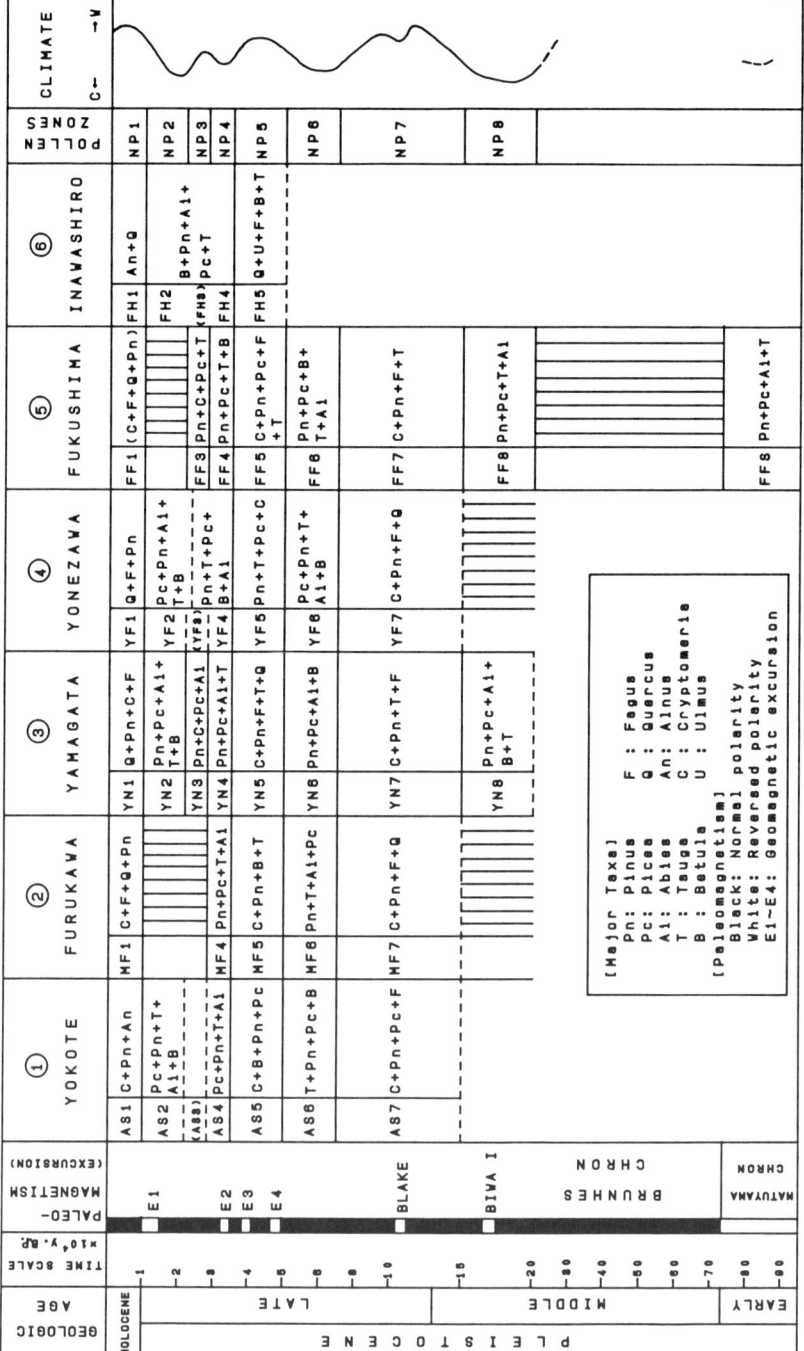

Figure 4. Paleoenvironmental history of the northeastern Japan. Labels in left-hand column of each studied area denote the pollen assemblage zones. Location numbers of the studied area are shown in Figure 1. Data for Yokote from Tohoku Regional Agrical. Administ. [4]; for Yamagata from Matsuoka and others [1]; for Yonezawa from Nagae and others [5]; for Fukushima from Manabe and others [2]; for Inawashiro from Suzuki and others [3]; Furukawa in this paper.

CONCLUSION

Geochronological, palynological and paleomagnetic studies have been made on sedimentary cores taken from six inland basins in Northeast Honshu Japan.
As a result of paleomagnetic measurements, four local geomagnetic excursions (E1, E2, E3 and E4) and two short events (Blake and Biwa I) within the Brunhes Chron were confirmed. The youngest E1 excursion occurred between 14,000 and 17,000 y.B.P., and the older E2, E3 and E4 excursions between about 35,000 and 50,000 y.B.P.
From the result of pollen analyses, the following eight pollen zones were established, during a time interval from the Middle Pleistocene to Holocene.
NP8 (Termination: ca 200,000 - 150,000 y.B.P.) The late Middle Pleistocene is characterized by a boreal forest dominated by *Pinus*, *Picea*, *Tsuga* and *Abies*, which indicates a cold and dry climate.
NP7 (ca 150,000 - 70,000 y.B.P.) This period represents the Last Interglaciation. The temperate forest mainly composed of *Cryptomeria*, *Alnus* and *Pinus*, accompanied by the deciduous broad leaved trees, replaced the preceding boreal forest under warm and moist climates. About the middle of this period, a relatively cool climatic condition appeared.
NP6 (ca 70,000 - 50,000 y.B.P.) The dominance of *Pinus*, *Picea*, *Tsuga*, *Abies* and *Betula* suggests cold and dry climates.
NP5 (ca 50,000 - 35,000 y.B.P.) *Cryptomeria* accompanied by deciduous broad leaved trees increased under a relatively warm and moist climate.
NP4 (ca 35,000 - 28,000 y.B.P.) The appearance of the boreal coniferous forest with deciduous broad leaved trees suggests that climate deteriorated again.
NP3 (ca 28,000 - 24,000 y.B.P.) The boreal elements slightly decreased and the temperate elements increased, though it is not clear in some areas. It is assumed that this vegetational change was caused by a climatic amelioration.
NP2 (ca 24,000 - 10,000 y.B.P.) This period is characterized by the boreal coniferous forest, indicating the onset of the coldest climatic condition through the Last Glaciation.
NP1 (after about 10,000 y.B.P.) This period represents the Holocene. The deciduous broad leaved temperate forest, mainly composed of *Quercus* and *Fagus*, flourished under warm and moist climates.

ACKNOWLEDGMENT

We are much indebted to Professor Tsunemasa Saito of the Institute of Geology and Paleontology, Tohoku University, for correcting the English text.

REFERENCES

1. I. Matsuoka, J. Akutsu, K. Manabe and S. Takeuti. Quaternary deposits of the Yamagata Basin, Northeast Honshu, Japan, *Jour. Geol. Soc. Japan.* **90**, 531-549 (1984). (*in Japanese*).
2. K. Manabe, S. Takeuti and F. Yabe. Quaternary deposits of Fukushima Basin, Northeast Honshu, Japan, *Ann. Rep. Synthetic Study of Fukushima Univ. The Nature and Man.* **3**, 1-11 (1992). (*in Japanese*).
3. K. Suzuki, K. Sohma, T. Kashimura and K. Manabe. Vegetation around the Hoshojiri moor and plant fossil assemblages from the Hoshojiri Formation, *Ann. Rep. Synthetic Study of Fukushima Univ. The Nature of Lake Inawashiro.* **3**, 51-64 (1982). (*in Japanese*).
4. Tohoku Regional Agricultural Administration Bureau. *Hydrogeology of the southern part of*

the Yokote Basin. Tohoku Regional Agricultural Administration Bureau, Ministry of Agriculture, Forestry and Fisheries, Sendai (1989). (in Japanese).
5. R. Nagae, H. Nakazato and T. Takami. Hydrogeology of the Yonezawa Basin, In: *Essays in Geology, Professor Hisao Nakagawa Commemorative Volume.* Prof. H. Nakagawa Taikan Kinenjigyo-kai (Ed.). 342. 177-184. Institute of Geology and Paleontology, Tohoku University, Sendai (1991). (in Japanese).
6. J.D. Smith and J.H. Foster. Geomagnetic reversal in Brunhes normal polarity epoch, *Science.* **163**, 565-567 (1969).
7. N. Kawai, T. Nakajima, K. Yasukawa, M. Torii and N. Natsuhara. Palaeomagnetism of Lake Biwa sediment, *Rock Magnetism and Paleo-geophysics.* **3**, 24-31 (1975).
8. K. Manabe. Geomagnetic variation during the last glacial age, *Mem. Geol. Soc. Japan.* **29**, 269-280 (1988). (in Japanese).
9. K. Sohma. Two Late-Quaternary pollen diagram from northeast Japan, *Sci. Rep. Tohoku Univ. 4th ser. (Biology).* **38**, 351-369 (1984).
10. Y. Yasuda. The cold climate of the Last Glacial age in Japan - A comparison with southern Europe -, *The Qut. Res.* **25**, 277-294 (1987). (in Japanese).
11. Japan Association for Quaternary Reseach (Ed.). *Explanatory text for Quaternary maps of Japan.* University of Tokyo Press, Tokyo (1987). (in Japanese).
12. K. Hibino, Y. Morita, T. Miyagi and H. Yagi. Palynological study on the vegetational change during the last 120,000 years B.P. in Kawadoi basin, Yamagata Prefecture, Japan, *Sci. Rep. Miyagi Agri. Col.* **39**, 35-49 (1991). (in Japanese).
13. S. Takeuti. The climatic change during the Last Interglaciation in Northeast Honshu, Japan, *Saito Ho-on Kai Mus. Nat. Hist. Res. Bull.* **53**, 13-19 (1985).

BEGINNING OF A NEW PERIOD -- THE TECHNOGENE

G. TER-STEPANIAN

Laboratory of Geomechanics, Institute of Geophysics and Engineering Seismology, Armenian Academy of Sciences, Yerevan, Armenia

ABSTRACT

The Holocene should be considered as the first epoch of a new geological period -- the Technogene. It came after the Pleistocene as a result of a new, previously unknown geological agent -- the reasonable and unreasonable human activity. Paleolithic Man's activity was comparable and did not differ in principle from action of other living beings as plants or animals: they extract from the environment no more substances that is necessary for their existence. The situation has changed in the course of the Neolithic revolution consisting in transition from food appropriation to food production. Food gathering and hunting changed to agriculture and cattle-breeding. This made it possible to extract more substances that is necessary and accumulate the wealth. Virgin dense forests with a rich animal life were cut down and changed into ploughed fields and pastures; later on some of them turned into semideserts and deserts. Due to agriculture, industry, and construction the face of the Earth changed cardinally. Man is able to affect, reproduce, imitate, prevent and impede many exogenous, a number of endogenous, even some extraterrestrial processes, and several biological ones as well. No genuine Quaternary feature has been conserved even by now at the very beginning of the Technogene. The new geological period will take shape fully by the end of the forthcoming millennium. Conception of the role and place of the Technogene in the system of our mentality is necessary for true appraisal of modern trends of development of the Earth's face and possibility of geological forecasts which are the most related to the future of our civilization. The unique distinguishing feature of the Technogene -- the ability of living beings to affect purposefully the course of geological and biological processes should be used to make the Earth a prosperous and reasonable planet. For the first time Geology is not so much concerned with the past as with the future.

INTRODUCTION

A new geological period is formed recently before our eyes which in its full development will differ sharply from all previous periods the Quaternary included. This period began with the Holocene.

The Holocene holds a special place in the history of the Earth. Being very short in the geological time scale, about ten thousand years only, the Holocene surprises us by enormous rapid changes in environment which are taking place all over the world in a tectonically calm period and without any cosmic catastrophe as collision with an asteroid. During this short time interval the face of the Earth has changed completely and no genuine Quaternary feature unaffected by human activity has been preserved; acid rains fall even in Greenland and ozone hole has been formed over Antarctica. The Holocene amazes researchers by impetuosity of changes in fauna content, extinction of forms, dynamism of geographical distribution of species and the most intricate interaction of different tendencies and factors of fauna formation [1].

MAN REPRODUCES NATURAL PROCESSES

The only cause of these changes is Man. Man's technical, agricultural and industrial activity became a powerful geological agent influencing the environment. The Holocene is characterized by constantly increasing human interference into the course of geological processes occurring on the Earth and in the life's harmony in Nature. This influence started with the beginning of the Holocene, increased during last millenniums and became quite evident at present. These changes are of prime importance not only in biological, ecological, economic, or social respects but are of geological significance as well.

Results of Man's technical activity are comparable by their variety and intensity with results of numerous natural processes occurring on the Earth (Table). On the left side of the Table the action of various geological, cosmic, and biological factors in natural conditions is shown while on the right side results of corresponding technical actions of the Neolithic and modern Man are displayed.

Table. CORRELATION BETWEEN NATURAL AND TECHNOGENOUS PROCESSES

NATURAL PROCESSES (action of geological, cosmic and biological factors)	TECHNOGENOUS PROCESSES (action of Neolithic and modern Man)
EXOGENOUS PROCESSES (in Nature take place on the Earth under action of reaching the Earth surface Sun's electromagnetic radiation, water, wind, ice, meteoritic bombing and organisms, Paleolithic Man included).	
Gradual decrease of the greenhouse effect beginning from the Proterozoic in consequence of CO_2 and water vapor content decrease and oxygen role increase	Extremely rapid increase of the greenhouse effect beginning from the 19th century in consequence of fossil fuel burning and CO_2 content increase
Transgressions in the interglacials in consequence of getting warmer and thawing the glaciers	Transgression due to thawing the continental glaciers in consequence of the technogenous heating of the Earth's atmosphere
Physical weathering	Rock grinding in mining and construction
Chemical weathering	Change of substances in chemical engineering
Forming the relief	Changing the relief at forming the mining landscapes and construction of cities, roads, canals, etc.
Denudation	Soil cutting in construction and soil transfer by ploughing on slopes
Subaerial accumulation	Soil filling at grading

Soil formation	Changing the soils by land tilling and fertilizing, deforestation, forming the artificial soils by land recultivation
Development of longitudinal profile of rivers	Stopping the longitudinal profile development of rivers by hydraulic engineering
Stream erosion and subaqueous accumulation	Changing the distribution of fluvial deposits by hydraulic engineering
Formation of meanders	Straightening of rivers
Coastal processes -- abrasion and travel of littoral deposits under action of longshore currents	Protection of coasts and changing the distribution of littoral deposits by port engineering
Changing the position of the coastline under action of waves, upswashes and longshore currents	Changing the position of coastlines at coast protection and development of new lands on shelves (draining the sea bottoms and filling up the shoals)
Earth surface subsidence in consequence of karst processes (poljes)	Earth surface subsidence in consequence of mining, gas production and pumping out the water and oil
Ground water level oscillation in consequence of changing the climatic conditions	Ground water level change in consequence of draining, irrigation, underflooding, pumping out the groundwater and leakage from networks
Formation of karst	Origination of technogenous karst
Formation of underground streams in karst areas	Construction of hydraulic tunnels
Formation of rock falls, landslides and debris flows in denudation processes on slopes	Formation of rock falls, landslides and debris flows due to pore and seepage pressures, slope undercutting, loading the slopes and deforestation
Formation of rock falls and landslides at earthquakes	Formation of rock falls and landslides due to slope vibration
Sedimentation	Formation of technogenous deposits

Formation of permafrost	Soil freezing
Degradation of permafrost	Formation of technogenous thermokarst
Formation of meteoritic craters	Formation of deep quarries at opencast mining
Meteoric winters	Thermonuclear winter

ENDOGENOUS PROCESSES (in Nature take place on the Earth under action of radioactive decay and internal heat of the Earth)

Formation of native metals	Production of metals in metallurgy
Heat transfer in the earth's crust	Use of underground heat in geothermal power
Heating the earth's crust as a result of radioactive decay	Use of fission energy of radioactive elements in atomic power
Seismicity caused by tectonic processes	Technogenous earthquakes at filling up the water reservoirs
Demolition caused by lava flows	Control of lava flows and protection from them
Contact metamorphism of clay rocks	Production of brick, ceramics, pottery and porcelain
Formation of conglomerates with carbonate cement	Production of concrete
Formation of dikes	Construction of impervious curtains in water-development works
Formation of obsidian	Production of glass
Origination of diamonds, rubies and corundum	Production of artificial diamonds, rubies and corundum
Metal concentration in consequence of magma differentiation and hydrothermal alteration	Ore benefaction

HELIOGENOUS PROCESSES (in Nature take place in interior of the Sun and stars under action of low-temperature nuclear reactions)

Synthesis of helium from hydrogen	Production of thermonuclear energy by blasting the hydrogen bombs and in thermonuclear reactors

COSMOGENOUS PROCESSES (in Nature take place during supernova outbursts under action of high-temperature nuclear reactions)

Synthesis of heavy metals	Production of plutonium in atomic reactors

BIOLOGICAL PROCESSES (in Nature take place on the Earth in consequence of ability of open dynamic systems with probabilistic bonds to self-organization)

Natural selection of the most favorable for the given species organic forms accompanied by their specialization	Artificial selection of the most profitable for Man organic forms accompanied by their domestication
Wide variety of numerous close forms leading to filling up of great quantity of ecological niches with small areas	Wide prevalence of few, very distinguished species all over the world leading to desertion of many ecological niches
Extinction of specialized organisms in tectonically active epochs as a result of impossibility to fit up for changing conditions	Extinction of many species of organisms in tectonically calm epoch as a result of direct or indirect extermination by Man
Origination of new species, genera, families and higher taxons of organisms	Gene engineering

MAN AS A NEW GEOLOGICAL AGENT

It is evident from this rather instructive Table that Man is able to reproduce, imitate, affect, prevent, or impede many exogenous, a number of endogenous, even some extraterrestrial and biological processes. Thus Man, due to his reasonable and unreasonable activity from the beginning of the Holocene appears as a new, previously unknown geological agent.

An objection concerning this statement may be made since human beings appeared on the Earth about six or seven million and not ten thousand years ago. However, there is a great difference between action of ancient and Paleolithic men on one hand and Neolithic and modern men on the other. The formers as all other living beings as plants and animals extract from the environment no more substances than is necessary for their existence; they enter into complicated and economical self-regulating trophic chains with other living beings and, therefore, they are in harmony with Nature. Paleolithic Man led the same predatory mode of life as his environment. This very long time interval covering the considerable part of the anthropogenesis was characterized by food appropriation. In the beginning of the Holocene in the course of the Neolithic revolution Man went over from hunting and food gathering to agriculture and cattle-breeding, i.e. to food production [2]. This process was accompanied by population growth, creating the family and private property, accumulation of wealth, and transition from matriarchy to patriarchy. Self-regulation of trophic chains was disturbed and harmony with Nature began to relax.

The idea to consider the Man as an independent geological agent was put forward earlier [3, 4] but the distinguishing features for the separation from the common group, organisms, were not specified clearly and thus the time of starting point remained vague.

There are serious reasons to separate the Holocene from the Quaternary and consider it as the first epoch of a new geological period -- the Technogene, which came after the Pleistocene [5, 6, 7].

Enormous changes took place during the past time. In the geological time scale the transition to the Technogene occurred extremely rapidly.

Virgin dense forests with rich animal life were cut down and changed into ploughed fields and pastures; later on some of them turned into semideserts and deserts. Deforestation and desertification processes continue with increasing rate. A menacing situation developed in Africa, Siberia and Brazil where the lungs of our planet are being destroyed; due to sea pollution oxygen production by plankton has been oppressed badly.

As a consequence of changes induced by realization of water-development projects on rivers such as construction of dams and embankments, formation of water reservoirs, their silting, river regulation, bank protection, diversion of flow in neighboring systems, etc. the river drainage networks are often being completely transformed. These technogenous processes proceed considerably faster than formation of longitudinal river profile in normal geological conditions.

Coastal processes as abrasion and formation of littoral deposits under action of longshore currents is prevented due to construction of quay walls, piers, breakwaters, etc. Physical and chemical weathering of rocks is reproduced by Man in mining, crushing of stones in construction and in chemical technology. It is impossible to describe here even the most important kinds of technogenous interference in the course of exogenous processes.

Among the endogenous processes reproduced by Man, the most important one is the mass production of seemingly native metals in metallurgy -- iron, aluminum, copper, zinc, tin, etc.

Human activity accelerates two kinds of endogenous processes which occur in Nature very slowly: radioactive decay of uranium-238 in atomic reactors and convective heat transfer of the planet through the use of the underground heat in geothermal power plants. Man induces earthquakes by construction of big water reservoirs which increase the load on the ground surface and the pore pressure in deep strata. Man can reduce damages caused by volcanic eruptions by bombing small lava flows, watering them or diverting these flows.

In this century Man has created extraterrestrial processes previously nonexistent on the Earth; he receives thermonuclear energy by the helium synthesis from hydrogen in H-bombs as it occurs in the Sun and star interiors in consequence of low temperature nuclear reactions, and even produces transuranian metals as plutonium in atomic reactors as it takes place during supernova outbursts from high temperature nuclear reactions.

The Holocene interglacial we live in should have been terminated in this millennium [8]; indeed the Little Ice Age began in 16th and 17th centuries and culminated 200 years ago. Owing to Man's technogenous activity (the greenhouse effect) instead of being threatened by growing the Greenland and Antarctic ice sheets we are facing quite an opposite danger -- overheating the planet, thawing the existing continental ice sheets and rise of the ocean level.

Man influences the biological processes too. In natural conditions selection of the most favorable for the given species organic forms takes place; this process was extended for millions years and accompanied by specialization of species. In technogenous conditions it is substituted by artificial selection of organic forms which are the most profitable for Man; the process takes very short time and is accompanied by domestication. In Nature wide variety of numerous close forms originate; they result in filling up many ecological niches. In technogenous conditions wide prevalence of few, very distinctive species led to desertion of many ecological niches. And finally, gene engineering is menacing with creation of new unexpected and dangerous effects which may be fatal for the very specialized high mammals and Man himself. It should be kept in mind that the rate of technogenous changes exceeds that of natural ones by several orders.

Island-like districts of human influence were formed on the Earth during the Holocene. This influence was at first weak but it intensified with time. These "Technogene" districts increased on the common "Quaternary" background, merged together, turned into vast "Technogene" regions enveloping the greater part of the Earth's surface and becoming more and more transformed. Now they compose megapolies -- extremely urbanized strips of continuous highly concentrated population. The biggest megapolies are in the USA (Atlantic shore and southern California), Japan

(Tokyo-Osaka), Germany (Rhein-Ruhr) and England (London-Liverpool).

Deep and fundamental changes of environment will continue to increase undoubtedly in the future. The new geological period will take shape fully by the end of the forthcoming millennium.

Conceptions of the role and place of the Technogene in the system of our mentality is necessary for true appraisal of modern trends of development of the Earth's face and possibility of geological forecasts which are mostly related to the future of our civilization.

The unique distinguishing feature of the Technogene -- the ability of living beings to affect purposefully the course of geological and biological processes, should be used to make the Earth a prosperous and reasonable planet. For the first time Geology is not so much concerned with the past as with the future.

List of References:

1. K. Paaver, Forming the Terriofauna and Changeability of Mammals of Baltic Region in the Holocene, Botanical Institute, Estonian Academy of Sciences, Tartu (1965) (in Russian).
2. V.G. Childe, Man Makes Himself, London (1948).
3. R.F. Legget, Věstnik ústředního ústau geologickeho, 18, 4, 241-248 (1968).
4. R.L. Sherlock, Man as a Geological Agent, An Account on his Action on Unanimate Nature, London (1922).
5. G. Ter-Stepanian, Abstracts, XI Congress, Intern. Union for Quaternary Research, Moscow (1982) p.260.
6. G. Ter-Stepanian, Bull., Intern. Ass. Eng. Geol., Paris 30, pp. 133-142 (1988).
7. G. Ter-Stepanian, Problems of Geomechanics, 10, pp. 45-57, Yerevan (1988).
8. J.T. Andrews, in: The Winters of the World. Earth under the Ice Age, B.S. John, (Ed.), David and Charles, London (1979).

THE ESSENTIAL CHARACTERISTICS AND FACIES MODEL OF WINDBLOWN SAND DEPOSITS DURING THE LAST GLACIAL PERIOD IN EASTERN CHINA

Zhang Mingshu & Liu Jian

Institute of Marine Geology, Ministry of Geology & Mineral resources, Qingdao, China, 266071

ABSTRACT

On the basis of two different types of Last-Glacial sedimentary sections of windblown sands found in eastern China, this paper not only gives the facies model of windblwon sand deposits, but also suggests their nine characteristics as follows: limit in spacial distribution, diversity of types, cyclicity of sequence, relative constancy of preveiling wind directions, proximal components, the same associated materials, typicalness of beddings, special hierarchies of bounding surfaces and uniform controlling factors.

Keywords: China, Pleistocene, last glacial, stratigraphy, eolian sand, dunes

1. INTRODUCTION

Eolian deposits are among the sediments which are first internationally studied and used as an object for the ancient and modern comparative research, but the sedimentary model has not been established yet. In recent years, the eolian deposits, formed since the Last Glacial Period, are playing an increasingly important role in global change studies.

People usually associate eolian deposits with vast deserts, but not all eolian deposits are formed in desert environments. If leaving rock deserts and gravel deserts out of consideration, we may classify eolian sediments into two main types: windblown sands and windblown dusts. The former results from transportation of windblown sand current by monsoon systems during relative cold terms, while the latter is attributed to west wind circulations which transported them in the air and then dropped them. As a typical case of Quaternary winblown dust deposits, loess has been widely and profoundly studied throughout the world, and China loesses have occupied the leading place in international research. But, presently, the research on Quaternary windblown sand deposits of China is in its infancy.

In eastern China, as is known at present, the Last Glacial sedimentary sections of windblown sand deposits, verified by dating results, have been found at three places, namely: the eolian carbonates in Shidao Island (1), "Liukuang Red Beds" in Shandong Peninsula (4) and the sandstones in Hongguangshashan Mountain, Pengzhe, Jiujiang, Jiangxi Province (3). These sediments are distributed in China over areas across as many as 20 degrees of latitude and 10 degrees of longitude. In our opinion, to summarize the essential characteristics and facies model of Last Glacial windblown sand deposits in eastern China should not be meaningless for correctly understanding the burried windblown sediments in the shelves of China, truly appreciating the paleogeography of China shelves during the Last Glacial Period and furthering windblown sand deposit studies.

2. ESSENTIAL CHARACTERISTICS

The Last–Glacial windblown sand deposits in eastern China have nine characteristics as follows:

2.1., Limit and zonation of spacial distribution

During the Last Glacial Period, the shelves of eastern China emerged out of sea water a few times due to climatic and sea –level changes. When the shelves were exposed above sea level, the climate was arid and monsoon active throughout the areas, windblown sand deposits resulting. This kind of sediments is widespread in the world. For example, the Last–Glacial eolian dunes are widely distributed in India, Australia, Africa, America and other countries, and researches on these dunes have been carried out since more than 100 years ago. In China, however, aside from the west deserts, it was not until early 1980′s that on modern coast regions and islands were found the windblown deposits, which are not at all as vast and majestic as the west deserts. These sediments appear on a small scale on initial negative landforms, each outcrop being less than 2 km^2; but on the whole, the sediments are located along the northeast monsoon zone.

The eastern–China coast regions and adjacent shelves, lying in the margin of West Pacific Ocean, were subjected to the strong NE monsoon during the Last Glacial Period because there were no very high mountains to stop the wind. In the low sea–level terms, the regions and shelves had aboundant sediment source and also a lot of suitable negative landforms, which should have built so much more windblown sand sections in the monsoon zone. The fact that only a few sites with the Last–Glacial windblown sand deposits are found presently is ascribed to the following two problems:

(a) The exposed shelves during the Last Glacial Period, covered at present with sea water and younger sediments, are not at all investigated systematically and extensively during the shallow stratigraphic survey. Nevertheless, on the northern shelves were reported the windblown sand deposits, and inference was made about the desertification of the shelves [4]

(b)Some sedimentary sections of windblown sand deposits, already determined as Holocene sediments by ^{14}C dating, are in fact different in features from true Holocene windblown sand, and may have been mistakenly dealt with according to the low ^{14}C dating age values.

With the further research on China coast zones and continuous understanding of windblown sand deposits, the authors hold that there are gradually increasing chances to find new Last–Glacial sedimentary records of windblown sands year by year, and the more found sections in future would further make clear that the northeast–oriented windblown sand deposits are characterized by zoning distribution.

2.2., Diversity of types

Situated in different geographic units, the known Last–Glacial windblown sand deposits are greatly different in their environment conditions and provenances, so they may be divided into two main types: one is terrigenous clastic sedimentary type, which scarcely contains carbonate components and marine fossils; the other is partly even completely made up of carbonate components bearing a large amount of marine fossils and, for this reason, called carbonate type. If further divided, the eolian sediments have another transitional type between the two main types. The two types are evidently relative to geographic divisions with 30° N as the boundary in between. The north division, corresponding geographically to the region north of the Changjiang River, is dominated by terrigenous clastic windblown sands; the south division to the south of Changjiang River is characterized by the windblown sands with increasing carbonate content toward low latitudes.

The difference between the two types is caused mostly by two factors. The first is the geologic settling: the north division covers Liaodong Peninsula and Korea Peninsula where granites and metamorphic rocks are well developed, which produced quantites of terrigenous clastics after being weathered; while the south division is adjacent to the shelves covered by younger sediments. The second is the climate condition: the north division is in temperate zone, so during the Last Interglacial Period, on the surrounding shelves were deposited megafossils–bearing sediments with very low content of carbonates; the south division is largely in subtropic and tropic zones, the carbonate saturability in adjacent sea water being higher, and increased organic reefs in the sea toward the south relevantly provided more abundant carbonate materials.

2.3., Cyclicity of sequences

The Last Glacial Period is not a monoglacial age. During last 60000 years, the frequent alternations of cold and warm terms and changes of climate–environment conditions caused alternate development and stagnation of windblown sands, the cyclic windblown deposits being formed. The number of cycles depends on the retention potentiality of sediments. The time spans of every cycle are not equal but the corresponding cold and warm terms are more or less the same. The cycle numbers in the known sections are not identical, for instance, the carbonate rock section in Shidao Island, Xisha, has five cycles, each consisting of windblown sands in the lower part and fossil soils in the upper; the "Liukuang Red Beds" section in Shandong Province is composed of nine cycles, each comprising windblown sands in the lower part and soil layers and alluvial sand–gravel beds in the upper.

The cyclic deposits, with windblown sands in cold terms directly overlain by soils or alluviums or erosive hiatuses in warm terms, may often be regarded as a sign to distinguish windblown sand deposits of the Last Glacial Period from those of Holocene. The Holocene windblown sand deposits, distributed on modern coast zone of China, are mostly monocyclic, loose sediments, and no soil layers occur, because of the unstable sediments and not long enough time to finish pedogenesis. Comparatively, the Last Glacial windblown sand sections, after frequent and tremendous changes in the climate and environment through tens of thousand years, can at least preserve two or more cycles of semiconsolidated rocks and soil layers.

Each cycle of the windblown sand deposits has a definite sequence. A complete cycle begins from the basal erosive surface or hiatus upwards to dune sediments, then to seceond–order bounding surface and interdune sediments and finally to upper erosive hiatus or soil layer or other sediments formed in a warm term. On field investigations, however, such complete sequences are rarely seen, and commonly seen are the sequences short of dune sediments and second–order bouding surface or short of second–order bounding surface and interdune sediments. Nevertheless, these incomplete sequences can also be capable of analysising the climate–environment conditions. If the windblown sand deposits occupy a larger area, deposits of the same stage may show different sequences in the distribute scope.

2.4., Relative constancy of prevailing wind directions

The prevailing wind directions during the Last Glacial Period, recorded by all cycles of windblown sand deposits in the known sections, are northeast without exception, because all foreset beddings kept dominantly in the deposits dip southwestly. This suggests that during the Last Glacial Period, the interaction between the northeast monsoon and arid loose sediments on Earth's surfaces is the principal mechanism responsible for the windblown sand deposits. The prevailing wind direction in a relatively cold term is basically unchanged, but changed greatly in different cold terms, for example, from N 20 ° E to N 80 ° E.

According to the prevailing wind directions, the authors have preliminarily done the correlation of five stages of windblown sand deposits in Shidao Island, Xisha with nine stages in Liukuang, Shandong Province, and come to the following conclusion: the second, fourth and fifth stages of windblown sand in Shidao correspond to the fifth, sixth and eighth or ninth stages in Liukuang respectively. By comparision, two inferences are obtained: (a) the fourth stage in Shidao, Xisha and the sixth stage in Liukuang were probably in the Last—Glacial Maximum; (b) all the five stages in Shidao Island occurred within stronger monsoon activity terms of Last Glacial Period.

2.5., The same associated materials

In two types of windblown sequences, namely the terrigenous clastic and carbonate sequences, the associated materials are basically the same, and they are also an important mark to judge environment conditions. Main associated materials in the windblown sand sequences include Rhizoliths, calcretes, deformation structures, mollusk fossils, lag deposits, ripple structures and so on. These associated materials, except deformation structures, lag deposits and ripple structures, never appear on the active sedimentary terms of windblown sands but always on the stagnant terms of wind action and thus considered as the products formed after the cold terms turned into warm terms.

2.6., Typicalness of beddings

Of windblown sand deposits, the most characteristic is the bedding types, which are also one of the important marks to distinguish eolian origin from other environments.

Dune subenvironments are characterized by high—angle foreset beddings. The forsets in modern windblown dunes dip in all directions due to veer of the winds resulting in the dunes, but only unidirectional cross—beddings are dominantly preserved in the forsets of ancient dunes. The foreset beddings vary in forms with the dune types and chiefly include the following: (a) tabular cross—beddings; (b) trough cross—beddings; (c) convex cross—beddings; (d) Seif beddings (pyramid—like beddings). On areas with windblown sand deposits, trough cross—beddings, convex cross—beddings and Seif beddings are useful for us to judge dune types; but if only tabular cross—beddings are discovered, it would be difficult to recognize dune types and to infer ancient prevailing wind directions.

Interdune subenvironments don't occur until dunes reach certain heights. As dunes move, windblown sands in interdunes climb upwards and accreate along dune flanks. In result, the bedding type developed with interdunes is a simple, shallow-trough-shaped bedding which climbs upwards with the wind and gently dips against the wind. Superimposed each other with low angles, the beddings look like horizontal beddings, and only when the sets are traced with the wind direction, can be found the climbing-up and dissection between them.

Within the same stage of windblown sand deposits, both modern and ancient, subhorizontal shallow-trough-shaped beddings of interdunes invariably overlie high-angle forset beddings of dunes. Therefore, eolian beddings are evidently diffeent from water-formed ones.

2.7., Proximal components and unidirectional dominant preservation

Windblown sand sedimentation originates with windblown sand currents, which are air currents moving near earth's surfaces and carrying only non-adhesive grains. Windblown sands are generally transported within a short distance, and made up of the same components as those of the surrounding bedrocks, indicating a proximal origin. The windblown grains are related in size to the wind force strength and their specific gravities, as a result of which the same wind force is capable of carrying away coarser organic skeletons than grains of rock-forming minerals and lithoclasts. In particular, microfossils including foraminifera, having light specific gravities due to larger coeloms, are carried by the wind even in the case of larger individuals.

The foresets of the Last-Glacial and earlier windblown sand deposits always have unidirectional dip. The windblown sand deposits formed by the northeast monsoon, for example, have the foresets invariably dipping southwestward. During the relative cold terms of Last Glacial Period, the northeast monsoon is not the solely prevailing wind, as in the summers the southwest wind prevailed and is supposed to have formed eolian foresets showing northeasterly dips. But the Last-Glacial windblown sand deposits are found to have only preserved the effect of dominant northeast monsoon, because only the northeast monsoon was strong and lasting enough to have the records preserved in the deposits.

2.8., Characteristic hierarchies of bounding surfaces

Based on the bounding surface studies of foreign scholars [5,6], the authors and others demarcated more clearly the three-order bounding surfaces [7, 8]. These interfaces are clear, definite, and suitable for any windblwon sand deposits.

The first-order bounding surface: it is an isochronous and restricted interface, located on the top and bottom of any stage of windblown sand deposits. As a hiatus, the interface is formed before or after the deposition of windblown sands.

The second-order bounding surface: it is a diachronous interface separating the interdune from the dune. In the section parallel to the paleowind direction, the interface seems to extend horizontally at a distant glance, but with close observation the surface is found to be made up of local curvatures, slightly convex or concave, caused by the seperimposition of different cosets, as a result of sets overlaping one by one in response to interdune sediments climbing upwards with the wind; in addition, this bounding surface is easy to be mistaken for an angular unconformity interface by the obvious fact that subhorizontal interdune beds discordantly overlie high-angle forset beddings of dunes.

The third-order bounding surface: it is an instantaneously isochronous interface. Controlled by the scale and shape of the cosets, this kind of interface may be horizontal or curved and ranges in declivity from 0 to the angle of repose. The surface, just associated with eolian sediments, has not direct environmental implication.

The three hierarchies of bounding surfaces as above are peculiar to windblown sand deposits and thereupon the most direct evidence of eolian origin.

2.9., Uniform controlling factors

Three essencial prerequisities, that is, three main controlling factors, are necessary to form any type of windblown sand deposits, modern and ancient. First, certain wind action strong enough to produce windblown sand currents; second, the provenance rich in non-cohesive loose materials; third, geomorphic units suitable for accumulation of windblown sands. Windblown sand currents, moving near the earth's surfaces, deposit sands due to being hindered or reduction of the wind velocity. Places favorable for windblown sand accumulations include: piedmont plains or open fields, bedrock coasts, low lands, basins, intermontane depressions, dry lagoons, dry lakes and so on. Therefore, wind-acted zones are not covered everywhere by windblown sand deposits. That is why during the Last Glacial Period, the windblown sand deposits on the NE trade-wind zone of eastern China were in band or point distribution.

3. SEDIMENTARY FACIES MODEL OF WINDBLOWN SANDS

In the light of the eolian section of Shidao Island, Xisha, the authors and others raised a regional windblown sand facies model [7], [8], this model has been proved to be of universal implication by means of testing and verifying in windblown sand deposits of different ages. By improving the previous model, we propose a united facies model of windblown sand deposits in this paper in order that it can be further tested and then accepted.

The complete facies sequence (or essential facies components) of windblown sand facies model is described as follows (Fig.2):

(1) The basal first-order bounding surface, an isochronous erosive interface, called Facies Component A. This surface represents a long or short

non-sedimentation term. Below the interface may be windblown sand deposits or deposits in other environments which lie outside the model and thereby have no genetic link between them.

(2) above the first-order bounding surface are dune sediments, called Facies Component B. This unit develops high-angle foreset beddings including: tabular, trough-shaped, convex, pyramid-shaped and drape-like cross-beddings. The bedding types depend on dune types. Generally the dune foresets dip by 20-35°, but locally by less than 20° or more than 35° because of sets climbing or sediments slumping.

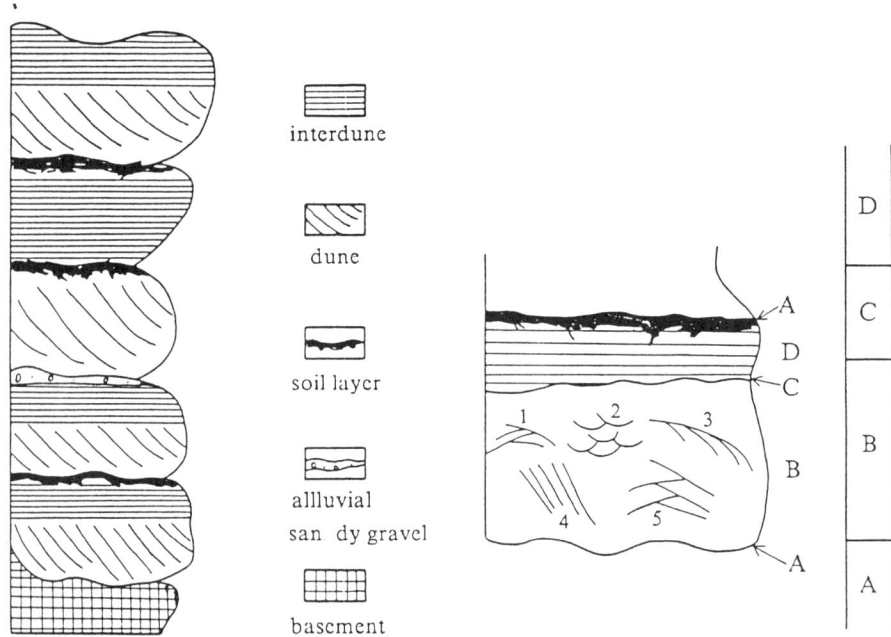

Figure 1

The sedimentary facies model of windblown sand deposits.

Legend: A, first order bounding surface; B, dune deposits; C, second order bounding surface; D, interdune deposits.

1, Convex cross-bedding; 2, trough cross-bedding; 3, drape-like cross-bedding; 4, tabular cross-bedding; 5, Seif (pyramidal) cross-bedding.

(3) The second-order bounding surface, above unit B, is known as Facies Component C, a diachronous interface as mentioned above.

(4) The upper interdune sediments are called Facies Component D, consisting of subhorinzontal, shallow trough cross-beddings. The sets dips at 0-15° and onlap themselves one by one.

Above the Facies Component D is a pedogenic layer, or an erosive hiatus, or sediments formed in water or in other environments, they are not essential components of the model.

The complete facies model from Component A to B to C to D can be seen in some windblown sand deposits, but more common are the sequence A-B and A-D. If the Component B and D appear concurrently in sections, the Component A and C are sure to exist.

REFERENCES

1. Ye Zhigheng, Zhang Mingshu, Han Chunrui, Wu Jianzheng, Li Hao and Ju Lianjun, Marine Geology & Quaternary Geology(in Chinese), Vol.4, No.1, 1—10(1984).
2. Zhang Mingshu and Liu Jian, Marine Geology & Quaternary Geology(in Chinese), Vol.12, NO. 1, 73—83& NO.2, 53—64(1992).
3. Wu Xihao, Xu Heling, Yin Weide and Deng Jiwen, in: Correlation of Onshore and Offshore Quaternary in China(in Chinese), Liang Mingsheng and Zhang Jilin, (Ed.), Science Press of China , Beijing (1991)p.262—269.
4. Qin Yunshan & Zhao Songling, in: Correlation of Onshore and Offshore Quaternary in China(in Chinese), Liang Mingsheng and Zhang Jilin, (Ed.), Science Press of China , Beijing (1991)p.23—29.
5. S.L. stokes, Sedim. Petrol.,38, 501—515(1968).
6. M.E. Brookfield, Sedimentology,24, 303—332(1977).
7. Zhang Mingshu, He Qixiang, Han Chunrui, Ye Zhigheng, Li Hao, Ju Lianjun and Wu Jiangzeng, Memoirs of institute of Marine Geology (1)(in Chinese), Shandong Press of Science and Technology, Jinan(1987)pp.189—341.
8. Ye Zhizheng, He Qixiang, Zhang Mingshu, Han Chunrui, Li Hao, Wu Jianzheng and Jui Lianjun, Acta sedimentdogica sinica,vol. 3, NO. 1, 1—15.